Human Nature & Biocultural Evolution

Human nature & Biocultural evolution

JOSEPH LOPREATO
University of Texas at Austin

Boston
ALLEN & UNWIN
London   Sydney

© Joseph Lopreato, 1984
This book is copyright under the Berne Convention. No reproduction without permission. All rights reserved.

Allen & Unwin, Inc.,
9 Winchester Terrace, Winchester, Mass. 01890, USA

George Allen & Unwin (Publishers) Ltd,
40 Museum Street, London WC1A 1LU, UK

George Allen & Unwin (Publishers) Ltd,
Park Lane, Hemel Hempstead, Herts HP2 4TE, UK

George Allen & Unwin Australia Pty Ltd,
8 Napier Street, North Sydney, NSW 2060, Australia

First published in 1984

Library of Congress Cataloging in Publication Data

Lopreato, Joseph.
   Human nature and biocultural evolution.
Includes bibliographical references and index.
1. Sociobiology. 2. Human behavior. 3. Social
evolution. 4. Behavior evolution. I. Title.
GN365.9.L66 1984   304.5   84–402
ISBN 0–04–573017–2

British Library Cataloguing in Publication Data

Lopreato, Joseph
   Human nature and biocultural evolution.
1. Nature and nurture   2. Psychology
I. Title
155.9   BF341
ISBN 0–04–573017–2

Set in 10 on 12 point Bembo by
Inforum Ltd, Portsmouth
and printed in Great Britain
by Butler & Tanner Ltd, Frome
and London

*To my sisters, Grazia and Francesca,
and to my children, Gregory and Marisa*

The meaning of human life and the destiny of man cannot be separable from the meaning and destiny of life in general. "What is man?" is a special case of "What is life?" Probably the human species is not intelligent enough to answer either question fully, but even such glimmerings as are within our powers must be precious to us. (George Gaylord Simpson, *The Meaning of Evolution*)

Developments in biological theory and in the social sciences have created firm grounds for accepting the fundamental continuity of society and culture as part of a more general theory of the evolution of living systems. (Talcott Parsons, *The System of Modern Societies*)

Contents

Preface   *page* xi

1 Introduction
    1.1 From Copernicus to the Idea of Natural Selection   1
    1.2 From Darwin to Sociobiology   5
    1.3 The Debate over Sociobiology   20
    1.4 Human Nature   33

2 Behavioral Predispositions, Cultural Universals, and Cultural Variants
    2.1 Isolating Behavioral Predispositions   37
    2.2 Behavioral Predispositions   49
    2.3 Universals and Variants   56
    2.4 The Problem of Analogy and Homology   65
    2.5 Nature and Nurture   68

3 The Interplay of Biology and Culture
    3.1 Intensity of Causation   80
    3.2 The Mutual Influence of Culture and Natural Selection   92
    3.3 The Modifiability of Predispositions   95
    3.4 The Form of Sociocultural Change   102

4 Predispositions of Self-Enhancement
    4.1 Introduction: The Problem of Inclusive Fitness   106
    4.2 The Climbing Maneuver   110
    4.3 Territoriality   120
    4.4 The Urge to Victimize   129
    4.5 The Need for Vengeance   141
    4.6 The Need for Recognition   146

5 Predispositions of Sociality, I
    5.1 Reciprocation   151
    5.2 Predispositions of Domination and Deference   161
    5.3 The Need for Conformity   177
    5.4 The Need for Social Approval   186
    5.5 The Need for Self-Purification   188

6 Sociality, II: Ascetic Altruism
    6.1 On the Conceptualization of Altruism   196
    6.2 Is "Homo sapiens" Ascetically Altruistic?   207
    6.3 Asceticism   211
    6.4 Ascetic Altruism   216

6.5	An Explanation of Ascetic Altruism	page 222
6.6	Definition of the Soul	228
6.7	Saving the Soul: Supporting Facts	229
6.8	Conclusion	233

7 A Model of Sociocultural Variation and Selective Retention

7.1	Biocultural Evolution Defined	237
7.2	Forces of Variation and Selection: a Conceptual Introduction	238
7.3	Criteria of Selection	251
7.4	Intra-Societal and Intersocietal Selection	262
7.5	Conclusion	264

8 Behavioral Predispositions and Religious Behavior

8.1	Explanatory Urge; Relational Self-Identification	267
8.2	The Denial of Death	274
8.3	Symbolization; Reification	281
8.4	The Susceptibility to Charisma	294
8.5	The Imperative to Act; the Need for Ritual Consumption	298

9 Evolutionary Foundations of Family and Ethnicity

9.1	Homologous Affiliation and Heterologous Affiliation	304
9.2	Incest Avoidance	314
9.3	Women, Men; Marriage and Children	321
9.4	Heterologous Contraposition	330
9.5	Conclusion	337

Glossary	341
References	348
Index of Names	381
Index of Subjects	389

Preface

The beginnings of this volume go back several years to a time when I found myself reviewing the sociological literature on various aspects of stratification theory. The task was both awesome and dispiriting—awesome because clearly it revealed collective tenacity of inquisitiveness; dispiriting because, though in many respects creative, our endeavors seemed to be hopelessly entangled in a maze of confusion and frustration.

The literature on status inconsistency (or incongruity) and its correlates was a case in point. Some scholars argued and purported to show, for example, that persons characterized by mutually misaligned statuses (for example, ethnic, educational, occupational, religious) were also inclined to be "liberal" in their political orientations. Other students of the subject failed to support this finding or altogether contended that the opposite was true. In principle, there was no harm in that. But the quarrel—and much of the dispute was just that—was often based, not always with full awareness, on such problems as different conceptions of status inconsistency, irreconcilable statistical techniques, conflicting measures of the correlates, especially those of liberalism and conservatism, and the failure to account for fluctuations in the intensity with which the correlates manifested themselves as "dependent variables."

More fundamentally, however, the controversy was an inevitable product of the fact that there was inadequate consideration of the question as to why the independent variable should have a given effect in the first place. Why, for example, should individuals who scored low on ethnicity but high on eduction show an inclination to favor liberal political candidates to conservative ones? Tenacity and cleverness of inquiry sometimes gave us a glimpse of the path to clarity. It made some sense, for instance, to read that status inconsistency was "stressful." It made even more sense to find attempts at a causal specification of the stress variable. Status-inconsistent persons, for example, might experience stress, and prefer one political formula over another, *because* they experienced a sense of distributive (in)justice in relation to the established system of rewards.

But such glimmerings of the causal nexuses and the logic of scientific inquiry did not penetrate deeply enough. The human animal is motivated by much more than a sense of distributive justice—assuming even that this is real strategy rather than mere stratagem. The widespread idea of a sociological "crisis" hit home hard, and I became convinced that the root of the crisis was our lack of a general principle with which to organize our research and guide our theoretical efforts. At about that same time, Edward O. Wilson published his now famous *Sociobiology: The New Synthesis*. A careful reading of it deeply impressed me and reminded me of some old scientific lessons that the "crisis" had managed to obfuscate. First, there is no science without some concept of cause, and cause entails the transfer of some form of energy from one agent (variable) to another. Second, causality operates at various levels of prowess, all of which are essential to more nearly complete theories; but the predilection of the scientific enterprise is the search for ever-deeper levels of causality; that is to say that science is by definition regressionist in nature. Third, all living things and their emergent as well as aggregate organizations are deeply historical in nature; the historical perspective is more or less farsighted—the farther the vision, the less trivial the findings. Fourth, scientific inquiry abhors artificial, administrative boundaries; that means that it pursues with relish the comparative and interdisciplinary method of analysis.

All of this strengthened my suspicion that the fathers of social science at the turn of the century and earlier may have been justified in their evolutionary perspective on sociocultural phenomena and in their companion notion that the human animal has a biological history, a "human nature," whose constituting forces may continue to be at work even today in our highly symbolized existence. Moreover, such forces might be under the control of a fundamental process, and since the general principle epitomizing it has not yet been found in the sociocultural world, the one basic to behavioral evolutionary biology might well deserve examination and even prove useful to sociocultural science.

Accordingly, this book aims at achieving two principal and related goals. One concerns the delineation of a theory of human nature—of what is sometimes termed the human "biogram." While the constituent items of the biogram will be viewed as products of natural selection, the latter, in turn, will be conceptualized as being under the enduring influence of culture. It follows that my fundamental strategy is to establish in some systematic fashion the intimate interplay of biology and culture. This is a work in coevolution.

The other aim is to make a contribution to the emerging but still embryonic theory of biocultural evolution. My commitment is to the idea that such a theory can be constructed along the lines, roughly, of the variation-and-selective-retention model found in Darwin's theory of evolution by natural selection.

More specifically, Chapter 1 will discuss certain basic developments in evolutionary biology with a view to highlighting the theoretical connection existing between evolutionary biology and social science, and getting in the process at the concept of human nature. Chapter 2 will discuss methods of isolating the *behavioral predispositions* constituting human nature. It will also contain a conceptual analysis of such predispositions and of two types of their cultural manifestations, namely, *universals* and *variants*. It will conclude with a treatment of the etiological interplay existing among the three levels of analysis. Chapter 3 will introduce the properties of *intensity* and *distribution* that relate to predispositions and cultural manifestations; and, through a critical analysis of certain classical theories in social science, it will continue to demonstrate biocultural interplay from the perspective of such properties. This will conclude what may be considered the first part of the volume.

Chapters 4 and 5 will discuss behavioral predispositions of *self-enhancement* and *sociality*, respectively, from the viewpoint of the cultural manifestations thereof. The forces of self-enhancement are intended to express in the most direct fashion the so-called selfishness of behavior implicit in the principle of natural selection. The predispositions of sociality, on the other hand, convey the idea of the attenuation of selfish orientation through group living. Chapter 6, which closes the middle part of the book, will continue along this line and seek to demonstrate and explain the evolution of genuine altruism, a phenomenon which dramatizes the influence of culture on genetics and which human sociobiologists have great difficulty conceptualizing.

Chapter 7, the first unit of the last part of the book, will make an effort to conceptualize the nature of sociocultural variation and offer a model of variation and selective retention. Chapters 8 and 9, finally, will present and discuss specific predispositions of variation and of selective retention; in the process they will seek to illustrate the relevance of such predispositions to an understanding of certain phenomena connected with religion, the family, and ethnicity.

It is an author's privilege to acknowledge at least the most immediate debt incurred in preparing a book. Various persons have read and helpfully commented on various parts of this book at several stages of its preparation. I especially wish to thank Richard N. Adams, Joe R. Feagin, Janice Hullum Edwards, Steven Lyng, Richard S. Machalek, Marilyn Ross, David A. Snow, Annette B. Weiner, and Edward O. Wilson. My departmental chair, Frank D. Bean, has always been generous with advice and aid that facilitate both research and clerical tasks. Several years ago the Dean of my College, Robert D. King, provided funds for summer research assistance that ended by intensifying my interest in evolutionary science. Two fellow workers and friends, Sharon Moon and Shannon Short, have with professional excellence and great patience gone through several drafts of the manuscript. My greatest debt is due to two persons. Sue Motzer spent a year in my office providing precious help as secretary, research assistant, and critic. Delbert D. Thiessen, my friend and colleague in Psychology, read two different drafts of the volume and opened my eyes to many difficulties and not a few promises.

Finally, many of the ideas contained in this volume have been tried more or less explicitly and thoroughly on several groups of students, especially those in my theory course and seminars. Perhaps unbeknown to them, even as I was peddling hard my wares, they taught me invaluable lessons by forcing me to practice self-criticism with sharp questioning if not downright skepticism. It is always edifying to feel oneself a member of the congregation even as one pontificates from the pulpit.

JOSEPH LOPREATO
Austin
February 1982

1 Introduction

1.1 From Copernicus to the Idea of Natural Selection

It is conventional to view the innovations of Copernicus—and even more of Kepler, Galileo, and Newton, among others—as both foundation and catalyst of a scientific process, "the Copernican Revolution," that has gone through several major stages, has left untouched no field of intellectual and moral endeavor, and is even now blossoming forth as if still in the spring of its prodigious potential.

Specifics aside, the general proposition of the Copernican Revolution, fully established by the eighteenth century, states simply that the universe is a system of matter in motion obeying natural, immanent laws. Its entrenchment in the history of ideas was no mean feat, for it entailed subscribing to propositions that had sweeping moral as well as intellectual implications (Kuhn, 1957; Gillespie, 1960). One proposition proposed that, far from being the center of the universe, the earth was but a speck rotating around one of the legion of stars constituting the universe. Another asserted that the motions of the planets around the sun conformed to the same laws that were obeyed by moving objects on our planet and by bodies in motion elsewhere in the universe.

As astronomy, physics, mathematics, modern philosophy, chemistry, and soon enough geology and biology were inevitably forged by the Copernican Revolution, certain intellectual implications came to permeate the scientific mentality. Motion, change, time, and the demystification of geocentricity and its major corollary, anthropocentrism, became the cornerstone upon which the modern scientific edifice was erected. Associated with them were certain methodological orientations: for example, the conception of phenomena as systems of interdependent parts, and of knowledge itself as a system of disciplines professing interdependent theories. Indeed, according to an axiom of the emerging scientific enterprise, the theoretical success or failure of a discipline was to a measurable extent a function of the degree to which it acknowledged the wider scientific boundaries and the orienting principles conjoined with them.

Unfortunately, the axiom has a poor standing in social science, and predictably social thought has of late progressed but slowly if at all toward a general theory of the sociocultural system, as many scholars in various disciplines maintain. That is ironic, for at one time social science was not at all sluggish in entering the stream of the Copernican Revolution. By the end of the seventeenth century, Europe, for example, was deep in the throes of the "quarrel of the ancients and the moderns" (Barzun, 1937; Brinton, 1959; Kuhn, 1957; Dobzhansky, 1962). What represented the apogee of human history, one expression of the dispute inquired, the classical achievements of the ancient Greeks and Romans, or the accomplishments and promise of the Copernican Revolution? The moderns carried the day. Giambattista Vico, and later Montesquieu, cogently put forth the proposition that human affairs are the results of human activity, and subject to a gradual development in time. Then, as the eighteenth century was approaching the dawn of "the evolutionist century," Condorcet prepared the ground for the elaborate historical theories of Auguste Comte and Herbert Spencer by advancing the idea that human history unfolded with inherent direction and could even be divided into a number of fairly identifiable stages.

Three features above all characterized the intellectual posture of the great social thinkers who brought nearly three centuries of exhilarating secular philosophy to a conclusion toward the end of the nineteenth century: the ideas of progress and reason (Bury, 1932) along with an inveterate commitment to the notion that the staple of the mind was *scientsia*, that is, knowledge which celebrates the wealth of channels along which it unfolds, yet cherishes the unity of the swelling origin and the common force behind its course.

Little wonder that in the academy of moral and social studies few have matched the theoretical span and the universal appeal of a Karl Marx. For more than a century, Marx has been a monumental figure on the stage of a worldwide intellectual as well as political drama. It is not merely his humane prescriptions for social ills that have gripped the minds still driven by the ideas of reason and progress. It is certainly not sufficient that Marx, perhaps better than anyone else since Machiavelli, tore down the ideological veil with which overly powerful governing classes cloak their work of plunder, pillage, and expropriation. The most impressive of Marx's virtues was the breadth of his knowledge. With a verve possessed by few, he sought to establish the theoretical link between social thought and physical science, on the one hand, and economic sociology and the emerging evolutionary paradigm in biology, on the other. A careful reader of Marx's work is necessarily exhilarated by an erudition so vast that philosophy goes into gear with elements of mathematics, physics, and biology just as smoothly as it links up with economics and sociology. What is truly engaging about Marx's scholarship is that his vision of science was not fragmentary, and his writing of history had a predilection for the capital H.

Moreover, as it is sometimes noted, Marx was a "sincere admirer" of Darwin's work in biology, seeing in it "the basis in natural history" for his own theory of human society. Without a doubt, Marx's search for the laws of social and economic development was viewed by him as an inquiry in social evolution or the transformation of society and culture according to definite and specifiable mechanisms (e.g. Engels, 1883). Unfortunately, intentions and erudition notwithstanding, he must be faulted for having after all relied a bit too much on industrial-capitalist society for an identification of the universal laws of social development. In the clutch—when it came to explanation—his vision narrowed. Moreover, he was an ideologist and utopist as well as a scientist, and having glimpsed the proletarian revolution, he abandoned the scientific method and lost sight of the enduring mechanism, "the struggle," that had lent his theorizing an evolutionary flavor.

To an extent, it is precisely this slight aberration of vision—this excessive emphasis of the current fact coupled with the nineteenth-century philosophical penchant for "perfectibility"—that has given Karl Marx, within the moral and social sciences, a certain preeminence in relation to many another scholar of his time and since. Such biases of temporecentrism ("today is the key to all time") and millenarianism not only touch the strings of prevailing "relevance," but, what is basically the same thing from another perspective, they also spare by default the Judeo-Christian assumptions about the moral order, the pre-Copernican premises of our species's incomparably unique place in the universal order of things, and the utopists' credo of the preordained sanctity of the individual human being.

While social thinkers were among the first to think in evolutionary terms, a theory of sociocultural evolution deserving of the label "scientific" and wide consensus has failed to materialize. The reason for this unfortunate circumstance, as this volume will suggest, is probably to be found in the fact that social thinkers have failed to utilize the scientific discoveries of the last 300 years, including the theoretical offshoots of their own labors. We may have lost little for our failure to draw a lesson from the cosmogonies of Kant and Laplace, according to which the solar system began as a gaseous cloud and gradually differentiated into its various bodies (Dobzhansky, 1962). But we probably would have gained much from a strict attention to Lyell's geological theory of "uniformitarianism," according to which the earth's features (mountains, seas, and so forth) have acquired their present shapes and locations gradually and under the operation of ancient forces that may still be observed at work today.

We could have learned even more from a scholar who avoided neither Lyell's lesson nor the insights of "evolutionizing" social thinkers of his time and earlier. The publication in 1859 of *On the Origin of Species* by Charles Darwin opened up a new era in intellectual history and marked a major stage in the Copernican Revolution by offering the conception of the organic realm as a system of matter in motion governed by the natural laws of variation and natural selection. Accordingly, it will be worth our while to begin the present volume with a brief consideration of Darwin's theory and some of the developments undergone by it since its inception 125 years ago.

Darwin's theory met with great resistance. After all, it challenged such long-revered postulates as that God had created all living species; that they had been created independently; and that they were basically immutable—that they were in 1859 what they had been at the time of the Creation and what they would be at the End. It also questioned the none too modest assumption that humans were the result of a special divine experiment wherein they were created in the very image of the Creator.

Still, the major social scientists of the time were hardly shocked by Darwin's theory of evolution. The basic elements of the theory had long been in the air, and in any case the Fathers were polymaths, scholars splendidly trained in various disciplines, who were hardly troubled by "the animal connection." What disturbed many of them was Darwin's failure to provide the link between the organic and the superorganic. Unfortunately, none was able or willing to fill the lacuna. They sought new directions and could not find them; or if they did, we have not seen them. Then, as the wave of the Fathers passed away, subsequent generations of social scientists came to equate evolutionism with biologism and this, in turn, with a rash of social prejudices passing for social theory. We even fell victim to the popular garbling of Darwin's proposition that "man and apes have descended from common ancestors" into the nonsensical statement that "man has descended from apes."

There is a sense in which it may be conjectured that civilization in general, and the religions that begin with and radiate from Judaism in particular, have at best been ambivalent about the development of behavioral science. The ideas of independent creations and the immutability of species are mostly fictions of the Judeo-Christian-Islamic culture, whose roots may be found in the book of Genesis. Ironically, in many circles they became basic tenets of natural history in the seventeenth and eighteenth centuries (Dobzhansky *et al.*, 1977), even as Copernicus's innovations were becoming the focus of an immense debate rising in dynamic tension with them (Kuhn, 1957).

In some respects, the more ancient religions known to us collectively under the label of animism or totemism reveal a remarkable natural philosophy that verges on fundamental evolutionary principles. For example, the idea was widespread among totemic peoples that all living things are related. So, to be a Crow was not to conceive of oneself in fact as the animal called crow. It was rather to believe that the same animating principle or force was found in people and crows alike (Durkheim, 1912).

Other totemic beliefs are even more striking. One concerns the idea that the soul exists prior to the formation of the body and survives the latter's decomposition. Closely associated is the idea of reincarnation, common also to Buddhism and Indian Brahmanism, whereby the dead are believed to return to life and once again have the feelings, the needs, the general characteristics of the living (Swanson, 1960). According to Durkheim (1912: 127f., 281), among the preagrarian peoples of Australia, along with many American Indians, conception and birth resulted from the act of a soul's entering a woman's womb; and "each individual is considered as a new appearance of a determined ancestor: it is the ancestor himself, come back in a new body and with new features."

Ideas that were at least partly consonant with modern evolutionary biology abounded in the ancient world. Among Greek philosophers, Anaximander, Democritus, and Empedocles, among others, taught that organisms could be transformed into different types. Much of Greek mythology concerns the transformation of mortals and immortals alike into animal form, and vice versa. The same may be said of the mythology of many other peoples and cultures. This ancient belief persisted to some extent as a form of paganism throughout the European Middle Ages. St Thomas Aquinas even argued that Christian philosophy was compatible with the belief that animals could rise from inorganic matter (Dobzhansky et al., 1977).

The first scholar to present a major theory of evolution was the French naturalist and philosopher Jean-Baptiste de Lamarck (1809). He was first to offer a coherent argument against the ideas of independent and invariable creations. Central to the theory are two theorems. One is the proposition, later to be adopted in a more general key by Herbert Spencer (1857a: Vol. I, 1857b) as the basic law of social evolution, that nature "has given to the acts of organization the faculty of making the organization itself more and more complex"—that, in short, there is in evolution a progression from simpler to more complex organisms. The other states that organisms undergo biological changes in response to environmental pressure, and that such transformations are heritable or transmissible by biological inheritance. Both propositions have been discredited by biologists. The second, however, is correct when translated to refer to sociocultural innovation and transmission. Accordingly, the approach to what in this volume is termed biocultural evolution requires a combined Lamarckian-Darwinian orientation.

1.2 From Darwin to Sociobiology

The father of modern evolutionary theory is Charles Darwin (1859). So great has his influence been that, despite the considerable confusion present in his theory and the essential role that genetic science has played in specifying it, Darwinian theory is today largely synonymous with evolutionary theory. Its centerpiece is the principle of natural selection.*

1.2.1 NATURAL SELECTION

Like many another great idea, the one behind the principle of natural selection was simple and yet hard to come by. Social scientists can take pride in the fact that, if scholars like Lyell provided the geological and paleontological thrust for Darwin's breakthrough, the notion of natural selection came to Darwin in 1838 while reading "for amusement" (Darwin, 1859: ch. 3) Malthus's (1798) *Essay on the Principle of Population*. On a five-year voyage around the world aboard the *Beagle*, Darwin had been struck by the richness and diversity of life both in the present and in the fossil past. In particular, he had become intensely aware of horizontal evolution or the significance of environmental differences for the differentiation of populations. He sought an all-encompassing explanation of these facts, and Malthus suggested it to him. This scholar had argued that human populations tend to reproduce beyond the limits permitted by the means of subsistence. "This implies a strong and constantly operating check on population from the difficulty of subsistence. This difficulty must fall some where; and must necessarily be severely felt by a large portion of mankind."

Darwin reflected that the tendency to overreproduce was a characteristic of the entire family of living things. Indeed, it could hardly be otherwise in view of the ever-present pressures from the environment and the related fact that the species serve as food for one another. Yet, in most cases, and over the long haul, populations tend to remain remarkably stable in numbers. Why? One partial answer, Darwin thought, concerned the fact of emigration, which helped to explain horizontal evolution. Another broadened the issue by in effect asking whether death within a given population took its toll at random or through some systematic or directive mechanism. The answer provided by Darwin was that in any given species some individuals were more successful than others in reaching the age of reproduction and in begetting offspring that were themselves better adjusted, or adapted, to their environment—whether through better health,

---

* Historically, however, for the theory of evolution by natural selection, Darwin must share credit with a number of scholars, notably with Alfred Russell Wallace who, working independently in the East Indies, produced a theory remarkably like Darwin's own. Both scholars presented *in absentia* their respective theories in 1858 at a meeting of the Linnean Society in London. Indeed, the third edition of *The Origin of Species* opens with a historical account of the natural selection idea and acknowledges that as early as 1813 a Mr W. C. Wells had held a fairly clear conception of the principle. Darwin's historical success is largely due to his naming of the principle, combined with his genius at synthesizing available information and a devotion to observation and experimentation over a sustained period of time.

greater strength, keener vision, or whatever. The numbers–food imbalance implied a process of competition, or "struggle for existence," in which the better adapted would be represented by the future generations more than the less adapted. They were in this sense the "fittest." This, basically, is Darwinian natural selection: the differential reproduction and contribution of offspring to the next generation by biologically different individuals belonging to the same population.

But the full meaning of this principle cannot be grasped without taking into account the concept of "variation." Darwin observed that the species tend to reveal variations in time. It was these variations, he argued, that in any given species underlay the differential reproduction and contribution of offspring to future generations. He (1859) put the matter succinctly as follows:

> As more individuals are produced than can possibly survive, there must in every case be a struggle for existence, either one individual with another of the same species, or with the individuals of distant species, or with the physical conditions of life. . . . Can it, then, be thought improbable, seeing the variations useful to man have undoubtedly occurred, that other variations useful in some way to each being in the great and complex battle of life, should sometimes occur in the course of thousands of generations? If such do occur, can we doubt . . . that individuals having any advantage, however slight, over others, would have the best chance of surviving and of procreating their kind? On the other hand, we may feel sure that any variation in the least degree injurious would be rigidly destroyed. This preservation of favorable variations and the rejection of injurious variations, I call Natural Selection.

Darwinian evolution, therefore, consists basically of a process whereby certain unspecified variations confer a reproductive advantage to certain members of a given species and a reproductive disadvantage to certain others. Because any given generation produces a large amount of variations, the species are constantly changing in one direction or another and gradually giving rise to new species, while old ones on the other hand become extinct.

A crucial implication of the Darwinian theory is that the multitude of species are all related, though in various degrees. They represent different offshoots and stages of the same evolutionary mechanism of speciation. This postulate, along with observed similarities in anatomy, physiology, and behavior across species and groupings thereof, as well as the more recent discovery of differential degrees of genetic similarities between species, eventually leads, as we shall see, to behavioral biology and the nascent science of human sociobiology.

On the face of it, the principle of natural selection was straightforward enough. Yet, it long continued to be a most problematic notion. The fundamental reason lies in the indefinite meaning of the companion concept of variation. "What is it that varies?" is a question that Darwin himself, and biologists in general, were for a long time unable to answer. And that problem, in turn, had to do with the fact that another great discovery of the nineteenth century remained unknown until 1900. I am referring to Gregor Mendel's 1865 laws of inheritance and the science of genetics that they helped to give rise to. Without knowledge of the unit of heredity, which on the basis of Mendel's work on "particulate" heredity, was termed the "gene" by Johannsen in 1909, Darwin could fancy but not know definitely the source of variation (Dobzhansky, 1937; Huxley, 1942; Mayr, 1942; Simpson, 1944, 1953).

His ignorance of the laws of inheritance had various deleterious effects. For example, he was never able to reject Lamarck's theory of biological inheritance of acquired traits. More important for us is the fact that, in his confusion about what natural selection acted upon, Darwin found Herbert Spencer's concept of "the survival of the fittest" as a "more accurate" expression than natural selection. Thus, in the fifth edition of *Origin of Species* he changed the title of the chapter on natural selection to read: "Natural Selection; or the Survival of the Fittest." The change was theoretically disastrous for social science and was made worse by the combined use of the metaphor "struggle for existence" which for many came to imply a cutthroat competition. Thus, a concept that gradually became the cornerstone of a major stage in the Copernican Revolution served for a while as the handmaid of a theory, Social Darwinism, which, while claiming the status of evolutionary science, was to a large extent an ideology and an apologia of the worst form of capitalism, ethnocentrism, and racism.

1.2.2 SOCIAL DARWINISM

Many critics have placed the nineteenth-century English sociologist Herbert Spencer and his school at the center of that ideology. The charge is not entirely warranted, and Spencer will probably be proven to have made useful contributions to the emerging theory of sociocultural evolution. But the Darwinian flavor of Spencerian concepts was consumed by connotations of a self-serving capitalism (see Hofstadter, 1955).

To a considerable degree, Social Darwinism was the unofficial core of capitalist ideology. Every major technological revolution begets an ideology, a set of political and moral formulas, that sings the praises of the new order and the social class that was most instrumental in bringing it about. In its heyday, the capitalist class required a free hand in the production of profit, its loftiest goal. It thus produced an ethic that to an inordinate degree steered political institutions and working masses to its active service. The roots of the ethic, of course, antedate both Darwin and Spencer. If we can lend any credence to Max Weber's (1904–5) thesis on the economic effects of the Protestant Ethic, the origin of Social Darwinism may be found in the seventeenth-century Puritan belief that economic success was not only a measure of personal worth but also nothing less than a sign of divine election. The leap from divine "fitness" to Darwinian or genetic fitness may be one of those peculiar instances of theological hocus-pocus that further exalt the soul of those who are holier than thou. Whatever the roots, however, there is little doubt that, strewn in Spencer's otherwise brilliant work, are the elements of a capitalist ethic.

The basic axiom of his system of ethics, allegedly the first clearly evolutionary such system (Quinton, 1966), was what he termed "quantity of life in breadth and depth," combined with the apparent qualification of "surplus of agreeable feelings" (Spencer, 1895–8). Just what he meant by such notions is not entirely clear. What is quite beyond doubt is that they somehow flow from his position that human will should not allow institutions to interfere with the struggle for existence. More specifically, Spencer viewed as socially deleterious any intervention by the state in favor of the indigent, the sick, or any particular category of the population for that matter. Such intervention could only interfere with the natural weeding out of feeble elements from the population. "The whole effort of nature is to get rid of such, to clear the world of them, and to make room for better." "If [people] are sufficiently complete to live, they *do* live, and it is well they should live. If they are not sufficiently complete to live, they die, and it is best they should die" (Spencer, 1864: 414–15). Accordingly, Spencer deplored any form of state intervention in the daily affairs of the citizenry: for example, public education, supervision of health measures, regulation of housing conditions, state banking, tariffs, postal service, and of course poor laws, although he believed that private charity to the poor had positive effects on the benefactors and hastened the entrenchment of altruism in society (Spencer, 1864).

Little wonder that Spencerianism became most deeply lodged in a country that at the time was inheriting the capitalist initiative. As Richard Hofstadter (1955) notes, Spencer was far more popular in the United States of America than in his native England. The most influential American Spencerian and Social Darwinist was William G. Sumner. Like the master, he was opposed to anything that interfered with the struggle for existence and natural selection as he understood them. "Poverty belongs to the struggle for existence, and we are all born into the struggle" (Sumner, 1914: 57). By the same token, huge fortunes were the result of natural selection and a measure of the managerial efficiency that replaced waste in production (Sumner, 1883). "The millionaires are a product of natural selection, acting on the whole body of men to pick out those who can meet the requirement of certain work to be done" (Sumner, 1914: 90).

Still, contrary to widespread belief and the most vulgar form of Social Darwinism, Sumner's hero was not the millionaire but the "Forgotten Man," the prototype of a popular conception that is deep-rooted in American populism. A few years ago, the Forgotten Man was synonymous with the Silent Majority. In Sumner's own time he was represented by his own father, a hard-working and self-taught English laborer who went quietly and self-sufficiently about the business of providing for himself and his family. The Forgotten Man is

> the simple, honest laborer, ready to earn his living by productive work. We pass him by because he is independent, self-supporting, and asks no favor. . . . Every particle of capital which is wasted on the vicious, the idle, and the shiftless is so much taken from the capital available to reward the independent and productive laborer. . . . He is a commonplace man. . . . Therefore, he is forgotten. All the burdens fall on him, or on her, for it is time to remember that the Forgotten Man is not seldom a woman. . . . The Forgotten Man is weighted down with the cost and burden of the schemes for making everybody happy, with the cost of public beneficence, with the support of all the loafers, with the loss of all the economic quackery, with the cost of all the jobs . . . (Sumner, 1919; also Sumner, 1963: ch. VII).

In general, it may be argued that the more scholarly Social Darwinists were more devoted to, and keener about, an extreme principle of laissez-faire economics than to the nuances of Darwinian theory. They really owed more to Adam Smith than to Charles Darwin. Not surprisingly, therefore, the extreme implication of what they advocated was what came to be known as the principle of "nature red in tooth and claw"—what T.H. Huxley termed the "gladiatorial theory of existence," wherein "struggle for existence" and "survival of the fittest" represented the principal slogans of an economic ideology.

On the other hand, the less scholarly Social Darwinists gleefully claimed that Darwinian theory justified unrestrained and destructive competition between peoples, classes, and races, for only in this way "the fittest" would inherit the earth. They further equated affluence, high location, political might, and Anglo-Saxon ethnicity with biological fitness, blissfully ignorant of the fact that Darwinian fitness referred to reproductive success. Accordingly, by virtue of one of the most amusing ironies of history, the fittest were likely to be those whom the worst Social Darwinists denigrated, namely, the prolific people of the lower social classes, although, as we shall see in Chapter 4, the upper classes always tend to practice various forms of victimization whose effect is typically a reduction of genetic fitness in the lower classes.

Naturally, apologias for wholesale rape of natural resources and for prejudice of all sorts multiplied like weeds in the springtime. According to Rockefeller, for example, "the growth of a large business is merely the survival of the fittest. . . . This is not an evil tendency in business. It is merely the working out of a law of nature and a law of God" (quoted in Flew, 1967a, 1967b). That had an echo in church sermons and hymns with benevolences like the following: "The rich man in his castle, the poor man at his gate. God made them high and lowly. He ordered their estate." Adolf Hitler was less pious and more fearful lest a slackening of the will of the strongest destroy the human race altogether. Thus, Christianity, one of his favorite targets, was "a rebellion against natural law, a protest against nature." Taken to its logical extreme, Christianity would mean the systematic cult of human failure (quoted in Flew, 1967b).

1.2.3 THE REACTION

In time, as we shall presently see, the gaucheries, errors, and arrogance of various strains of Social Darwinism managed to generate a defensive reaction in social science and a costly detour in the development of social theory. We were led to throw away the proverbial baby with the bathwater. But the eminent dissenters of eighty and more years ago had little or no quarrel with Darwinian theory per se. For example, scholars like sociologist Lester Ward and the pragmatists in philosophy sought to discredit only the ideological or "anti-social" interpretations of the Social Darwinists. Moreover, Ward, a most articulate and influential critic, drew a partly justified distinction between animal evolution, which was understood to be purposeless, and human evolution, which was allegedly subject to modification and direction through purposive action. The possibility of directed evolution, or "telesis," was presumably inherent in the very nature of science, whose aim was "the artificial control of natural phenomena" in order to "minister to human needs" (Ward, 1913–18: Vol. II). This excessive faith in the power of science, a sort of inverted image of Social Darwinist excess, was accompanied by a powerful argument against the laissez-faire axiom. Free competition, it was granted, was indeed a good to aim for, but it was feasible only along with some measure of regulation; for laissez-faire encourages mergers and thus tilts economic action toward monopoly (Ward, 1913–18: Vol. IV; also 1893: ch. 33).

Finally, Ward (1913–18, Vol. III: 303–4), one of the most learned men of social science, took issue with the Social Darwinists' tendency to misinterpret Darwinian fitness. Thus, in his review of Sumner's famed volume on *What Social Classes Owe to Each Other* he wrote:

> The whole book is based on the fundamental error that the favors of this world are distributed entirely according to merit. . . . Those who have survived simply prove their fitness to survive, and the fact which all

biologists understand, viz., that fitness to survive is something wholly distinct from real superiority, is, of course, ignored by the author because he is not a biologist, as all sociologists should be.

The "fundamental error" was attributed to Sumner without full justification, as the careful reader of this scholar's volume (1883) can attest. Sumner did err by minimizing the effect of what may be properly called artificial selection (for example, tampering with natural selection through inheritance of privilege and wealth, military and legal coercion of the masses, and so forth); but he was not unaware of artificial inequities, as the portrait of the Forgotten Man, for example, has shown. Ward, nevertheless, was justified in calling attention to the discrepancy existing between the Darwinian and the Social Darwinian conceptions of fitness. In the former case, the metaphor suggests a variable capacity or chance to make the best of biological and environmental factors and to contribute descendants to future generations. In the latter case, the metaphor was often stretched, ironically enough, to unwittingly equate fitness with social abuse against the fittest, that is, with power that is exploitative and not necessarily associated with reproductive success.

Whatever else may be said of Social Darwinian philosophy, it was rather poor evolutionism. It did not have an adequate grasp of natural selection. It also revealed faulty knowledge of the history of human society. Ethnography and evolutionary biology have shown that mutual aid has been a basic element of human, indeed, of much animal, evolution. In fact, as I shall argue in Chapter 6, natural selection has probably favored individuals living in societies featuring a high degree of public spiritedness. Mutual aid both reflects and reinforces conformity and solidarity. These social traits, in turn, grant a selective advantage to those societies, and their constituent members, that are rich in them.

While in Siberia, Petr Kropotkin (1902), an anarchist and naturalist, was much impressed by the amount of cooperation and mutual assistance existing among the rodents, birds, deer, and wild cattle of that region. Delving then in the literature on insects, birds, mammals, and humans, Kropotkin, like Darwin, found a strong element of cooperation within social species. "Happily enough," he argued (1902: 74), "competition is not the rule either in the animal world or in mankind. It is limited among animals to exceptional periods, and natural selection finds better fields for its activity. Better conditions are created by the *elimination of competition* by means of mutual aid and mutual support. . . ." There is abundant reason to believe that Kropotkin reached conclusions quite at odds with the facts. Still, the emphasis on mutual aid laid bare one of the worst deficiencies of Social Darwinism.

To a large extent, the reaction against Social Darwinism was associated with the historical forces that were developing *pari passu* with capitalism itself. Their weight was keenly grasped by the great revolutionary Karl Marx. On both sides of the Atlantic great fortunes were being made within a context of economic laissez-faire and cutthroat competition. Concomitantly, class exploitation was vicious. Incredible poverty, squalor, and moral degradation were the bloated birthmarks of an economic system gone berserk with the greed inherent in an unrestrained profit motive.

Reading *The Origin of Species* in 1860, Karl Marx was favorably impressed by it and reported to his friend and collaborator Friedrich Engels: "Darwin's book is very important and serves me as a basis in natural science for the class struggle in history" (Marx and Engels, 1935). Likewise, orthodox Marxian theorists at the turn of the century felt quite at home with Darwin's theory (Hofstadter, 1955: 115). But the same may not be said of their reaction to the socio-economic conservatism of the Social Darwinists. How could it be otherwise? The bourgeoisie or capitalist class, which received the great admiration of the Social Darwinists, and the proletariat, the epochal heroes of the Marxists, were after all in contention for the same tokens of what the Social Darwinists accepted as the epitome of fitness: the economic and political power released by the Industrial Revolution. The misery and suffering of the one was, of course, the malfeasance and malevolence of the other. As the socialist and labor movement picked up momentum, so did social reform. The movement was a revolutionary phenomenon in its own right. It also swept along the moral conscience of the new intellectuals, the well-bred children of hard-working proletarians.

In the United States of America, it was the Great Depression that dramatized the excesses of laissez-faire "Darwinism" and to an extent vindicated Marxist ideology and the "telic" or purposive sociology of Lester Ward, although the moral rejection of Social Darwinism was older, and included powerful spokesmen. Henry George (1879), for example, had accepted competition as a law of nature, but he had forcefully rejected both the fatalism inherent in the Malthusian explanation of poverty and the Social Darwinist apologia that went with it. Edward Bellamy's (1889) famed utopia attacked the principle of competition and the institution of private property itself. Awakening in the year 2,000, Julian West, the hero of *Looking Backward*, muses: "Human nature itself must have changed very much." Dr Leete, his host, is more cautious but no less optimistic. "Not at all," he replies, "but the conditions of human life have changed, and with them the motives of human action."

At the turn of the century, the robber barons of the United States of America had an insatiable hunger for hired hands, and most of these came from countries and religious groups that, according to some Social Darwinists and to race supremacists like Houston Chamberlain and the composer Wagner, were decidedly inferior to the master race of the

Nordics. Many among the second-generation children of those "inferior" peoples grew rich inferiority complexes. Many also went to college, and some became sociologists, psychologists, and anthropologists. Their humiliations and their skills led them to sense that ideas, like rocks, often hide a serpent under their weight. And indeed the thesis of the master race was moving toward brutal application in the nefarious laboratory of the Nazi German dictatorship. How could evolutionary theory receive a fair hearing under such circumstances?

Thomas Kuhn (1957: 190) has noted that most scholars in the sixteenth and seventeenth centuries learned about the universe, not from astronomers but from poets and popularizers. Since the publication of Darwin's famous volume, most critics outside biology have likewise "learned" evolutionary biology from an analogous assortment of poorly qualified spokesmen. And so it came to pass that, in the social sciences at least, evolutionary theory gradually came to be synonymous with the worst form of Social Darwinism. That was an error unworthy of our scholarly tradition.

To make matters worse, while it is true that evolutionary theory, when it glances in the human direction, contains in-principle sensitivity to the interdependence of cultural and biological factors, it is equally undeniable that in practice it all too often emphasizes the biological input to the total exclusion of the cultural. Such a theoretical posture offends not only our sense of anthropological worth but also two long-revered principles of European metaphysics. One concerns the ancient belief that we have a soul, an *anima*, without being animals. Like the good people of old who were appalled by the idea that the earth was neither flat nor at the center of the universe, today we are still a bit troubled, even if at the subconscious level, by the suggestion that we are not the work of the transcendental intervention whereby we were made in the image of the Creator.

The other principle refers to the assumption of free will. There is an old Christian saying to the effect that, to put it in Leo Tolstoy's words, not a hair can fall from a man's head without God's will. The substance of the saying probably precedes Christianity. But as the slightly irreverent religions of the Greeks and the Romans gave way to Christianity and a monotheistic belief in a just and loving God, the substance of the proverb ran into some serious trouble. If nothing happens that is not God's will, does it mean that God is responsible for the evil as well as the good of this world? We got out of the pickle by turning to the fiction of "free will." Theologically, that means in effect that all that is good is the result of God's will, and all that is evil is the result of free will fallen prey to Satanic forces. Sociologically, the solution of the inherited dilemma has entailed a stance that is excessively rationalistic and fundamentally ahistorical—a stance that, on the one hand, glorifies the power of will and, on the other, undermines the capacity to think of the sociocultural existence in terms of natural forces that have been at work for millions of years.

The factors that help to explain the reaction against Social Darwinism, and through it against evolutionary theory in general are too numerous to treat exhaustively. The fact remains that by the 1930s evolution of any sort was all but dead among social scientists. That was especially the case in the United States of America where the various aversions generated by evolutionism in general and Social Darwinism in particular helped to prompt a position no more tenable than many Social Darwinist conceptions. Central to it was the assumption, cloaked in multicolored garb, that at birth the mind was a *tabula rasa*, a blank slate, and that consequently all sociocultural behavior was the result of forces acting on the mind through processes of socialization or enculturation disconnected from any biological endowment.

I shall touch again on this issue. For now we may conclude the present section by noting that finally it is dawning on us that socialization so viewed does not tell the whole story. If it did, we would have to admit as a minimum that socialization has a perverse way of doing much of what it is not supposed to be doing. The evidence is staggering. We are presumably trained to love peace, and yet we are constantly moving toward war under the Caesarian assumption that *si vis pacem, para bellum* (prepare for war if you wish peace). The world has become an arms bazaar. We celebrate the sanctity of the individual, but everywhere practice genocide and holocaust. We produce and teach brilliant ideologies of government of, for, and by the people, not to mention those that promise total freedom and the classless society, but we come up with governments of, for, and by *a few people*; and with societies that make a mockery of freedom and classlessness. We set out to wage war on poverty, and the poor not only remain but also lose whatever dignity they ever possessed under the burden of a burgeoning bureaucracy whose self-perpetuation becomes its principal reason for being. And so we could go on endlessly.

According to Alvin Gouldner (1970: 440), the assumption that "society and culture shape men served at one point to liberate men from biological or supernatural conceptions of their destiny." Today, that same assumption increasingly would reveal a world of our own making that baffles and frightens us at the same time that the assumption encourages narrow perspectives and vacuous intellectual inquiry. The time has come for the freedom to enjoy the examination of the ways in which mind, culture, and biology play and interplay with each other to produce the complexities of the human condition.

1.2.4 THE MODERN SYNTHESIS (OR THE SYNTHETIC THEORY)

For at least a century Darwin's theory did little to advance the cause of social science. What is ironic is that for more than fifty years Darwin's discovery produced research that was very limited in scope within biology itself. It consisted largely of attempts to construct phylogenetic trees of various plants and animals, of the search for and description of fossil forms that aided in such endeavors, of the quest for examples of natural selection, and of the application of certain techniques and concepts of comparative anatomy and embryology: for example, homology and recapitulation. The problem lay in the failure to recognize the importance of Mendel's laws of inheritance. Without these tools, a thorough understanding of the basis of natural selection was not possible (Dobzhansky, 1937; Huxley, 1942; Mayr, 1942; Dobzhansky et al., 1977).

Mendel's laws were rediscovered in the spring of 1900, but for more than two decades they were of little help to Darwinian science. The early Mendelians were not naturalists; thus, they poorly understood the basic concept of adaptation, and did not accept the theory of natural selection (Mayr, 1978: 52). Or they were, basically, typologists interested in mutation to the almost total exclusion of changes in gene frequency as the basis of evolution. Many, moreover, rejected the "gradualism" of Darwinian evolution in favor of "stepwise mutational changes of large extent" (Huxley, 1958: xii). This idea in its essence has of late been rekindled by certain paleontologists and geneticists who oppose Darwinian gradualism or microevolution with a thesis of macroevolution or "punctuated equilibrium." Surprisingly, such scholars show little awareness of the vintage of their basic argument. Nevertheless, they may be on to an important development in evolutionary science. Setting out from the observation that the gradual history of life is punctuated with periods of mass extinction and rapid speciation, they pose the argument, among other interesting ones, that certain genes or gene complexes may play a "regulatory" role in evolutionary history. Changes in such regulators are responsible for relatively rapid rates of extinction and speciation (e.g. Gould, 1977: pt IV; Stanley, 1981). Thus, the famous "missing link" may be missing, for example, simply because it was never there.

The birth of population genetics brought about the marriage of Mendelian and Darwinian laws, known as the Modern Synthesis (or Synthetic Theory), and established the principle of natural selection as the principal, non-random, guide in evolution. R. A. Fisher (1930), S. Wright (1931), and J. B. S. Haldane (1932) are the names of the earliest scholars best remembered in this area among English-speaking people. A second wave of geneticists and evolutionists gave momentum to and refined the works of the masters (see, e.g., Dobzhansky, 1937; J. Huxley, 1942; Mayr, 1942; Simpson, 1944, 1953).

Evolutionary biology has come a long way since Darwin's time. The most important difference between Darwin's theory of natural selection and the Synthetic Theory consists in the addition of the genetic laws of

particulate heredity. These help to specify the principle of na...
and to provide a precise measure of its intensity by showing t...
is determined by specific genes located on chromosomes. The...
based on the fundamental proposition that "all biological orga...
down to the level of molecules, has evolved as a result of natural se...
acting upon genetic variation" (Dobzhansky et al., 1977: 18). The mo...
theory of evolution is often termed the Synthetic Theory not only becau...
of the synthesis of Mendelian and Darwinian laws but also because it integrates the contributions of many fields of knowledge, for example, zoology, botany, anthropology, microbiology, biochemistry, physiology, ecology and systematics, genetics, and paleontology.

1.2.5 THE NEW SYNTHESIS: SOCIOBIOLOGY

The Modern Synthesis has been a powerful, encompassing, and luring theory. One of the latest developments to claim inclusion under its paradigmatic umbrella is what has come to be known as sociobiology, or the New Synthesis (Wilson, 1975b, 1978; see also Alcock, 1975; Brown, 1975; Barash, 1977; Lumsden and Wilson, 1981). Fundamentally, sociobiology may be viewed as an alliance of disciplines tied together by the principles of genetics and modern population biology (Waddington, 1975). These include ethology, evolutionary ecology, evolutionary demography, behavior genetics, primatology, neurophysiology, and cellular biology, among others.

In his widely acclaimed volume, Wilson (1975b: 4), the foremost spokesman of sociobiology, defined it briefly as "the systematic study of the biological basis of all social behavior." The definition was intended to include *Homo sapiens* under the purview of sociobiology. Indeed, "one of the functions of sociobiology . . . is to reformulate the foundations of the social sciences in a way that draws these subjects into the Modern Synthesis." Wilson (1975b: 4) then wisely cautioned: "*Whether the social sciences can be truly biologized in this fashion* [reformulated to weigh phenomena for their adaptive significance and then relate them to the basic principles of population genetics] *remains to be seen*" (emphasis provided).

All summed, the principal, immediate instigator of sociobiology has been ethology, or behavioral biology. Pioneered by such scholars as Julian Huxley, Karl von Frisch, Konrad Lorenz, and Nikolaas Tinbergen, ethology has shown that certain patterns of animal behavior—for example, family planning and signaling displays of courtship—can designate taxonomic relationships as well as morphological traits do. This discovery, in turn, has encouraged the hypothesis that there is a fairly close correlation between structure and behavior: the more morphologically similar two species are, the more similar are their respective behavioral repertories likely to be, *and* the greater the probability that such similarities are homologous—that is, that they have a common genetic origin. Just as genetic endowment, then, explains anatomy and physiology, so it may, to a degree yet to be determined, explain social behavior as well.

This, of course, is the point of maximum friction between sociobiology and the social sciences. Fortunately for the promise of a gradual rapprochement, evolutionary biology has also shown that behavior often, if not invariably, serves as a pacemaker in evolution. A change in behavior, such as migration to a new habitat, gives rise to selection pressures that, in turn, result in genetic changes (see, e.g., Hamilton, 1964; Williams, 1966; Wilson, 1975b; Dawkins, 1976; Maynard Smith, 1978; Washburn and McCown, 1978; Lumsden and Wilson, 1981). Accordingly, the door is now open not only for a close examination of the provocative hypothesis that sociocultural behavior is at least in part causally rooted in biology, but also for research on the effects of behavior, including *cultural* behavior, on biology. This branch of sociobiology is known as human sociobiology. Biocultural science might be a better term for it.

In any case, the human view from a distance that is suggested by human sociobiology to social scientists has the advantage of harnessing the excessive anthropocentrism that is typical of their craft. The decades to come may very well show that it amounts to introducing into the social sciences the comparative method on a grandiose and extraordinarily fruitful scale. In this sense, biocultural science may truly represent a high point in the continuing Copernican-Darwinian revolution. It may in the long run help to show that at least some behavioral and sociocultural patterns have long roots, are to a degree still determined by ancient forces at work, and represent proof positive that the creation is a large family of life forms none of which was a favorite with the Creator. Whatever the results and the promises, however, sociobiology is levelling a formidable challenge at the social sciences (see, e.g., Alexander, 1971, 1974, 1975; Alcock, 1975; Brown, 1975; Wilson, 1975b, 1978; Dawkins, 1976; Barash, 1977; Lumsden and Wilson, 1981; also the excellent collection by Chagnon and Irons, 1979).

In proffering a theoretical invitation to social scientists, sociobiologists have entered morally hypersensitive areas and kindled some old passions. To make matters worse, they have not always been explicit about the necessity of viewing culture and biology as being in a state of mutual dependence. As biologists, they have tended to emphasize the genetic input on culture and behavior. This is a gauchery, and a limitation neither less nor more excusable than the tendency among sociocultural scientists to peddle their own wares and to wear lead shoes as they approach evolutionary biology. In part, it reflects an incomplete grasp of sociocultural behavior and the rules of its transmission. But it also flows from a polemical or defensive stance; namely, it reflects a reaction to the tendency in social science to deny the importance of phylogenetic phenomena.

Gradually, however, two defensive reactions, like two minus signs, will come to equal one positive, cooperative action. The present book is intended as a modest step toward the achievement of that goal.

Indeed, the rapprochement between the social sciences and sociobiology is proceeding at a faster rate than most scholars realize. There are various indications of this development, not least of which is the intense interest taken in the relationship by numerous scholars from disparate disciplines outside both areas. One senses something analogous to a marriage contract drawn with the help of the go-between. Alternatively, one is reminded of a peace treaty, even an alliance, reached through the good offices of a diplomatic third party. An excellent case in point is the recent book coauthored by E. O. Wilson and a colleague trained in theoretical physics (Lumsden and Wilson, 1981).

The Lumsden–Wilson book appeared when the present volume was essentially complete, and I regret the inability to take it fully into account. Still, an occasional word about it will not be amiss. Lumsden and Wilson claim to have "developed a renewed concern and abiding respect for social theory, the neurosciences, and psychology"; and they show it, despite the inevitable limitations inherent in working at the frontier. The book makes a creative effort (1) to connect genes to culture through an examination of mental processes and (2) to link mental development to culture and this to genetic evolution.

The subject is gene-culture coevolution, namely, a process or "interaction in which culture is generated and shaped by biological imperatives while biological traits are simultaneously altered by genetic evolution in response to cultural innovation" (Lumsden and Wilson, 1981: 1). Like Wilson's 1975 book, the more recent volume is widely interspecific in reach. Its focus, however, falls explicitly on *Homo sapiens*, and the authors frankly criticize some of the prior positions of ethology and sociobiology: for example, the categorical argument that genes prescribe behavior and the tendency to treat the mind as a mere replica of behavioral traits. An important statement is worth reporting at some length:

> In this book we propose a very different view in which the genes prescribe a set of biological processes, which we call epigenetic rules, that direct the assembly of the mind. This assembly is context dependent, with the epigenetic rules feeding on information derived from culture and physical environment. Such information is forged into cognitive schemata that are the raw materials of thought and decision. Emitted behavior is just one product of the dynamics of the mind, and culture is the translation of the epigenetic rules into mass patterns of mental activity and behavior. In contrast to the approaches of traditional ethology and sociobiology, including previous approaches to gene-culture coevolution, we take account of the free-ranging activities of the

mind and of the diversity of cultures created by them. Genes are indeed linked to culture, but in a deep and subtle manner. (Lumsden and Wilson, 1981: 2)

The Central Principle According to E. O. Wilson (1975b: 3), the natural selection of altruism is "the central theoretical problem of sociobiology," when altruism is understood as behavior that reduces the genetic fitness of ego for the benefit of alter (Wilson, 1975b; Barash, 1977). The problem is indeed crucial for a variety of reasons. It speaks to the nature of interpersonal relations and ultimately, therefore, of social order, which is necessarily a primary, if not the elemental, concern of any science of social behavior. At a more specifically theoretical level, furthermore, the conceptualization of altruism would seem to violate the logic of natural selection. Inasmuch as natural selection may be understood as a process that, within the struggle for existence, favors certain individuals (genotypes) as against others, the process may be specified behaviorally to imply that, when organisms are under the influence of natural selection, they tend to behave in such a way as to maximize their own fitness. This is known as the *maximization principle*, and would seem to deny the possibility of the evolution of altruism by natural selection. For, as Wilson (1975b: 3) puts it, "how can altruism, which by definition reduces personal fitness, possibly evolve by natural selection?" We shall return to this context in Chapter 6.

1.3 The Debate over Sociobiology

1.3.1 ADAPTATION

The publication of Wilson's (1975b) first book on sociobiology created a tempest that has spilled over quite beyond the proverbial teapot. To my mind, the theoretical issue that is at the center of the controversy concerns precisely the maximization principle. More specifically it concerns the related concept of *adaptation*, which is fairly widely defined in evolutionary biology as "any structure, physiological process, or behavioral pattern that makes an organism more fit to survive and to reproduce *in comparison with other members of the same species*. Also, the evolutionary process leading to the formation of such a trait" (Wilson 1975b: 577—emphasis provided).

Lewontin (1979: 6; see also 1978), a biologist and one of the bitterest critics of sociobiology, considers this discipline as an extreme example of the "adaptationist program," namely, an "approach to evolutionary studies which assumes without further proof that all aspects of morphology, physiology and behavior of organisms are adaptive optimal solutions to [environmental] problems." In other words, sociobiology tends *to show*

how organisms may attain optimal adaptation *rather than to test if* the assumption is valid (for other criticisms of sociobiology's use of adaptation, see Blute, 1976; Sociobiology Study Group of Science for the People, 1976; Quadagno, 1979). Lewontin (1978) claims, further, that natural selection does not lead inevitably to an adaptation, which is sometimes hard to define in the first place; and he (1979: 13–14) argues that there are many evolutionary alternatives to "direct adaptations." He is absolutely right.

There are many impediments to adaptation. One concerns the fact of mutation. Mutations are errors in genetic transmission that are random, fairly frequent, and usually detrimental to fitness, as when a child is born with poorly developed sex organs. Another is due to the phenomenon known as pleiotropy, namely, the condition whereby genes have multiple phenotypical expressions. For example, beauty, a high fertility potential, and a dullness of wits that incline toward trouble with the law might conceivably be rooted in the same genetic structure. The first two features would seem to be adaptive enough; the third, however, may be so maladaptive as to cancel entirely the adaptive value of the first two. Again, the irony with adaptation is that organisms are more likely to be adapted to past environmental conditions than to present and future ones. Indeed, the more rapid environmental changes are, the more problematic the assessment of adaptation becomes. The smokestack, for example, may have been very adaptive at one time in ridding the home environment of irritants and pollutants; but when it becomes commercial and multiplies to the point where it poisons the atmosphere, it ceases to be adaptive.

There are many other bases on which to reject a categorical adaptationist hypothesis. The crucial one, however, is suggested by the principle of natural selection itself, which is quite clearly accounted for in the above definition of adaptation by E. O. Wilson. The definition is properly couched in a comparative key, referring to something that lends *a survival advantage to an organism in comparison to another*. Thus, as we shall see in a later chapter, the haplodiploid method of reproduction and sex determination among eusocial insects may, from one perspective, be considered a mechanism of optimal adaptation: because of it, the caste of workers attain the highest rate of fitness in their society. But the fitness of their mothers, the queens, is not as high. And that of their half-brothers, the drones, is lower still. To even imply that Wilson, a distinguished entomologist, is unaware of such facts and their implications for adaptation is utterly unreasonable. Indeed, Wilson (1975b: e.g. 123–4, 326–7) has all along shown keen awareness of the problematics of an adaptationist strategy: for example, in relation to the phenomena of warning calls among birds and parental desertion in various species (see also Wilson, 1976; Oster and Wilson, 1978: ch. 8). It is surprising, therefore, to note that the favorite target of Lewontin's criticism is precisely Wilson's work.

Still, it is fair to note that in part the controversy in question arises from a failure on the part of some sociobiologists to stress that, *stricto sensu*, given the theoretical underpinnings of the craft, their statements concern populations of organisms not individual organisms. Indeed, some write as if they had lost sight of this fact, and the result is most unfortunate. It is compounded by the practical failure to recognize that population traits have averages, ranges, and the like; and consequently they should be viewed in strictly variable and probabilistic terms. But I would stress that the problem is at bottom one of communication. Thus, to say, for example, that culture is adaptive, as I shall myself maintain in this volume, is not to say that it is adaptive for all human beings; nor that *all* culture is adaptive; nor yet that what is adaptive today will be adaptive tomorrow. It is, strictly speaking, a way of saying that culture, too, is a mechanism whereby competition for fitness in a population takes place. To say otherwise is to speak nonsense. To leave room for a faulty interpretation is careless, though often understandable, for how often can one say or even clearly imply things without becoming tedious if not altogether obnoxious? To understand otherwise is to misunderstand evolution by natural selection at the very heart of the matter.

1.3.2 DETERMINISM AND CONSERVATISM

A second area of debate is two-pronged and exceedingly complex. It is epitomized by a statement of the Sociobiology Study Group of Science for the People (SSGSP) (1976), a circle of Boston area critics from various disciplines who see human sociobiology as just another Social Darwinism or "biological determinism" that favors the "legitimation of past and present social institutions such as aggression, competition, domination of women by men, defense of natural territory, individualism, and the appearance of a status and wealth hierarchy."

This charge cannot be taken seriously. It is less the result of careful inquiry than the product of collective and ideologically inspired criticism of Wilson's (1975b) work. In particular, as sociologist Eckland (1976) notes, Wilson's position has been grossly distorted. Wilson (1976) himself discerns a "self-righteous vigilantism" behind the SSGSP affinity for improper quoting and maladroit misrepresentation.

Indeed, the SSGSP broadside is sometimes startlingly erratic. For example, Wilson (1975b) suggests the search for an "anthropological genetics," by which he means research into genetically based behavioral predispositions that, as results of both natural selection and sociocultural action, may help to explain the recurrence in time and place of what have been termed cultural universals. But in the original statement that inspires, and is elaborated into, the SSGSP article, the numerous authors (Allen *et al.*, 1975) reinterpret Wilson's argument as follows: "In other words, we must study the process by which culture is inherited through genes." The error is elementary. No one in biology has ever held such a position, and Wilson (1975b, 1978) himself has recognized from the beginning that sociocultural evolution is largely Lamarckian in nature; that is, cultural transmission takes place at the ontogenetic, or life cycle, level through complex processes of socialization.

In the more recent work, Lumsden and Wilson have specifically focused on what they term "culturgens," namely, such transmissible units as behaviors, mentifacts, and artifacts. The culturgens are processed and transmitted through what have been called "epigenetic rules." Such rules are genetically determined and constitute behavioral procedures whose function is to direct the mental assembly and processing of culturgens. They are termed epigenetic because they comprise the restraints that genes place on development. But it bears stressing that they are subject to evolution under cultural pressure (Lumsden and Wilson, 1981: especially, chs 1–3). In short, culture is not inherited through genes. Rather, it is inherited through a process of gene-culture coevolution in which the mind is the direct conveyor of culturgens.

The charge of biological determinism has been stimulated by the strong doubts raised by human sociobiologists against the widely held assumption in social science that sociocultural behavior is shaped wholly by the environment. The alternative offered has been the hypothesis that such behavior is to an extent *conditioned* genetically. To say that X is conditioned genetically is to say that, however wide and numerous the degrees of cultural freedom, culture remains tied to a "leash," although, to be sure, "the leash is very long" (Wilson, 1978: 167). Still, to some extent genes cannot but "constrain human values" because, after all, behavior issues from the brain and "the brain is a product of [biological as well as cultural] evolution." Certainly, while Wilson, the foremost spokesman of human sociobiology, has not always been unambiguous, he has shown remarkable awareness of the special nature of cultural science and the lessons that it can teach sociobiologists. The message of the Lumsden–Wilson volume cannot be mistaken in this respect. A little earlier, in a brief dialogue with anthropologists, Wilson (1979: 519—emphasis mine) had stated: "Sociobiology is itself a malleable, growing subject rather than an immutable catalyst. Its future depends on the challenges created by its most difficult subjects and most especially anthropology. Both areas of science can expect to be enriched by a *mutual transformation*." Earlier still, he (1978: 13) had noted: "Biology is the key to human nature, and social scientists cannot afford to ignore its rapidly tightening principles. But the social sciences are potentially far richer in content. Eventually they will absorb the relevant ideas of biology and go on to beggar them."

To a degree, the controversy and the misunderstandings surrounding the issue of alleged biological determinism rests on the fact that systematic efforts to establish the interplay of biology and culture have been few and slow in coming. The deficiency is, in turn, due to the circumstance that the dialogue between the cultural sciences and evolutionary biology has barely begun. Still, Wilson deserves much of the credit for the fact that the dialogue has at least started. It is, therefore, mystifying to find so much criticism and so much "disappointment" over his (1975b) chapter on "Man: from sociobiology to sociology" (e.g. Eckland, 1976; Gould, 1976; Mazur, 1976; Tiryakian, 1976). It would be more edifying to receive from sociological sources Wilson's own clarification that the chapter "was intended to be a beginning rather than a conclusion."

To be sure, Wilson's prose at times sounds a bit categorical and imperial. So, he seems to have no doubt that, in the mind of a hypothetical zoologist from another planet, the social sciences would "shrink to specialized branches of biology" (Wilson, 1975b: 547). But substance is more important than form. The search is for a cooperation of cultural and biological sciences. In the process of comprising cultural science, biology would be gratefully transformed.

There is a widespread and absurd concern among social scientists that sociobiologists are intent in absorbing social science. To be related, however, is not to be siamese twins. Furthermore, if one can think of the Synthetic Theory as representing the life sciences, what would be lost if sociology and anthropology were to become specialized branches of such a broad scientific program? Note that we do speak freely of physical science and thereby refer inclusively to such disciplines as physics, chemistry, astronomy, and geology. Who cares whether astronomy is a branch of physics, or vice versa? Who worries about the fact that chemistry bestrides the inorganic and organic realms? Proprietary positions, inadvertent or otherwise, are unbecoming in the community of sciences, whosoever the incumbents may be.

There is something prepossessing about a charge of unjustified biological determinism when it comes from evolutionary biologists themselves. Stephen Gould (1976), a paleontologist, has argued for the distinction between "biological potential" and "biological determinism," and categorically attributed the latter to Wilson. "Biological determinism is the primary theme in Wilson's discussion of human behavior; chapter 27 [of Wilson, 1975b] makes no sense in any other context." Nevertheless, Gould does not succeed either in demonstrating his charge or in proving Wilson's position wrong, whatever his interpretation of it. He argues, for example, that "the statement that humans are animals does not imply that our specific patterns of behavior and social arrangements are in any way directly determined by our genes." To my knowledge, however, neither Wilson nor any other behavioral evolutionist has ever implied the statement that sociocultural behavior is determined *directly* by genes. Indeed, as Dawkins (1976) has made clear in his forceful discussion of the issue in question, what genes determine directly is the synthesis of protein. One is on firmer grounds in arguing that the statement "humans are animals" *could* imply that sociocultural behavior is to some degree connected causally, though indirectly, to genetic structures and processes. The point is to inquire about the possibility. It is no great feat to recognize that between the following two hypotheses, (1) X is *not* associated with Y, and (2) X *may* be associated with Y, the psychology of scientific inquiry counsels in favor of the latter. There are really many ways of being "conservative"; one entails emphasizing the obstacles rather than the means to the achievement of scientific ends.

Gould connects his critique to the issue of homology versus analogy, a topic that will be confronted in more detail in the next chapter. Here we may note that his position restates some of the criticism typical of social scientists. Consider the case of aged Eskimos who, instead of endangering their whole family group by slowing down a difficult and dangerous migration, choose to be forsaken, and thus to die a frightful and lonely death. One explanation considered by Gould is as follows: "Family groups with no altruist genes have succumbed to natural selection as migration hindered by the old and sick lead to the death of entire families. Grandparents with altruist genes increase their own fitness by their sacrifice, for they insure the survival of close relatives sharing their genes."

Such a sociobiological explanation is alleged to be plausible but "scarcely conclusive since an eminently simple, *nongenetic* explanation also exists:. . . . The sacrifice of grandparents is an *adaptive, but nongenetic*, cultural trait. Families with no tradition for sacrifice do not survive for many generations. In other families, sacrifice is celebrated in song and story; aged grandparents who stay behind become the greatest heroes of the clan. Children are socialized from their earliest memories to the glory and honor of such sacrifice." Which explanation is more acceptable? Both are plausible, according to Gould, but neither can be proven, and therefore neither is superior to the other.

The problem here is to an extent a question of what we want from science. Different scholars apparently expect different things. Nevertheless, the corpus of the scientific enterprise has for several centuries proceeded on the assumption that scientific inquiry is most productive when it is tied to general laws or principles, such as universal gravitation and natural selection, that come to the service of scientists with great regularity in one problem after another. Such principles, sometimes termed principles of uniform causation, are not only great heuristic devices but also eminently helpful organizers of facts and theories, so that researchers know at all times the logical structure within which their work is situated. That makes for cumulative theoretical development.

Now, in the above problem, the sociobiological explanation specifies that the mechanism accounting for the presence or absence of the tendency toward self-sacrifice is natural selection operating on gene complexes associated with given behavioral predispositions. In the evolution of Eskimo society, families with genes predisposing to self-sacrificial behavior have been favored as against those families without such genes—or even against families possessing such genes at a low intensity of potential behavioral expression. The explanation is but the bottom line of a fuller argument, and does not deny the influence of environmental factors. Indeed these are among the variables that account for the "scaling," or variability of genetically relevant behavior. For example, individual A has a genetic predisposition toward self-sacrificial behavior, but he has been socialized to save his dear life at all costs; accordingly, the "altruistic" predisposition has been scaled down, weakened, by environmental factors, and whether the predisposition will ever manifest itself to the degree suggested by the Eskimo case is highly problematic. Thus, biocultural science would begin by making recourse to a general principle and would end, if sufficient time and help were given it, by considering those conditions under which the explanation is most likely to hold and those under which it is least likely to do so. More precisely, the general principle invites predictions and research into conditions that help confirm, or alternatively confute, those predictions. The fact that the practitioners of a science are not always explicit in the pursuit of this strategy does not gainsay the fundamental logic of their enterprise.

The question now arises: what mechanism does Gould offer, polemically, as an alternative to account for the presence of the proclivity toward self-sacrifice among Eskimo elders? The answer is, "tradition for sacrifice," which is further specified in terms of (a) "celebration of sacrifice in song and story" and (b) the socialization of children to the glory and honor of such sacrifice. But—and that is precisely the big But behind the necessity to redirect social science toward a biologically informed evolutionary paradigm—why such tradition in the first place? Why the need for glory and honor? The implicit answer may be found in the sentence: "The sacrifice of grandparents is an adaptive, but nongenetic, cultural trait." But that is hardly informative at all. What is an adaptive, but nongenetic, cultural trait? It would be helpful to know in order to be able to decide whether the two "explanations" in question have equal scientific merit.

Suppose that we give Gould the benefit of the doubt and understand an adaptive, but nongenetic, cultural trait to be a cultural trait with biologically adaptive value. There is a respectable theoretical context for such a position (e.g. Campbell, 1965; Durham, 1978), and there is some warrant for this interpretation in Gould's own statement that "Families with no tradition for sacrifice do not survive for many generations." We have now reached the point where a causal connection has been made, implicitly, between *tradition for sacrifice* and *biological fitness*. But why not make it explicit? Is explicitness of etiological structure not one of the fundamental aims of science?

Gould's "eminently simple" explanation is no explanation at all. Apart from the broader theoretical context that Gould only unwittingly hints at, the explanation is too simple. Again giving the benefit of the doubt, we may perhaps more fairly say that his alternative explanation is already comprised by the sociobiological explanation, which holds that culture tends to conform to the maximization principle. Human sociobiology, in short, takes the position that sociobiological and cultural explanations are probably complementary rather than alternative or antithetical (e.g. Trivers, 1971; Campbell, 1972, 1975; Wilson, 1975b, 1978: Lopreato, 1981; Lumsden and Wilson, 1981).

Most of the genetic evolution on which sociocultural behavior is based took place during the 5 million years that led to the agricultural revolution. Since then, genetic evolution "cannot have fashioned more than a tiny fraction of the traits of human nature." Genetically, then, we may have changed but little in the last 10,000 years or so, while by far the greater part of sociocultural evolution has occurred precisely during this short space of time (Wilson, 1978). These facts invite some skepticism about the hypothesis of biological "determinism."

Still, is it reasonable to assume that the genetic endowment which made culture possible in the first place, and which accompanied its evolution for millions of years, has been completely jettisoned by 10,000 years of sociocultural evolution? To be sure, we may discover that the assumption, widespread but waning in sociology and cultural anthropology, is after all correct. But in the meantime it is proper scientific procedure to question it and to test the grounds on which it is based. To behave so is to seek new horizons, and that is the cherished virtue of the scientific mentality. Human sociobiologists are encouraged in their hypothesis of partial genetic causation by a variety of facts, some more tenable than others. To begin with, there is a great similarity in the basic institutions and the basic classes of behaviors found across human societies (Murdock, 1945). The present volume will stress this fact, and accordingly we may leave the matter as stated for the time being. Further, there are many striking and basic similarities between human beings and the other primates, especially the great apes and monkeys of the Old World, not to mention animals phylogenetically further removed from our species. They include the size of intimate social groupings, the protracted period of socialization of the young, social play, competitive games, mock aggression, a highly developed curiosity—really a host of fundamental characteristics. On the basis of carefully drawn comparative portraits of chimpanzees and hunting-gathering peoples, Wilson (1978: 24f.) is led to speak of chimpanzees as "a little-brother species," and to hypothesize that the similarities between the two species "are based at least in part on the possession of identical genes" (on chimpanzees, see especially van Lawick-Goodall, 1971).

Before closing this section, it may be fruitful to return to the charge of conservatism and especially to a critique fashioned by the anthropologist Marshall Sahlins. This scholar raises various strictures, some quite tenable and constructive. Ultimately, however, his argument is vitiated by the contention that "the theory of sociobiology has an intrinsic ideological dimension, in fact a profound historical relation to Western competitive capitalism. . ." (Sahlins, 1976: xii). Perhaps so, but there is little or no basis for the assumption, apparent in Sahlins's argument, that Marxism, or "the Left," has a greater claim than sociobiology to the arbitration of an expanding science of *Homo sapiens*, or to the higher virtues of the human condition for that matter.

My own debt to Marx's great synthesis and to his brilliant insights into the historical conditions and dynamics of industrializing society is very great (e.g. Lopreato and Hazelrigg, 1972). Still, *de omnibus dubitandum* was Karl Marx's own motto. And so let us consider a more or less hypothetical case. A revolution is consummated by a Marxist ideology on the presupposition that the state—the exploiting ruling class—will soon disappear and the people will benefit from the moral as well as economic law: "from each according to his ability, to each according to his needs." Suppose now that the state grows larger rather than smaller. The people ill suffer the betrayed promise of the lofty formula. Their leaders, conversely, are likely to engage in psychological fabrications that would fashion them in the molds that they are not constituted to fit in. And so it can come to pass, as it has indeed far too many times, that people by the millions will have to pay for the temerity of their natural constitution with endless forms of abuse, including forced labor, torture, exile, the Gulag, and the insane asylum.

I for one shall not condemn Marxism for Stalinism; nor Christianity for the Inquisition. The truth is that there is probably no theory which cannot be tied to any number of ideologies, often mutually contradictory, and which does not lend itself to the uses and abuses that people of good or evil hearts and minds are wont to devise. That goes also for sociobiology, of course. But returning to the hypothesized revolution above, let us substitute its ideological assumptions with the following assumptions: people tend to maximize their own interests, and any extrinsic attempt to impede the free challenge of the dominated against the dominant is merely capricious, unnatural, and in the long run bound to fail. The former scenario suggests the concentration camp. The latter, which ironically offends by making assumptions about human nature that we would wish away, conveys the image of a society in a constant process of revolution and challenges to the powers-that-be.

There is great danger in trying to debilitate unpleasant facts or hypotheses possibly grounded in human nature with humanitarian and rationalistic wishful-thinking. The peril is transformed into a tragedy, often of worldwide proportions, when the aversion to the ideological dimensions, real or imagined, of scientific theories leads us to muster our animus toward the theories and to ignore their technical principles. For then, how can we decide which human policies are theoretically warranted and which are unwarranted? More crucially, how can we even begin to prevent the unwarranted ones? One necessarily wonders, for example, whether the human conscience would today have to bear the ignominious catastrophe of the Holocaust if more people of science and goodwill had not *yielded* their right to interpret the logical structure and the political implications of evolutionary theory to those who were bent on using it only as a weapon. The story of humane intentions gone astray, and of the readiness, almost the relish, with which men and women of the pen have allowed themselves to be intellectually exploited by dominant classes, shall one day be written, much to our self-righteous mortification.

The irony is that the development of science ill suffers straitjackets. Nothing has been able to halt the Copernican-Darwinian revolution. It is an incontrovertible fact, further, that systems of knowledge thrive when they are linked by open lines of communication. One wonders whether the long-standing isolation of sociology and cultural anthropology from the community of scholars does not account in large measure for the lack of a systematic theoretical advance that few fail to recognize and decry. Social scientists cannot ignore sociobiology, nor dismiss it out of hand, any more than physics could ignore astronomy, chemistry could disregard physics, and biology, in turn, could proceed with theoretical unawareness of chemistry. The development of scientific thought has an immanent logic. As it marches on, it increasingly reveals its unitary character by enveloping any discipline that deals with empirically verifiable observations—whatever its perspective on natural order and whatever the nature of its subject matter. In any case, sociologists at least should be no more disposed to value gratuitous attacks on sociobiology, however immature and rash this may be in some respects, than we are to condone the vitriolic charges levelled at sociology still today and more so when this discipline first dared raise its head in the community of scholarship barely 150 years ago.

1.3.3 THE QUESTIONS OF MEANING AND CULTURAL VARIABILITY

There are several other types of controversy constituting the sociobiology debate, and some will be dealt with directly or indirectly throughout this volume. One deserves somber if brief attention here, before we close this brief treatment of the debate, because it is crucial to the aims of this volume and to an eventual rapprochement between evolutionary biology and the cultural sciences.

Viewed as criticism, the controversy is actually a two-pronged assault. It asserts that in its focus on biological, or naturalistic, non-mental principles, the New Synthesis: (1) disregards the question of the meaning that phenomena have for the people implicated in them; and (2) sacrifices the complexity, richness, subtlety, and variability of human facts in favor of an emphasis on the universals of the human condition, which may be contrived in the first place, and do not, in the second place, justify the accompanying assumption that, as cultural universals, they are rooted in genetic forces.

The charges are not altogether fair, first, because the study of meaning and cultural variability are not extraneous to the theoretical program of the sociobiological alliance, and second, because these are precisely the sorts of skills for which sociobiologists are especially eager to have the participation of social science in their agenda. Anyhow, on the question of meaning, one of the clearest statements is provided in Clifford Geertz's semiotic concept of culture and his (1973: 5) argument: "Believing, with Max Weber, that man is an animal suspended in webs of significance he himself has spun, I take culture to be those webs, and the analysis of it to be therefore not an experimental science in search of law but an interpretative one in search of meaning."

This statement predates the onset of the major debate on sociobiology, but it exemplifies the general criticism better than any other known to me. Its rejection of the sociobiological overture seems categorical and is explicitly based on an exclusion of cultural science from the family of experimental sciences. I suspect that future development in behavioral science will prove Geertz to have been either too modest or too isolationist. For the time being, the reference to Max Weber is deplorable. This scholar offered various definitions of sociology, and one of them did indeed emphasize "the interpretive understanding of social action" (Weber, 1922, I: pt 1). Moreover, his most famous volume (1904–5) is a direct application of this definition. Weber, however, was keenly sensitive to the potential intractability of the concept of meaning insofar as a systematic study of cultural action was concerned. Accordingly, he recommended "ideal types" as heuristic devices for the methodic classification and comparison of meaningful subject matter in terms of some crucial property (for example, efficiency). They were especially intended as tools with which to tie the possibly infinite number of meanings to particular anchorage points which, themselves "objective," underscored the most salient, general, and recurring features of meaning (Weber, 1949). For Weber, a science of meaning—experimental in the broad sense—was both possible and desirable. His aim was the "causal explanation" of the "course and effects" of social action.

What is even more remarkable is the fact that Max Weber was both aware of Darwinian theory and committed to the applicability of some concept of selection to the task of sociocultural analysis. For example, "all changes of natural and social conditions have some sort of effect on the differential probabilities of survival of social relationships," and it is accordingly both possible to speak of "a process of 'selection' of social relationships" and necessary to inquire into the causes behind the changes that, in turn, have an effect on the differential "survival" of "competing forms" (Weber, 1922, I: 38–40).

Yes, the web of meanings is wondrous to behold. In the history of life on earth there has probably never been a being who could construct so marvelous a web and live so intricate, diverse, and rich an existence in it, with it, for it, and against it. But what a pity that we have so little appreciation for the antiquity of the human spider who, in building a web, follows so many stupendous variations of determinate patterns. We cannot fully understand the web without a comparable understanding of the possibly ancient forces that lead the builder to construct the web according to fairly predictable themes. And, of course, we surely cannot fully understand the builder without fully understanding also the beauty of the web and the wealth of experiences that inhere in the variations from the main patterns or themes.

On the question of cultural variability, the complaint is fairly typically represented by Sahlins's (1976: 2) assertion that sociobiology is "completely unable to specify the cultural properties of human behavior or their variations from one human group to another." An analogous, but less categorical, statement was made by pre-evolutionist Parsons (1954: 199), though specifically in connection with his dated discussion of "instinct theory." It might be easier to cope with Sahlins's stricture if one did not also have to cope with the vagueness ironically inherent in such seemingly straightforward terms as "specify" and "properties." If the term that strains to emerge here is "explain," I am afraid that the same criticism may be just as easily made of cultural science. We may to a degree count cultural traits and compare their variations in time and place (a feat that human sociobiologists, following social science techniques, could accomplish just as easily); but as to their explanation, our good intentions have simply been richer than our accomplishments. If, on the other hand, the issue in question is the absolute distinction between the biological fact and the cultural, we are back in the old lemon orchard vociferously disclaiming any relationship to old orange fellows on the other side of the proprietary fence.

Ultimately, both variants of the criticism here under review are based on a conception of science that is extraneous to established practice. The traditional conception, shared by biologists and all other "natural" scientists, holds that the first goal of science is precisely what Geertz eschews, that is, the search for laws or uniformities of fact. Is there, for example, a general principle under which we may classify the motion of celestial bodies, and what are the conditions that must be fulfilled for it to hold? Once the law has been isolated, we proceed to an explanation of the exceptions, that is, the variable conditions.

The other conception of science is common to what may be termed the metaphysical branches of philosophy, history, literature, and obviously social science itself. It emphasizes uniqueness, nuance, richness, variability. The irony, mind you, is that sociologists at least repudiate this orientation in their innumerable textbooks. The students are solemnly informed that ours, unlike history, for example, is a law-like science. What happens between lofty statements of intention and earthy practice is no mystery at all. Exceptions, variations, uniqueness appeal to our urgent desire to know the here and the now. The problem is that this uncontrolled curiosity is dreadfully frustrating: when we think we know the cause of a fact, the fact becomes a new exception and requires a new explanation. In the long run, though not in the short, it typically turns out to have been a wasteful pursuit. The value of the law, the constancy, the uniformity of fact, lies precisely in the fact that it is an anchorage point for the rich and variable details so dear to the semiotically inclined. Weber was painfully aware of this crucial fact of intellectual progress.

In bringing this section to a close, let me affirm that, appearances notwithstanding, I have my own reserve of strictures against sociobiology; and they will be advanced in due course. Moreover—and that is the more important point—there is a sense in which it must be said that to complain against criticism, indeed to fail to celebrate it, is itself to perpetrate harm against the course of scientific development. As previously noted, science thrives under conditions of open and intense discourse. On the surface of it nothing would seem more absurd than attacks, some clearly wanton, against sociobiology for its alleged irrelevance to one thing or another, or for its ostensible overdeterminism, conservatism, and the like. Still, strange, absurd, and misguided as such criticisms may be, or may seem to be, they too are an intrinsic element in the procedure of scientific discovery. They frequently incite to clarification; they expedite the correction of hidden errors; they give valuable clues to needed elaboration; or they merely stimulate the defensive stance of the opposition that, after the normal period of pouting and obstinacy, reveals itself to be the staple of inquisitive perseverance. Dialogue is precious. Too bad that its structure frequently appears to negate this verity.

I have one reservation only about open debate as it has been practiced in relation to human sociobiology. My own, perhaps limited, examination of published materials indicates that, on balance, the reactions to Wilson's (1975b: especially ch. 27) path-breaking publication have been remarkably positive in the major journals of the social sciences. Ironically, some of the most virulent attacks have originated from within biology itself. My reading further shows that such opponents are certainly not better informed about the substance and the needs of social science than those whom they so vigorously criticize. I am inclined to think, as a sociologist proud of his craft and its potential, but concerned about its stubborn frailties as well, that social scientists could well dispense with self-appointed paladins who come galloping across the biological battlefield to sanctimoniously safeguard our chastity belts.

1.4 Human Nature

The anthropologist E. W. Count (1958, 1973) has argued that culture is "man's peculiarly elaborate way of expressing the vertebrate biogram." In human sociobiology, biogram, a term of rare usage nowadays, is roughly synonymous with "human nature," namely, what this volume will call a set of genetically based behavioral predispositions that have evolved by natural selection in part at least under the pressure of sociocultural evolution. In a related key, the biogram may be viewed as "a pattern of potentials built into the heredity of the species as a whole" (Wilson, 1977: xiv; also 1978; and Lumsden and Wilson, 1981).

Accordingly the search for the components of human nature has become a major task of biocultural science (Wilson, 1975b: 548; Lumsden and Wilson, 1981). The present volume is to a large extent an inquiry into this problem. It does not accept, however, Count's extreme loading of vertebrate phylogeny. It is my position that culture is a system of symbols that call a *much culturalized* biogram into action. That entails the hypothesis of biocultural interplay, or coevolution, *and* the assumption that human nature has been influenced by sociocultural pressures and is, therefore, significantly different from the vertebrate biogram.

We do not know the age of our biological givens, or behavioral predispositions, nor the rate at which, through various processes, they have undergone alteration, and are experiencing it still. That amounts to saying that we do not know where the peculiarly human, and thus recent, behavioral predispositions begin and where their ancestral forms or origins end. To the extent that they have theoretical relevance for us, however, they must be viewed as a set of behavioral forces that are ultimately the foundations of the sociocultural system and, in continuous interaction with this system, account for the manner in which sociocultural organization and behavior gravitate toward certain patterns recurring in time and place. These patterns will be termed "the universals."

Work on human nature by biologists is still sparse, although a fair literature could be cited of useful attempts at partial delineation. In general, too, the behavioral predispositions are brought to surface by way of their sociocultural manifestations. These are said to include the following, among others: incest taboo, bond formation, parent–child conflict, sex-biased infanticide, territoriality, hierarchy, parochialism and insularity, ethnocentrism, love of the homeland, indoctrinability, and the need for social approval (see, e.g., Count, 1958, 1973; Wynne-Edwards, 1972; Wilson, 1975b, 1978; Lumsden and Wilson, 1981). My principal aim is to improve on such preliminary work.

1.4.1 THE CONCEPT OF HUMAN NATURE IN SOCIAL SCIENCE

The area is old, deep-rooted, and in some respects richer in the social and moral disciplines than in the biological ones. One of the earliest statements of the modern era on human nature may be found in that monument to the power of reasoning that is Hume's (1739) *Treatise of Human Nature*. In the section on "moral distinctions not derived from reason," Book iii, Hume notes: "In every system of morality which I have hitherto met with I have always remarked that the author proceeds for some time in the ordinary way of reasoning, and establishes the being of a God, or makes observations concerning human affairs; when of a sudden I am surprised to find, that instead of the usual copulations of propositions, *is* and *is not*, I meet with no proposition that is not connected with an *ought* or *ought not*." It is one of the fundamental insights of modern philosophy and science. For Hume, the imperceptible shift from the descriptive to the prescriptive was of "the last consequence," and thus required an explanation. He found it in the hypothesis that norms or moral values were not intrinsic to human actions and events, but rather represented the blossoming out of innate human needs to which actions were somehow linked. Today, Hume's great insight is relevant to a considerable number of less synthetic, more theoretically reduced, traits of human nature. They seem to include what have been called the need for conformity and the instinct of vengeance, among others. The intuition may also help to explain the resistance to sociobiology, which is in part fueled by the feeling that cultural norms neither have other roots but themselves *nor* ought to have them.

The putative father of sociology, Auguste Comte, built an entire sociology around a theory of instincts, and the same applies to a lesser extent to the chief British sociologist of the nineteenth century, Herbert Spencer. Comte (1896), for example, had a rather clear conception that the course of human "progress" was tracked in part by a dialectical fluctuation in the intensity of "conservative and innovating instincts." An increase in population density, for instance, creates new problems of both a utilitarian and a moral character. Encroaching temporarily on the conservative instincts, the innovating instincts go to work on the solution of the newly arisen problems.

William F. Ogburn (1922), one of the first social scientists to show a profound and systematic grasp of the relationship existing between sociocultural change and "original nature," theorized that cultural aims and "instinctive tendencies" are often in conflict with each other and that, moreover, cultural changes are only to a slight degree "controlled and purposively directed by man."

W. I. Thomas (1923) identified four "wishes" or needs (new experience, security, response, recognition) tied to "organismic processes," and discussed with great cogency the tension existing between them and the demands of the individual's group. In the process of this enduring conflict between the individual and society, a moral code is continually forged whose function is above all to regulate the wishes and orient the individual meaningfully to social situations.

T. Veblen (1899), W. G. Sumner (1906), V. Pareto (1916), G. H. Mead (1934), B. Malinowski (1944), A. H. Maslow (1954), and a host of other social scientists have made remarkable contributions to the emerging theory of human nature. Collins (1975: 90) argues convincingly that Freud and Durkheim (not to mention Jung) both "drew their power" from "the perspective of animal ethology deriving from Darwin." Noting that Darwin himself (1859, 1871, 1872) was sensitive to this aspect of social theory, we can add that the same may be said of nearly all major figures in psychology, sociology, and cultural anthropology writing up to fifty years ago. The great minds who wrote in the nineteenth century and soon thereafter were bred to be keen about the emerging paradigm in behavioral evolution. Typically, they also had interdisciplinary training; and biology, despite the idiocies of some Social Darwinists, had first call on their scholarly attention.

It is true, however, that, with few exceptions, the susceptibility to biological theory was pregnant with ambivalence. In part, it was the result of ignorance about animal behavior. An excellent case in point is provided by the sociology of Emile Durkheim. In his fundamental work on the division of labor, this scholar (1893: 345–7), for example, argues not only that biology "predetermines" the existence of animals, but also that "collective life is very simple" among them. "Man, on the contrary, is dependent on social causes" and his organism is "spiritualized." Later, in what is perhaps his most creative work, he (1912: 264) goes entirely out of bounds by declaring that human "social life, in all its aspects and in every period of its history, is made possible only by a vast symbolism."

Today few are so ignorant of the complex social life of numerous animal species. Fewer still would be so rash as to argue that human social life is made possible only by symbolism. How could such a specious assumption linger in view of the fact that there are tens of thousands of species that are exceedingly social but unable to symbolize? Nor could we rescue Durkheim from his predicament by interpreting, as is sometimes done, his attribute "social" to mean "cultural." For, in that case, his faulty statement would be reduced to the vacuity that "cultural life is made possible by culture." But if Durkheim was confused, he was decidedly not hopelessly so. Thus, in this same book we find numerous references to the role that "instinctive sentiments" play in the religious life of human beings. Moreover, the 1893 volume offers much interesting evidence to the scholar who is inclined to accept Collins's characterization of Durkheim's basic theoretical thrust. For example, in the discussion of vengeance in the social order, Durkheim (1893: 87) argues that vengeance "consists of a mechanical and aimless reaction, in an emotional and irrational movement, in an unintelligent need to destroy; but in fact, what it tends to destroy was a menace to us. It consists, then, in a veritable act of defense, although an instinctive one. . . . The instinct of vengeance is, in sum, only the instinct of conservation exacerbated by peril."

There is little hope of reconciling such statements with those encountered above. But the latter brings us much closer to a position where our perspective on *Homo sapiens* is enriched by the courage to make naturalistic assumptions about ourselves and by an unambiguous awareness that the assumptions we make about human nature have theoretical implications that had better be made explicit.

2 Behavioral Predispositions, Cultural Universals, and Cultural Variants

If it is true that there is a causal connection between gene and sociocultural behavior,* then one of the important tasks of behavioral science is to determine the nature of such a link. But how? Ideally we would need direct genetic evidence, but that is still scant. So, for the time being we must rely largely on indirect proof. That amounts to saying that the problem of the probable link between biology and behavior rests to a high degree on inferential grounds. Inferences, basically, are surmises or guesses, and whatever we may conclude on their basis must always be accepted with a grain of salt. Consequently, it bears stressing here that I am not intent in this book in proving that sociocultural behavior is determined to any degree by biological forces, what I term *behavioral predispositions*. My aim is merely to buttress my inclination to accept as heuristically useful the *hypothesis* that to an unknown degree the circumstances of our phylogenetic history still have a causal bearing on our sociocultural behavior, just as the latter, in turn, influences our continually evolving genetic endowment. Should I, therefore, sound here and there as if hunch and curiosity had become certainty after all, I beg the reader to accept that as an expression of enthusiasm for the hypothesis, and nothing more.

2.1 Isolating Behavioral Predispositions

In absence of a direct genetic approach to the quest for the human biogram, there appear to be two major indirect methods open to us: (1) an intense examination of human behavior across time and place in search of recurring uniformities in its repertory; and (2) a comparison of other

---

* For convenience sake, the terms "culture" and "sociocultural behavior" (or simply "behavior") will be used interchangeably in this book.

species, especially the higher primates, with our own species, *Homo sapiens*, in an effort to see which behaviors are shared across related species and which are strictly intraspecific. Again ideally, in so reasoning we should like to have grounds for the assumption that the sociocultural uniformities observed in time and place, along with the inter-specific similarities, represent homologous developments, namely, behaviors based on common genetic origins, or common behavioral predispositions. I have little doubt that the future of evolutionary science will be partially accommodating in this respect. In the meantime, however, we should do what we can this side of the ideal goal and let curiosity prosper.

My emphasis in what follows will be on intra-human comparisons, although throughout the book I shall broach a few inter-specific subjects with little timidity. Further, the search for sociocultural uniformities clearly recommends the assumptions that (1) some regularities are theoretically more consequential and revealing than others; and (2) moderation and economy are well advised in an area of research that has barely begun.

One approach to cross-cultural uniformities, and, through them, to behavioral predispositions, calls language into action. Human beings not only perform certain acts; they also weave endless theories about them. If in examining such theories, we observe themes that tend to recur in time and place, we may in fact be confronted by basic classes of sociocultural behavior that represent sociocultural universals properly speaking. This approach to human nature is actually fairly salient in the history of social thought. One scholar gives the following rationale:

> Current in any given group of people are a number of propositions, descriptive, preceptive, or otherwise. For example: . . . "Love thy neighbour as thyself." Such propositions, combined by logical or pseudo-logical nexuses and amplified with factual narrations of various sorts, constitute theories, theologies, cosmogonies, systems of metaphysics, and so on. . . . all such propositions and theories are experimental facts, and . . . we are here obliged to consider and examine them. That examination is very useful to sociology; for the image of social activity is stamped on the majority of such propositions and theories, *and often it is through them alone that we manage to gain some knowledge of the forces which are at work in society—that is, of the tendencies and inclinations of human beings.* (Pareto, 1916: 7–8—emphasis provided)*

Consider, for example, the Latin maxim *Dulce et decorum est pro patria mori*—it is sweet and proper to die for one's homeland or group. The

---

* Reference to this scholar's work will follow convention, whereby number citation refers not to page numbers but to the section numbers into which his work is divided.

saying bears valuable information for behavioral scientists. We may begin by dividing it into two parts: one constitutes a principle, a moral injunction, to die for one's society or homeland; the other purports to provide a reason for the desired act. If we abstract the latter part ("it is sweet and proper"), we are left with a part that, as moral and ethnographic history shows, appears to be universal in time and place. As such, it is a uniformity of fact and thus falls under the special purview of scientific inquiry.

Two other sets of facts encourage the hypothesis that the cultural universal is not a universal accident of sociocultural history—that indeed it may be associated with the evolution of the biological substratum. First, human beings at all times and everywhere have in fact sacrificed their lives for their groups on special occasions. When, for example, Great Britain and the United States of America entered World War II, numberless individuals hastened to enlist in the military service, thereby offering their lives, if the occasion arose, to their country. Second, if we survey the animal kingdom, we discover that there are other social animals who exhibit a similar tendency (see, e.g., Wilson, 1975b; Barash, 1977). We are not obligated, of course, to focus only on cases of self-immolation; the actual sacrifice of one's life is in part contingent on the consequences of the risk that is taken. Indeed, most acts of self-sacrifice probably entail minor lasting cost to the actor.

The reliance on linguistic clues for inferences about human nature is encouraged by certain theoretical developments in modern linguistics. Noam Chomsky (1972, 1975), for example, has argued cogently that there are basic similarities in language structure and acquisition across the great diversity of human cultures. There appears to be a "deep structure" of grammar that is shared by different languages, and points to an innate linguistic capacity which is more directive than was once believed. Indeed, the study of language structure reveals properties of mind that underlie the exercise of mental capacities in various behavioral activities. Anyhow, we do not learn to speak *ex nihilo*. The capacity to learn a grammatical speech is somehow in the brain; it has been pre-programmed by millions of years of biocultural evolution.

Apropos of language, it is by now a truism in the human sciences that this basic *mos*, as Sumner termed it, is both our delectation and our affliction. Returning to the Roman saying above, we must recognize that if, on the one hand, it gives us a clue to a sociocultural uniformity that may have deep biocultural roots, on the other, it contains the means with which to lead us foolishly astray. Part of the rationale or explanation for the inclination toward self-sacrificial behavior makes good sense: the appeal to propriety or duty is both morally and theoretically compelling. But another part ("it is sweet") is morally superficial and theoretically specious.

Many reasons (attitudes, beliefs, opinions, meanings) associated by human beings with their actions are no better, and some are worse, than simple sophisms. To that extent, they are misleading if we accept them, as we often do, as *the* causes of behavior. Consider the attitudinal research on why people get married. The "reasons" uncovered are interesting and, on the face of it at least, theoretically appealing. For example, we seek happiness; we are in love; we prefer permanent companionship to haphazard relations or to loneliness; we wish to get away from intolerable parents; and it is the thing to do. To be sure, any one or more such "reasons" may have been (or may retrospectively become) salient in the minds of the actors. But there is a fundamental sense in which it may be said that with few if any exceptions we have no choice but, and therefore have no "mental" reason, to get married. The force behind the maximization principle is ultimately *the* reason. The proof lies in the terrific competition throughout the zoological world for reproductive mates. There is no evidence that human beings do not partake of that competition.

It seems reasonable to argue, therefore, that research on why people get married is not really research on causes of marriage (except perhaps at very superficial, "proximate," levels) but rather research on the *beliefs* that people give, often glibly and in creative retrospect, for getting married. There is nothing wrong with studying beliefs! And this book will argue that they can in some sense be treated causally even from an evolutionary perspective. The problem lies in attributing to them excessive causality or in loading them with any degree of causal influence when they have none.

The human mind is full of tricks, and therefore one fundamental impediment to an evolutionary social science resides precisely in the mind itself. I shall refer to it as the *ultra-rationalistic bias* and depict it simply as follows: B → A, in which a belief, B, is erroneously taken as *the* cause of a given action, A. This mistake is very common. We have encountered it just above. It is also committed when we say, for instance, that human beings sacrifice themselves for their country *because they believe it their duty* to do so. The explanation begs the question: Why the belief in the first place, and why is it universal in human society? More: why does the same act appear even in the absence of the belief, as among animals who engage in acts of self-sacrifice in defense of the group (e.g. Wilson, 1975b: ch. 5)? Thus, in attributing excessive causality to human belief in this and numberless other contexts, we may commit not only a basic scientific error but also an act of gross anthropocentrism that unnecessarily restricts the scope of our knowledge. More often we commit acts of temporecentrism, as when we accept as cause of an event a "reason" that is peculiar to a given point in time.

Often, not always, there is no direct connection between belief, B, and action, A. Belief and conduct both are determined by a third factor, C. The

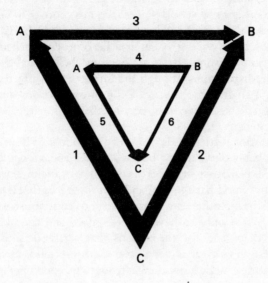

Figure 2.1 *Relationships among behavioral predispositions (C), beliefs (B), and actions (A).*

latter represents the evolutionary force that we have been intent in elucidating. There are times, too, when the force, C, gives rise to the conduct, A, and this, in turn, produces the belief, B. For example, a protracted series of national victories, based at least in part on intense "patriotic" predispositions in the soldiery, may lead a nation to the belief that it is invincible. Finally, there is the possibility that the action, A, is the result of the belief, B, which, in turn, is generated by the force, C (Pareto, 1916: 267–9). In all cases, the search for C widens the scope of the analysis, renders it more historical in the broad sense of the word, and, as I shall argue later, facilitates a parsimonious classification and explanation of human behavior without sacrificing weighty information.

In so stating the case, I have deliberately overstressed the causal import of factor C and understated, indeed failed to consider, the influence of A and B on C. The theoretical scenario that best fits the facts is probably represented more faithfully by Figure 2.1. The numbers as well as the thickness of the arrows in the figure represent together the order of magnitude among the various causal relationships depicted. Thus, the two triangles together indicate that behavioral predispositions (C) influence actions (A) and beliefs (B) more than A and B influence C. Furthermore, the outer triangle suggests that C has a greater, and more direct, influence on A than on B, with the result that, looking again at both triangles together, A is more causal in relation to B than vice versa. Action, as Marx among other great scholars noted, counts more than beliefs, words, consciousness, and so forth (all being represented here by B). The figure, finally, shows that, once a belief has become entrenched, it too can influence action. Further, both actions and beliefs have an input on behavioral predispositions, though the former more effectively than the latter. Later, I shall argue that, as a minimum, actions and beliefs call behavioral predispositions into action, they mitigate their force, or conversely intensify it.

The ultra-rationalistic bias is ironically a product of human evolution. Indeed, it probably flows from a behavioral predisposition that will later be termed the "explanatory urge." This is a force of major significance, and we shall in due course have to deal with it at some length. Here, suffice it to say that it is, among other things, the source of consciousness, and impels human beings not only to have beliefs, goals, and the like but also to account for them and for the actions that are associated with them. Consciousness, however, is time- and culture-bound. Being a characteristic of an ontogenetic individual, it tends to provide a perspective that is unique to a given time and place, and thus tends to hide more fundamental forces at work. So, human beings say that they sacrifice their lives for their country because they believe it their duty to do so, but in fact they may have little or no choice in the matter. When societies are at war, normally the number of individuals who are "willing" to sacrifice their lives for their country automatically increases manifold.

This first method of trying to get at behavioral predispositions thus sensitizes us to the fact that universals are heavily loaded with verbal fabrications, some of which verge on the sophistic. A second method casts light on the great variety of appendages, in the form of action procedures followed and means applied, that accompany a cultural universal. I choose for illustration a universal that several scholars have termed "assimilation" (e.g. Durkheim, 1912; Pareto, 1916). I favor the term "ritual consumption." As far as I know, it is peculiar to our species. It comprises a large class of facts. "Human beings have often believed that by eating certain substances one may come to partake of the properties of those substances. On rare occasions such phenomena may imply belief in some form of mysterious communion between a man and his totem or divinity; but more often those are different things" (Pareto, 1916: 937). That is, the force underlying ritual consumption is in principle independent of religious belief, the latter often being merely a carrier which the underlying predisposition attaches itself to. The substance in question may be any number of things, including human flesh.

Ritual consumption figures heavily in mythology, which may be viewed as a fertile field for the symbolization of behavioral predispositions, and plays an important role in Freud's (1913) theory of totemism and taboo, among other classics of behavioral science. The famous birth of Athena, the goddess of wisdom and of women's crafts in Greek mythology, was the result of an assimilative act. Zeus had received the prophecy that Metis, who was pregnant with Athena, would one day bear him a child who would become supreme in Heaven. He swallowed Metis. Hesiod explains in *Theogonia* that Zeus's act was perpetrated "to the end that the goddess might impart to him a knowledge of good and evil." Pindar relates in *Olympia* that Tantalus stole from Zeus the nectar and ambrosia that made him immortal, however pathetically so. Apollodorus informs in *Bibliotheca* that Achilles's strength, courage, and fleetness of foot were believed to be the results of his having been fed, as a child, on marrow from the bones of lions and stags.

One of the famous treatments of ritual consumption among people in the flesh belongs to one of the great masters of social science, Emile Durkheim (1912). His discussion will be dealt with in a later chapter. Here we may briefly note that, according to that scholar, under special circumstances the preagrarian peoples of Australia slew the totemic or sacred animal and solemnly ate sparingly of it as a form of communion with the totemic or sacred principle, which in Durkheim's (1912) interpretation was both God and a symbolic representation of the society itself. They also believed that at death the soul lingered near the corpse or continued to remain in the body altogether. Since in this state it was assumed to run any number of risks, ritual consumption of the dead's flesh, known technically as funeral anthropophagy, was practiced as a way of expediting the definitive departure of the soul.

In a close parallel, the Olo Ngaju and other peoples of Borneo practiced the custom of eating the flesh of their dead relatives in order to rid the dead body of its impurities and, thus, to speed up the soul's journey to the world of the ancestors. In the process, they also absorbed the good qualities of the departed (Hertz, 1907–9). Hogg (1966) reports as another example that Melanesians partook of deceased relatives as a gesture of affection, but ate enemies as a sign of contempt, although underneath it all "the transference of 'soul' " was presumably the crucial feature of the belief complex.*

Ethnographic literature abounds with accounts of peoples who have ritually eaten not only animals and plants but also their kin and neighbors "in order to" become possessed with their strength or other qualities (e.g. Hubert and Mauss, 1898; Swanson, 1960; Harris, 1977), although recently someone has unconvincingly tried to debunk the "colonialist myth" of cannibalism (Arens, 1979). Frazer (1890) relates in the ever-appealing *Golden Bough*:

---

* There is some evidence that anthropophagy has also been practiced simply as dietary necessity, as a result of a lack of protein (Harris, 1977).

> The savage commonly believes that by eating the flesh of an animal or man he acquires not only the physical but even the moral and intellectual qualities which were characteristic of that animal or man. . . . Thus, for example, the Creeks, the Cherokees, and kindred tribes of North American Indians, believe that nature is possest of such a property as to transfuse into men and animals the qualities either of the food they use or of those objects that are presented to their senses.

I have been writing in the past tense, and have noted that ritual consumption is heavily overlaid with heterogeneous beliefs, procedures, and explanatory reasonings. Can we speak also in the present? I think so, but we must be ready to encounter less "peculiar," and more insidious, sophistries associated with the basic behavior in question. Consider that from time to time scientific committees of the United Nations and maverick physicians everywhere inform that more than eight out of ten maladies are self-curing, and that indeed many of the drugs we take may have little or no therapeutic value, or may be more detrimental than beneficial. Yet every year we consume countless tons of the prescriptions granted by our esteemed colleagues, the physicians. Could it not be that in modern society the predisposition toward ritual consumption has reached a high level of intensity through means and rationalizations that the people of centuries to come will consider on an equal footing with the "barbarous" ways of yesterday's "savages"? Might it not be, further, that to an extent far beyond our present conception today's physicians correspond to the priests of an ancient and venerable art? That might help to explain the inordinately high esteem in which physicians have been generally held in recent decades.

It might also help to explain the remarkable phenomenon known as the placebo effect whereby sick individuals improve their condition or become altogether well after receiving inert medication. Until not long ago, medical science had not gone far beyond the mere temptation to attribute such a baffling effect to the mysterious workings of the mind. Today it is believed that at least in cases of pain relief, the placebo effect may be a result of mechanisms, still obscure, that release endorphins, or the body's own pain relievers. There is some experimental evidence for this hypothesis. In one study, for example, a placebo was first employed to relieve the pain of dental patients; immediately after, the patients were injected with a drug that is thought to inhibit the effect of endorphins; the pain returned (for a brief discussion of this topic and review of literature, see Jones, 1977: 204–14).

The part of special interest to me concerns, of course, the obscure mechanisms that under the form of, or through, the placebo does release the body's own pain relievers. Suppose that we are wired, as it were, to have a need to consume special substances which is associated with the belief that we thereby transform an existing physical or moral state. If the need is for the consumption, what is consumed may matter very little, provided that the belief exists that the appropriate substance is in fact assimilated. There is in fact some explicit proximate evidence for this hypothesis. Studies show, for example, that those who react profitably to placebos also fail to differentiate between analgesic drugs in terms of their efficacy. Conversely, those who do not react to placebos are significantly more likely to prefer one analgesic as against another (e.g. Jellinck, 1946; Beecher et al., 1953; Lasagna et al., 1954; Jones, 1977). Accordingly, some scholars have stated and confirmed the hypothesis that the patients' expectations, or beliefs in efficacy, are an important factor in the placebo effect (e.g. Jones, 1977).

The question naturally arises as to the evolution of such expectations or beliefs and of the need, as I have termed it, associated with them. Why the need for ritual consumption? The query makes an enormous demand on evolutionary science, and any speculation, therefore, is likely to verge on presumptuousness. Still, in a later chapter we shall return to ritual consumption in a manner that can briefly be anticipated here to provide one plausible answer to the query. In his theory of religion, Durkheim (1912) cogently suggests, more or less explicitly, that the historical viability of a society is sustained through the ritual absorption by its members of a symbol that represents and sanctifies the society. Since the symbol is religious in nature, religion may be viewed as an evolutionary adaptation of human societies. We may further hypothesize that ritual consumption of this sort is reflective of an intense socialization to the group's moral binders and that, therefore, individuals inclined toward such behavior were favored by natural selection through the support they received from other group members.

The hypothesis may be extended to suggest that the "idea" of nourishing oneself symbolically through ritual consumption was, in turn, a generalization of the idea that consumption of foodstuff was effective in nourishing the body. It is a well-known fact that actions and their functions go through complex historical vicissitudes during which they often radiate far afield from the original type. The process concerns such mechanisms as "differentiation," separation of form and content, and hypertrophy (or the extreme growth of an originally simpler phenomenon). They have been widely examined in behavioral science (e.g. Spencer, 1857a, 1897; Simmel, 1950; Parsons, 1966; Wilson, 1975b, 1978). In times long gone by, for instance, our ancestors chased animals only because they were driven by hunger. Today, we all too often hunt for sport alone and not infrequently in a highly ritualized manner, as in the English fox hunt.

A common expression of the need for assimilation is represented by the phenomenon of communion in the Christian tradition. Christians have believed that through communion they partake symbolically of the body of Christ. On the surface of it, such ritual consumption would seem to be a unique phenomenon, and so it has been treated by many a social scientist. In fact, the substance if not the form of the practice has been widespread in human society. Thus according to Durkheim (1912: 150), mystical repasts of totemic rituals "sometimes serve as veritable sacraments." The practice of some form of communion was widespread among the ancient peoples of the Mediterranean basin. Initiates in the Eleusinian mysteries drank the cyceon (a substance made of water, flour, grated mint leaves, and often wine) and recited a special formula for the occasion. In a related key, we note in the *Argonautica*, one of the Orphic compositions, that the Argonauts sealed their oath by drinking a cyceon made of flour, bull's blood, sea water, and olive oil. Sometimes, the mystical union takes a more startling form. Mau-Mau warriors, for example, had several stages of initiation ceremonies, progressively taking higher oaths in connection with more and more shocking acts, eating specified parts of a man or child, including brains and blood (Hogg, 1966).

In sum, cases of ritual consumption in its various forms have probably been universal in human society. Yet, it is fair to note that social science has had no end of difficulty and discord in explaining such phenomena. Our approach suggests that the problem may lie in not recognizing that, even if the form of the practice differs widely, the underlying substance is probably not at all different in time and place. The Christian communion, in short, is unique in a sense, but in another it may be but one version of a larger class of sociocultural facts. That, we may state in anticipation of a later discussion, is the fundamental reason for my emphasis on behavioral predispositions as explanatory variables; science looks for recurring elements in phenomena in order to get at uniformities of fact that invite scientific laws. The behavioral predispositions appear to be the fundamental causal units around which behavioral uniformities crystallize. The guiding hypothesis of this book concerns the biocultural interplay underlying behavior; hence cultural fact and the likely biological fact associated with it are both theoretically important to us. But, from a causal viewpoint, what matters most in the present context is not the great variety of rituals practiced; nor the great number of diverse substances consumed; nor yet the plethora of popular explanations advanced for such acts. Those are historically and spatially too fickle and, therefore, somewhat secondary to the scientific logic. What matters most is that the ritual practice of consuming substances to partake of their properties is universal in the human family. That encourages the hypothesis of a biological predisposition toward such a practice. Reference to ritual consumption, thus, helps to bring together a large number of seemingly disparate sociocultural facts and may help to explain such facts by reference to a possibly evolutionary cause. In this sense, the behavioral uniformity epitomizes a scientific uniformity.

Conversely, if we seek to explain as distinct entities the behavioral forms treated above and the multiplicity of those left out, we fail to recognize the denominator possibly common to them all. And that necessarily leads to theoretical frustrations that render scientific laws Chimera-like fictions. If we, for instance, focus cross-sectionally on the ceremonial consumption of human flesh among "primitive" peoples, and seek to explain the many rites in terms of the particular beliefs associated with them, we not only disjoin numerous facts from their behavioral class, but we may also find that our numerous explanations must vary greatly in time. For, beliefs and peculiarity of practice varied in time within a single society. The same applies to the Christian communion. In this connection I have spoken in the singular only for convenience sake; in fact, here too variations abound among the numerous denominations and subdivisions. A science following this course is a discipline that discards scientific laws as fast as it seeks to construct them.

But does that mean that particularistic explanations have no place in behavioral science? The answer is a definite no. Granting the hypothesis of ritual consumption as a uniformity anchored in an evolutionarily selected behavioral predisposition cannot gainsay the fact that people differ in time and place in the manner in which they express that predisposition. That is culture. Without it, people are not people. It is at least in part the different beliefs of given times and places that explain the different forms that the ritual consumption takes in time and place (Lorenz, 1977). These variations, in turn, organize to some extent the ways of life and the world-views of different peoples. Hence, it is legitimate scientific enterprise to inquire why some people satisfy, *and* explain, the predisposition in one form and others in another. It is useful, from another perspective, to have both long-range and short-range explanations—what are sometimes termed ultimate and proximate explanations (Pareto, 1916; Wilson, 1975b).

The emphasis on the ideas, meanings, beliefs, and practices of given times and places, however, is fraught with grave scientific dangers. The felt force of ideas is often a function of the importance of the person who holds them. Thus, such an emphasis often leads, at least implicitly, to Great Man theories in which the aggregate or demographic analysis that is a hallmark of social science research gives way to excessively individualistic, and necessarily specious, analysis. It was the great wisdom of Leo Tolstoy to say in *War and Peace* that "The king is the slave of history. In the events of history, so-called great men are merely tags that supply a name to the event, and have quite as little connection with the event itself as the tag."

Consider this example. In recent years, the US Supreme Court has issued a number of rulings that have reintroduced capital punishment in the nation. Likewise, in the spring of 1977 the high court ruled that the beatings administered to school pupils by school authorities are not prohibited by the constitution, even when they result in serious injury. A rash of publications have appeared in social science offering explanations that essentially revolve around the point that today's Supreme Court is "Richard Nixon's Court." But that is hardly an explanation. It does not, for example, account for the fact that Nixon after all was elected by a landslide. More importantly, it neglects to consider that to the extent that societies are historically viable, when they experience a breakdown in discipline, as it is generally conceded that American society did in the 1960s, they tend to react by reintroducing disciplinary measures, oftentimes in excessive doses. Normally the feat is accomplished through implementation of popular conceptions of what is morally proper and salutary. Scapegoats, even when they do deserve a skinning, make poor scientific reference points for rounding up the actions seemingly revolving around them.

The emphasis on behavioral predispositions and the sociocultural uniformities that may be their manifestations carries its own risks, of course. Most noteworthy perhaps is the danger that we classify together as one uniformity multiple phenomena that belong to two or more. Consequently, it may easily be that, for example, the fundamental cause of Olo Ngaju funeral cannibalism has little or nothing to do with the fundamental cause of the Christian communion. In part, however, this sort of problem probably implies that sociocultural complexes are compounds of lesser facts, just as chemical compounds are constituted by simple elements. That may mean, in turn, that to the extent that sociocultural facts are anchored in biological forces, as compounds they are the results of compound such forces, that is, of systems of behavioral predispositions. If this is the case, it may also be true that certain facts appear to be cross-culturally, or even cross-specifically, similar by virtue of the fact that they share parts of complexly bound biological forces.

In the meanwhile, and in conclusion of the present section, we should endeavor a preliminary sorting out of the three conceptual units or levels of analysis encountered so far. We have observed, or thought we have observed, certain widespread sociocultural complexes that may be termed self-sacrifice for others and ritual consumption, respectively. Depending in part on the method of inquiry—whether, for example, the focus was on verbal behavior or on other types of conduct—we have subsumed such complexes, or we could subsume them, under such general conceptual abstractions as "moral principles," "moral injunctions," "universals," "constancies," and "uniformities." All such labels are useful. My own preference is for *universals*, because of the wide diffusion of this concept in sociocultural science, and for *uniformities*, because of the implication in this term that a universal may be a locus of a possible scientific uniformity or law.

Closely connected with universals has been an assortment of phenomena, widely varying in time and place, in the form of linguistic expressions, beliefs, and practices that are, in turn, associated with the use of a variety of substances and techniques often of a ritual nature. So, for example, the universal of ritual consumption is sometimes combined with eating human flesh. Other times it entails eating part of a totemic object, a wafer, or any number of other substances. The rituals practiced in connection with such activities are themselves quite variable in time and place. In some, we observe people approaching an altar and a priest; in others, flesh is melted on fire. Finally, the beliefs, reasons, or explanations expressed for engaging in such rituals are also quite variable in time and place. For example, one people will say that eating the flesh of the departed will expedite the arrival of the latter's soul in the eternal abode of the ancestors. Another will explain that partaking of a given substance will impart the physical and/or moral virtues of the being who is materially represented or symbolized by the substance in question. It follows that a science of such a plethora of behaviors is difficult in the extreme if not altogether impossible to practice. Its constitution would by definition have to give theoretical primacy to an infinity of exceptions rather than to an economy of regularities.

A number of labels may be used to refer to the accretions of universals. As verbal facts, and on the assumption that they are more or less specious reasonings, they may be usefully termed *sophisms* or *sophistries*. As practices, means, and techniques, we shall refer to them simply as *variants* or *variations*. The latter terms will also be employed to be comprehensive of the two varieties when no distinction is intended between them.

The last of the three conceptual units has been referred to on several occasions with such labels as *behavioral predispositions, genetically based behavioral predispositions, behavioral forces*, or merely *predispositions*. I shall continue to use these concepts, all the more so because they are fairly common in behavioral science, including evolutionary biology.

2.2 Behavioral Predispositions

A more detailed discussion of the three concepts now follows. The gradual emergence of the evolutionary theory of human nature has various roots. For example, until not long ago psychology produced a number of what are sometimes termed "impulse energy theories" (Bandura, 1977). Such theories postulated an abundance of drives, instincts, or motivators so that for almost any given unit of behavior there was allegedly a unit of hereditary material. On the other hand, they failed to account for variation in such things as the frequency of a given behavior, its strength, its susceptibility to input from other factors in the social context, and its change in the course of time. Today impulse energy theories may be largely dismissed as being barely relevant to a biocultural perspective.

A second basis for a theory of human nature is provided by genetic science. The genetic approach for the time being relies largely on inferences from cross-species phenotypic similarities and on the hypothesis that phenotypic likeness may reflect underlying genetic similarities. The higher primates typically serve as the basis for comparison. We do share with them, especially the chimpanzees, a wide array of traits: for example, a highly developed curiosity, self-recognition or consciousness, social play associated with such things as mock aggression and highly stylized conciliatory behavior, an extended period of socialization in which female relatives and peers play a major role, the adoption of orphans, long-lasting bonding or friendship, incest avoidance usually associated with female exogamy, cooperative hunting, sharing of food, begging for food, and living in groups comparable in size to our hunting and gathering societies (e.g. Kummer, 1971; van Lawick-Goodall, 1971; G. G. Gallup, 1977; Wilson, 1978).

Some direct genetic evidence is not lacking, and considerable progress is being made (e.g. McKusick and Ruddle, 1977). A reasonably good example is provided by the case of individuals bearing XYY chromosomal combinations. They are all males and appear at the rate of about one in a thousand. Such persons have certain distinct physical characteristics: above average height, for example. But they also end up disproportionately in prisons and in hospitals for the criminally insane. One of the more thoughtful studies, based on massive data from Denmark, convincingly shows a rather peculiar type of biology-culture interplay at work to explain this unfortunate fact. Contrary to what was previously believed, XYY men are not likely to be more agressive than average. Rather, their relatively high criminal record seems to be a partial effect of a lower average intelligence quotient, the major behaviorally relevant genetic difference between them and the Danish population as a whole. This circumstance, in turn, sets the stage for relatively easy apprehension by penal agencies. In short, XYY men may run more frequently into problems with the law simply because they are less intelligent and thus less clever at dodging its enforcing agents (Witkin et al., 1976).

Some evidence, essentially direct in nature, in favor of the genetic hypothesis is provided by behavior geneticists in studies of identical, or monozygotic, twins. Because identical twins have identical genotypes, any behavioral differences between them must be attributable to different environmental experiences. The opposite unfortunately is not as true: behavioral similarities may be due to identical socialization processes and other environmental factors. Still, the array of behavioral similarities between identical twins is so startling that it would be rash to dismiss the idea of genotypical influence. Among the common traits that have a bearing on social behavior are word fluency, number ability, spelling, syntactical ability, homosexuality, perceptual skill, and mental illness (see E. O. Wilson, 1978: 45–6 for a brief review; also R. S. Wilson, 1978). Some studies, furthermore, have applied fairly refined techniques to control for environmental factors. The reasonable conclusion is that correspondence of socialization experiences is not sufficient to explain the sizable set of behavioral resemblances existing between identical twins, for these similarities are found also among most identical twins who have been subjected to dissimilar environmental factors (e.g. Loehlin and Nichols, 1976).

A related area of research concerns the adoption of very young children. As Joseph Horn and his associates (1979) note, despite certain problems of sampling and control, this practice affords one of the best opportunities for estimating the respective inputs of genetic and environmental influences on behavior. Children reared by their natural parents receive both genotype and socialization from them. Adopted children receive genotype from natural parents and socialization from adoptive parents. Consequently, resemblance to natural parents must be due to genetic kinship, while similarity to adoptive parents and other adoptive relatives is a result of shared environment.

In their study of 300 families who between 1973 and 1975 adopted children from a private, church-related agency for the residential care of unwed mothers in the Southwest of the United States of America, Horn and associates (1979), found that genetic variability contributes significantly to individual differences in intelligence. Common family environment was also important, indicating an interplay of biology and culture, "but its influence was less than that from genetic variation in almost every analysis" (for related studies, see Williams, 1975; Plomin *et al*., 1977, 1980; Scarr and Weinberg, 1978, among others).*

The main source of evidence in favor of the genetic hypothesis must come indirectly from cultural science. I have already hinted at this topic: it concerns the matter of cultural universals. Briefly, there appear to be several scores of major sociocultural similarities across human societies. Our choice of explanations of the fact is fairly restricted. One possibility invokes convergence or accidental similarity. This option may be dismissed on two grounds. First, it tends to impede scientific curiosity instead of stimulating it. Second, the similarities in question are too many and are shared by too many societies to lend much plausibility to the hypothesis of chance occurrence.

An alternative explanation makes appeal to diffusion and imitation. This hypothesis is a little more attractive than the first because culture, like people, does migrate. Again, however, it is hard to imagine why there

---

* For an excellent recent review of studies on human behavior and genetics, see Henderson (1982).

should have been so much diffusion and imitation across so many cultures, or along so many routes, without predispositions to behavior acting along those same channels. The fact is that cultural imitation is selective. We do not borrow everything. Why then do we borrow what we do borrow? Furthermore, major cultural similarities exist, or have existed, between societies—for example, between pre-Columbian American societies and European societies—in the likely absence of any contact at all, direct or indirect, for many thousands of years if not forever.

The third option, the best I believe, accepts, at least tentatively, the hypothesis of homology of common genetic origin. The acceptance of homology entails, as we shall see later in this chapter, a conception of human learning that is preferential, or that develops along certain largely obligatory routes. That is, cultures and socialization develop along certain channels, the universals, under the partial influence of predispositions that on the whole obey the maximization principle.

Interestingly, the universals-approach to a theory of human nature predates genetic science. With some oversimplification, the emerging theory may be viewed in large part as a by-product of the cultural relativism that was spawned by the intercontinental contact of peoples and the accompanying rise of sociology and anthropology during the last three centuries. Keen observers of the human scene were sometimes awe-struck by the fact that underneath the multitude of distinctive beliefs and practices representing the many peoples and cultures of the world there were striking similarities as well. It is this insight, perhaps more than anything else, that in early social science provided the context for the emergence of an evolutionary orientation.

Two questions were preeminent in the minds of early sociocultural evolutionists: What transformations were undergone in the course of time by presently known institutions—what had social institutions been like in primeval time? Why did human institutions develop in the first place? I shall briefly discuss the first query first.

What shall be done by one interested in the sociocultural conditions of times long past in the absence of archeological and other direct data? One answer—and it was fairly general—was the "comparative method." This method typically entailed the postulation of the principle of "contemporary ancestors" on the presupposition that the sociocultural forms of present time had differential degrees of resemblance to extinct societies and cultures. The following statement by A. Lane-Fox Pitt-Rivers (1906: 53) fairly expresses this position:

> the existing races, in their respective stages of progression, may be taken as the bona fide representatives of the races of antiquity. . . . They thus afford us living illustrations of the social customs, the forms of government, laws, and warlike practices, which belong to the ancient races

from which they remotely sprang, whose implements, resembling with but little difference, their own, are now found low down in the soil. . . .

By extension, this approach also tended to assume that presently living preagrarian societies were representative of the earlier evolutionary stages of European peoples.

The idea, however, that one contemporary sociocultural system corresponds to an earlier stage of another contemporary system is largely preposterous. It assumes "unilinear" or "unitary evolution," namely, development along a continuous line that is obligatory for all peoples. The probability, rather, is that living primitive societies did not resemble modern societies at an earlier stage for the reason, if none other, that the ancestors of present primitive societies were lacking in those biological and cultural qualities which resulted in modern civilization. In short, the differences existing between modern societies and living primitive societies were probably an indication of the differences existing between the direct ancestors of modern societies and those of other societies living in their same time. What were those differences? Eventually some scholars came up with an answer that is in effect consonant with the principles of population genetics: there were differences, however minor, in genotype or biological endowment as well as in cultural development.

The notion of contemporary ancestors was controversial from the very beginning (e.g. Morgan, 1877). Lowie (1937: 25) echoed many of his predecessors in arguing that "even the simplest recent group has a prolonged past, during which it has progressed very far indeed from the hypothetical stage" (but see Harris, 1968: 154–6, for an attempt at a partial salvaging of the notion).

The more persistent question was the second: Why did human institutions develop? In the absence of an adaptationist or functional approach, the question gradually petered out to a quest for "origins," and thus it faded into the first question. The method pursued here, however, was pure speculation, and the rationalistic fallacy had a field day. The search for origins entailed a variety of errors. A case in point is the famous theory of "animism." Spencer (1897: Vol. I), for example, assumed that religion was the result of primitive reasoning about such phenomena as dreams, trances, and death. Thus, dreams allegedly suggested to primitive people that they had doubles which were capable of separating themselves from the body. Like syncope and catalepsy, therefore, death must have seemed only temporary. Or if it was permanent, it must have been because the double was kept too long away from the body. Furthermore, according to Spencer, "belief in re-animation implies belief in a subsequent life." Hence arose the idea of mountain-tops or heavens populated by the departed souls, or doubles, of human beings. On the basis of such speculations, a fantastic theory of religion was woven together which, Spencer's evolutionism notwithstanding, attributed a most complex problem of evolutionary emergence to the human power of reasoning.

The fundamental problem with the old quest for origins is that it ends by explaining the known with the unknown. But the strategy of science is the very antithesis: it is the less well known that must be inferred from the better known. And what is the better known? In the religious sphere, it is the religion of today (Durkheim, 1912). More precisely, it is those behavioral forces that seem to underlie all basic religious behavior. We shall return to this notion in Chapter 8 in connection with a discussion of Durkheim's theory of religion.

Origin explanations often blended with another serious error, namely, the attempt to derive evolutionary explanations by interpreting historical facts as deviations from an original type or as constituting components of a definite rectilinear series with a limit. Explaining phenomena as deviations from a progenitive type requires a complex and probably impossible task. Even if we were able to empirically identify a first term, its inclusion in a rectilinear series would rarely be justifiable. The fact that A precedes B in time in no way provides grounds for concluding that A is the origin of B. Further, noting that B follows A and that G follows B does not allow one to insert, without evidence, phenomena C, D, E, F so that a series is completed, and thus, G "explained." In constructing sequences of this nature, early evolutionists fell prey to the "temptation to ask how things ought to have gone rather than how they actually went" (Pareto, 1916: 340–6).

Unitary theories of evolution often resulted from nearsighted conceptions engendered by excessive concern with the present. Thus, Spencer developed a theory on the basis of an assumed limit, "industrial peace," toward which societies and institutions were supposed to be progressing. The problem is that even when such theories could identify both an end point and a beginning; and even when a few successive stages could be observed, it simply would not do to draw a line through these few points and take this line as the path, and explanation, of evolution. Unless the equation for such a line could be determined and asserted as a testable proposition, nothing more than a sentimental hope could probably be expressed. Thus, if one was a pacifist under the spell of industrialism, as Spencer was, the facts obligingly showed that human society tended toward industrial peace.

The history of sociocultural forms neither moves in unilinear fashion nor constitutes a unitary continuum of deviations from a single type. Sociocultural evolution exhibits a complex divergence and redivergence of social forms. This conception of the pattern of sociocultural evolution is congruent with what social scientists now term multilinear or multi-branched evolution, and is by analogy consistent with modern biological conceptions of speciation and adaptive radiation (Harris, 1968).

All this, however, does not mean that all along the adaptive radiation of sociocultural forms there may not also be at work certain behavioral predispositions that are universal in time and place. It is these that would, then, account for major sociocultural similarities, or for the recurrence of sociocultural themes, in human society. One of the first scholars to forcefully express this notion was Sir J. G. Frazer in his controversial but still compelling study of comparative religion. The primary aim of *The Golden Bough* (Frazer, 1890) was to explain as a single class of facts such phenomena as the sacrificial doctrines of Christianity, the story of Incarnation and the Virgin Birth, the doctrine of the Resurrection, various forms of totemic worship, the beliefs and practices of the Greek pantheon, the sacrifices on the ancient altars of the Israelites, and of course the priesthood of Diana's shrine at Nemi, Italy, the starting point of Frazer's fascinating story. Here is his startling rationale:

> recent researchers into the early history of man have revealed the essential similarity with which, under many superficial differences, the human mind has elaborated its first crude philosophy of life. Accordingly, if we can shew that a barbarous custom, like that of the priesthood of Nemi, has existed elsewhere; if we can detect the motives which led to its institution; if we can prove that these motives have operated widely, perhaps universally, in human society, producing in varied circumstances a variety of institutions specifically different but generically alike; if we can shew, lastly, that these very motives, with some of their derivative institutions, were actually at work in classical antiquity; then we may fairly infer that at a remoter age the same motives gave birth to the priesthood of Nemi.

Frazer did not always treat his concept of "motive" unambiguously, but a careful examination of his reasoning leaves little doubt that his was a study of "independent invention" predicted on the assumption of the basic "psychic unity" of humankind. I understand psychic unity to refer not to what Lumsden and Wilson (1981: 2) call a "single promethean genotype" but to an indefinite number of behavioral predispositions that are shared across human society, and underlie various classes of cross-culturally shared behaviors, of which Frazer's particular focus may be one.

There are many illustrations of such universals, though some of them do not appear to be important enough to recommend inclusion, at this time, of the corresponding behavioral predispositions in a theory of human nature. Take, for example, a custom described by Petronius for Marseilles. When afflicted by the plague, Marsilians would choose a volunteer beggar and support him in the greatest luxury at public expense for a year. At the end of the period, he was clad in sacred vestments, decked in flowers, paraded through the streets of the city where he was hailed with imprecations that all the city's woes might fall upon him, and finally he was thrown into the sea.

The Aztecs in faraway and isolated Mexico had a similar practice (Sahagún, 1530). A young man among their prisoners was chosen for sacrifice a year in advance. In the meantime he was worshipped and showered with all sorts of luxuries and enjoyments. The Roman Saturnalia had a feature whereby slaves became for a time masters of the house; the king of the Saturnalia was sometimes some poor devil who for thirty days was pampered and treated as if he were Saturn; on the thirtieth day he was obliged to kill himself on Saturn's altar. In the *Arabian Nights* there is a story about a poor wretch who enjoys all the delights of king for one day and is severely beaten the next. The pages of social history and ethnography betoken a great variety of similar practices. The possibility cannot, therefore, be discounted that the same behavioral force underlies what appear to be variants of a single theme and that, as a sociologist put it, the various rites "are like the points A, B, C, D . . . on branches shooting off from a common course" (Pareto, 1916: 737).

Again, what is the alternative? The most tempting one, as already noted, concerns diffusion. At one time, diffusionist theories were popular, and proposed common geographic origins for cross-cultural similarities. A number of scholars (e.g. W. H. R. Rivers, 1911; W. J. Perry, 1923; G. E. Smith, 1928) were especially active in defending the thesis that cultural resemblances across societies were due mostly not to independent invention but to diffusion (Harris, 1968: ch. 14). Some (e.g. Smith, 1928) went as far as to argue that practically the entire cultural inventory of human society developed in, and diffused from, ancient Egypt. It would be rash to deny the importance of diffusion, but it is equally erroneous to try at all costs to generalize a fact that may be true only in the particular case. Besides, as far as we know, there had never been any contact, direct or indirect, between the ancient Marsilians and Romans, on the one hand, and the Aztecs, on the other. And if there ever was, it took place long before the civilizational development that the common cultural trait in question would seem to entail.

2.3 Universals and Variants

We need not place much emphasis here on universals. The discussion of methods of arriving at them has served our purposes for the time being; the rest of the volume will continue to focus on them as a way of inferring the possible operation of underlying behavioral predispositions. The basic idea behind universals is that beneath the forest of variants that point to the peculiarities of cultures across time and place there are certain major themes that suggest common cultural responses to the conditions of sociocultural evolution. The idea is well established in social science.

Among the best known students of the subject were the sociologist Vilfredo Pareto (1916) and the anthropologist G. P. Murdock (1945, 1949, 1957, 1967). Both scholars devoted much of their career to the search for species-wide cultural similarities or universals. We have already noted some. Others include age-grading, sports, community organization, cosmology, divination, ethics, food taboos, hospitality, kin groups, law, magic, propitiation of deities, ritual, sexual restrictions, status differentiation, weather control, defense of and/or attachment to territory, extreme curiosity or explorative behavior, symbolization, reification and personification, dominance–subordination, self-sacrifice for others, and vengeance, among others. Some of these will fit neatly within the context of the present volume. Others—for example, ethics and weather control—seem either too broad or too narrow to be reconcilable with my purposes.

The search for uniformities and the accompanying hypothesis of underlying biological tendencies have not gone unchallenged. As already noted, Geertz (1966, 1973), for example, has argued for a focus on variants rather than universals (see in part also Harris, 1979). But the skepticism, which is quite old, has also engendered powerful rebuttals. Lévi-Strauss (1950: xlii–xliii), for example, ably defended Emile Durkheim (1912) and Marcel Mauss (1950a, 1950b) against critics of their tendency to generalize across cultures. Thus, "conceptions of the mana type e.g., *wakan, orenda* [a preeminent power pervading all things] are so common throughout human society that we must unavoidably ask whether we have not come upon a universal form of thought that . . . is a function of a certain state of mind in a given situation and necessarily appears each time that situation is present."

Other scholars have helpfully discussed the value of Pareto's (1916) theory of "residues," his term for cultural universals (e.g. Homans and Curtis, 1934; Henderson, 1935; Parsons, 1937; Busino, 1967). More recently, Lumsden and Wilson (1981: 57–8, also 350–4) have cogently argued against the focus on the explanation of differences rather than similarities between cultures because such an approach never results in much more than surface description and correlation. Indeed, variants cannot be fully understood apart from the uniformities that underlie them. Thus, as these authors rightly note, an orbiting satellite, an airplane, and a falling meteor have three quite different trajectories. But all three are also subject to Newton's laws of motion. Accordingly, it is important to know something that is unique to each of those three phenomena, but it would be a pity to fail to subsume them all under a common, broader explanation.

The variants are the most variable elements of culture—or, what is the same thing, the most abstract and ephemeral manifestations of behavioral predispositions. By and large, variants have been taken seriously as basic units of sociological explanation more by the social scientists of recent decades than by the great scholars who opened the sociological frontier. Durkheim (1912: 202; also 1893: *passim*), for example, warned strenuously against superficial observations and "ingenious fabrications of the mind." His (1912) volume on religion, regarded by many as his greatest accomplishment, is a creative attempt to get at the fundamental uniformities of ritual, the sacred, the sanctification of the community known as the church, the conception of the transcendental or sacred principle, and of course the reification or objectification of the sacred principle through its equation with the sacred community by way of its totemic symbolization. In order to get at the substance of religion, at its "elementary forms," he had to choose a religion where, by virtue of its alleged simplicity, he could more easily "go underneath the symbol to the reality which it represents and which gives it its meaning" (Durkheim, 1912: 14).

The variants, including the sophistries or ingenious fabrications of the mind, are then the material that must be peeled away (though not thrown away) in order to get at the underlying substance. It is only to be decried that, while Durkheim was diligently engaged in this work of exposure, and while here and there he had no qualms about speaking of "instinctive sentiments" and arguing that the "most barbarous and the most fantastic rites and the strangest myths translate some human need" (Durkheim, 1912: 14), he was also inextricably entangled in a metaphysical, highly indefinite (238, 467–79) conception of "society," apparent *causa causarum* of all his sociology. He thus failed to systematically discuss the possibility that universals, units akin to his "elementary forms," might be anchored in ancient evolutionary forces that, in turn, give meaning to the universals and, as he put it just above, to the symbols which represent them.

It is apparent by now that, whatever else variants may be, they are sometimes products of the mind through which we are deceived by appearances, deceive ourselves, and deceive one another. We are with Hans Vaihinger's (1911) philosophy of "as if"—*als ob*. In part, *Homo sapiens* has evolved to live by fictions, to act *as if* illusion were Reality, *as if* a fleeting idea were the foundation of a whole Philosophy, *as if* preparing for war were an act of Peace, *as if* he had a Free Will, *as if* he were eternal by divine destiny, and so forth.

In sociology, few if any have matched the elegance and perspicacity of Vilfredo Pareto (1916) and Erving Goffman (e.g. 1959, 1961, 1969) in the study of deception. Much of human interaction is an intense exercise in appearance, pretense, deceit. Moreover, like actors on a stage, we are constantly engaged in constructing, reconstructing, and presenting our preferred images in our encounters with one another. In Goffman's dramaturgical model, the human animal changes roles or masks as often as he shifts position on a hard bench. When the work of a mind so constituted comes in a flood of words, clichés, and artifices, the subject matter that pertains to the domain of social *science* must be excavated as fossils are dug out by archeologists from under the material strata deposited by the inexorable forces of geological time. Thus, to the extent that variants are tools of deception, the emphasis on human nature and sociocultural uniformities represents a stubborn search for truth lying behind appearances.

At a structural level, variants may be a by-product of emergent features like hypertrophy, which may be a result of "the multiplier effect" (Spencer, 1857b; Wilson, 1975b). The growth of a phenomenon beyond its basic elements very often achieves enormously complex formations. Wilson (1978: 89, 218) defines hypertrophy as the "extreme development of a preexisting structure" and views it as the key to the emergence of civilization, arguing that most sorts of sociocultural behavior represent hypertrophic stages of simpler forms more directly associated with Darwinian adaptation. Religion affords a good example of hypertrophy. For example, the beginning of Christianity may be said with some oversimplification to be represented by "the humble Nazarene," his twelve disciples, and a series of sermons and parables. Since then, "the Church" has become many churches characterized by enormously complex organizations and endless theological variations and disputes.

Those marvelous masqueraders, the sophistries, probably have great adaptive value in a number of ways. For example, self-interest so often breaks through with such a force that the social order may be properly viewed as being ever in a precarious state of equilibrium. There are, as we shall see, various forces of sociality and persistence in human nature that help to account for the endurance of social orders. To some extent, however, social order on the large scale—for example, in modern societies—is ironically made possible by ruse, chicanery, fraud, stratagem, pretense, deception. One of the fundamental functions of variants, especially of the sophistic variety, is to clutter the mind, becloud the thrust of pure selfishness, and facilitate compromise, contracts, exchanges, mutual obligations beyond the narrow family group (Pareto, 1916: Vols III–IV, *passim*). In short, deception works at the service of the maximization principle by facilitating reciprocal helpful behavior, a subject that will be treated in Chapters 5 and 6.

An almost identical argument has been presented quite independently by E. O. Wilson in his theory of altruism. This scholar (1978: ch. 7) distinguishes two basic forms of cooperative behavior, termed "hard-core altruism" and "soft-core altruism," respectively. The first is mostly the result of natural selection at the level of the kin group and does not involve the expectation of reciprocation. The soft-core variety is found in the larger society. It entails the expectation of reciprocation and a number of behavioral strategies, including lying, pretense, and deceit, of self as well as of others. Hard-core cooperative behavior is "the enemy of civilization." It tends heavily toward nepotism, parochialism, extreme ethnocentrism, racism, blood revenge, feuds, vendettas. Soft-core cooperative behavior, on the other hand, is the key to human civilization. "In summary, soft-core altruism is characterized by strong emotion and protean allegiance. Human beings are consistent in their codes of honor but endlessly fickle with reference to whom the codes apply. The genius of human sociality is in fact the ease with which alliances are formed, broken, and reconstituted, always with strong emotional appeals to rules believed to be absolute." One must be careful not to read cynicism into such an argument. What we are dealing with here is an irony, a paradox. The more committed we are to a social order by unbridled personal interest, the narrower the circle of the social order. Deception and related traits seem to restrain, or at least restructure, self-interest enough to permit a broadening of social horizons. Thus, the variants are of capital importance for the social life, and for evolutionary theory.

In an etiological key, we have encountered three levels of causality: (1) the genetically coded predispositions, which are the ancient judges of last resort; (2) the uniformities or universals, which are their most direct and stable sociocultural manifestations; and (3) the variants, which are the less direct manifestations of the predispositions and may be thought of in part as accretions and embellishments of the universals. Of the three levels, the variants are also the most variable in time and place, and often do work of camouflage much as the forest, to use an old metaphor, tends to hide the trees.

The causal relationships existing between these three levels of analysis have essentially been noted in Figure 2.1. What we have done since that context is to subsume the factor B, previously termed a belief, under the broader term variant, which comprises a class of items under such labels as beliefs, opinions, symbols (including artifacts), ephemeral behaviors, mentifacts, particular procedures, and so forth. What in Figure 2.1 was termed the element A, or action, has now been replaced by the concept of universal. Furthermore, whereas in the previous context "A" referred conveniently to what may be called a simple unit act, the universal is much broader in scope, being viewed as a major component, or pattern, of the institutional framework. The element C of Figure 2.1 remains unchanged and now as then refers to behavioral predispositions.

From another perspective, following language introduced recently by Lumsden and Wilson (1981), we may speak of three strata of culturgenetic functioning. At the deepest level, the predisposition is a behavioral necessity or an organizational property of the genotype. It flowers into a major premise or channel of behavioral development and sociocultural formation, the universal, which constitutes the second level. But the human mind is satisfied neither with basic premises of behavior nor with implementation of such premises. It must also have an explanation of them and a rich repertory of means and practices with which to put them to use.

That tendency produces variants, which in turn help to organize complex systems of morality, philosophy, theology, metaphysics, and, in their most logical form, science itself, of course.

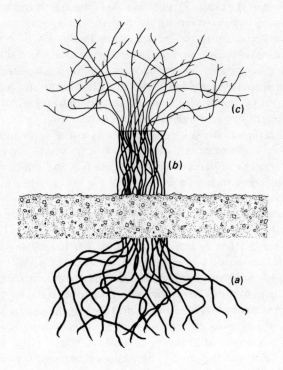

Figure 2.2 *Evolutionary relationships between behavioral predispositions, universals, and variants.*

Key:
(*a*) the hidden behavioral predispositions; (*b*) the universals; (*c*) the variants, conceived in various degrees of variability.

Figure 2.2 depicts these relationships. The Figure takes the form of a tree. Conceptualizing it as a bush, such as a European hazel tree (yielding the filbert nut), or certain species of the fig tree, is better still. The area marked (*a*), represented as being underground, and thus invisible, refers to

62   Human Nature and Biocultural Evolution

the behavioral predispositions. Some of these probably predate the onset of culture and, thus, are likely to be shared, to a degree, with our primate ancestors, from which sprang both our species and the living primate species like the chimpanzee and the gorilla. In short, they are the pure results of natural selection. Other predispositions are probably more recent, and have been fashioned under the influence of both natural selection and cultural pressures.

The bottom of the bush, (b), portrays the cultural universals. In addition to representing the basic sociocultural patterns that are common to human societies, they constitute the broad channels along which learning or socialization takes place in any given generation. Socialization may consequently be viewed largely as a process through which the mind develops the capacity to yield behavior that fits these broad channels under the pressure of the predispositions and according to rules and techniques that are inherent in the universals themselves.

The (c) area directs attention to the variants, namely, the multiplicity of variations that develop around the themes (the universals) across time and place. The concept of variants requires a clarification which is especially critical for it but which to a lesser degree is relevant to universals and predispositions as well.

The basic point is this. To come closer to the fact, further distinctions would be useful within each of the isolated levels of analysis. The rationale for this statement lies in the obvious fact that not all predispositions, for example, are equally generative of behavior or equally efficient in honoring what will later be treated as the master criterion of biocultural evolution, namely, the maximization principle. But the problem of internal differentiation is especially critical with respect to variants. Surely, there is a difference worthy of theoretical note between the baptismal practice, for example, of immersing a child in water, rather than merely wetting its head, and the custom of asking one type of person instead of another to be the godfather for the occasion. The former variants may be of little or no sociocultural or biological relevance; they may merely represent one of the outermost and most superficial consequences of theological divergence between two religious groups. One of the latter, on the other hand, may refer to a significant adaptation by strengthening the bridge between religious ritual and family relations, or by promoting relations of reciprocity between one family and another. The latter type of variant, therefore, may be more durable and less culturally specific. Accordingly, Figure 2.2 depicts variants as occurring in various degrees of magnitude. The larger branches refer to the more complex and less variable variants. The twigs—and we could add the leaves—refer to less complex and more variable ones.

I should now add that, important as such distinctions unquestionably are for a continuing program of theoretical development, they will necessarily be disregarded in a first approximation such as the present study. My own emphasis will be on the fundamental, unrefined, relationships existing between behavioral predispositions, universals, and variants. The example of baptism, mentioned just above, may help to clarify those relationships. Some form of baptism may be found in many if not all human societies. This recurrent phenomenon, when viewed in its basic expression, is the universal. It is, in turn, based on "the feeling, the sentiment, that the integrity of an individual which has been altered by certain causes, real or imaginary, can be restored by certain rites." This is the predisposition. In addition to provoking acts of self-purification (baptism), through diverse procedures and substances, this predisposition and its direct manifestation, the universal, lead to the proliferation of various "theories" which allege to explain and bring about the restoration of individual integrity (Pareto, 1916). This multiplicity of theories, along with the multitude of procedures and substances employed in the practice of baptism, refers to the variants.

As previously noted, there is mutual influence among the three levels of analysis; we shall return to this problem in more detail in the next chapter. Here, suffice it to say that the need for self-purification gives rise to actual acts of purification and all manner of associated practices and "theories." But the implementation of the need and the various formulas employed, in turn, function to maintain the vitality of the underlying predisposition. Predispositions, universals, and variants do not, however, have equal etiological import. That may be seen in the fact that their degree of variability in time and space is not the same. Predispositions vary the least, while variants change the most. Put otherwise, the rate of cultural change is more rapid than its genetic counterpart. That means that predictions made by reference to the workings of the latter are more durable, even if less rich, than those made by resort to the cultural uniformities. And predictions based on the latter are more reliable than those suggested by the variants.

The issue in question may also be viewed as part of the logic of the system concept. Systems are typically constituted by asymmetric relationships. In their interdependences, the values of system variables do not change uniformly. Causally, the steadfastness of the value of a variable in a system is a good index of its importance. Consequently, "the least variable part of a system is always the best one to examine first; in the complex interaction of all parts, it must appear more frequently as cause and least frequently as an effect" (Lorenz, 1958: 72; see also Wilkie, 1950; Lopreato, 1971).

Before closing this section, a comment is in order about my approach as presented so far and the approach followed in the recent Lumsden–Wilson (1981) volume. As already noted, for these scholars the basic unit of culture is the *culturgen*. By analogy to the basic unit of archeology, the "artifact type" (e.g. Clarke, 1978), a culturgen is defined as "a relatively homogeneous set of artifacts, behaviors, or mentifacts (mental constructs having little or no direct correspondence with reality) that either share without exception one or more attribute states selected for their functional importance or at least share a consistently recurrent range of such attribute states within a given polythetic set" (Lumsden and Wilson, 1981: 27). Note that the emphasis is on the set rather than on the unit. Moreover, the definition makes reference to the polythetic set, a concept from mathematical taxonomy comprising entities that typically have a large number of attributes in common. Accordingly, a culturgen must be viewed as a cluster of cultural facts that are functionally homogeneous along one or more theoretically relevant dimensions. It is, thus, a class concept, and to that extent it is related to the cultural universal, which refers to a class of cultural units that coalesce around a nucleus.

Like us, Lumsden and Wilson (1981: 28–29) recognize that sets or clusters come in different degrees of magnitude and complexity, and that, in part, taxonomic boundaries are necessarily arbitrary. There is continuity and permeability in biocultural systems. To a degree, therefore, classifications impose boundaries where these are not entirely appropriate. On the other hand, systems do have landscapes, as it were, and, however harmonious may be the various features of a landscape, they do, to another degree, imply system heterogeneity and, thus, the utility of subdivisions. In general, then, I understand the Lumsden–Wilson concept of culturgen to correspond to a range that goes roughly from complex variants to what may likewise be considered minor universals.

A comparison of a different sort may be drawn between the Lumsden–Wilson concept of epigenetic rules and my concept of behavioral predispositions. The latter are conceived as genetic programs properly speaking whose function is to guide the development and expression of behavior along particular channels or routes; these are the historically and spatially stable universals along with the variants that attach to, and detach themselves from, them in their relatively fickle course of selection and extinction. Epigenetic rules are basically conceived in like manner. They are "genetically determined procedures" that guide the mind toward the use of certain culturgens as opposed to others (Lumsden and Wilson, 1981: 7, 370). Rooted in genetic structures, they are rules of behavior that develop during the whole process of interaction between genes and environment—"from protein assembly through the complex events of organ construction to learning." Epigenetic rules, therefore, are the developmental reflections of the behavioral predispositions. They say something about how genetic program is translated into behavior. Accordingly, Lumsden and Wilson focus on developmental processes and to the extent possible emphasize properties of the brain and the mind. By contrast, the present volume touches more superficially on epigenetic processes; my stress is on sociocultural data and on the inferences that may be hazarded about the genetically programmed predispositions that may underlie uniformities revealed by the data. On the whole, however, the two volumes are complementary in their pursuits.

2.4 The Problem of Analogy and Homology

There are several critical problems linked to the type of reasoning adopted in this book. I shall conclude the present chapter with a brief discussion of two. Both concern the credibility of the concept of behavioral predisposition and its postulated relationship to cultural universals. One, central to the debate between social science and evolutionary biology, refers to the old nature-versus-nurture question, which will be considered in the next section. The other, the subject of the present section, was previously alluded to, and concerns the concepts of analogy and homology.

Simply stated, the problem is as follows: Assuming that the concept of sociocultural universal is empirically justified, can we attribute the cross-cultural similarities represented by the universals to the operation across sociocultural systems of common behavioral predispositions? More broadly, to the extent that behavioral evolutionists find similarities between *Homo sapiens* and other species, are they justified to entertain the hypothesis of homology, and therefore of any degree of biological determinism in sociocultural behavior?

Let us first define again our basic terms. By general convention, two or more similar features may be said to be homologous if they are determined by common genetic ancestry. Conversely, they may be viewed as analogous if they are the results of chance occurrence or convergent evolution as opposed to common genetic ancestry.

The more controversial question, or in any case the issue that provokes the greatest interdisciplinary acrimony, concerns the hypothesis of interspecific homology. Sociobiologists who have dared venture into interspecific comparisons have of late been at the center of the debate. The Sociobiology Study Group of Science for the People (1976) has with some justification argued, for example, that sociobiologists are eager to establish evolutionary homology, rather than simple analogy, as the basis for human–animal similarities because that would provide direct evidence of the genetic determination of sociocultural behavior. But such evidence is missing. "A behavior that may be genetically coded in a higher primate may be purely learned and widely spread among human cultures as a consequence of the enormous flexibility of our brain" (SSGSP, 1976).

I have no intention of defending the hypothesis of human–nonhuman homology, and in any case it is not at all certain to me that sociobiologists are intent in doing anything beyond what the term hypothesis itself suggests, namely, using a hunch as a way of attempting the discovery of possibly hidden relationships. But I find curious the position of the opposition precisely because it denies the legitimacy of scientific curiosity. The reader will note that, on the one hand, the SSGSP statement asserts what many might consider the obvious, namely, that behavior X may be genetically determined in one species but learned in another; but, on the other, it does not demonstrate that X is not genetically coded in both species. Note, too, the peculiar argument that X may be widely spread among human cultures because of the enormous flexibility of the human brain. On logical grounds alone, one would on the contrary be led to hypothesize that the greater the flexibility of the brain, the lower the likelihood of a shared behavior across human cultures. For, without any genetic direction, cross-cultural similarities would be a result of chance alone (at least in the absence of diffusion), and such chance convergence (or analogy) would be improbable precisely because of high flexibility in the brain. Note, finally, the sharp division made between genetic coding and learning at a time when, as we shall see in the next section, the basis for such a dichotomy has been completely eliminated. Notions like "purely learned" behavior hark back to the *tabula rasa* assumption, which stands to modern behavioral science roughly as the astronomy of Aristotle stands to the astronomy of Newton.

The Sociobiology Study Group continues to note correctly that sociobiologist Wilson (1975b), their favorite target, argues more often from analogy than from homology. Further, they correctly concede that, while analogies cannot help to establish biological determination of sociocultural behavior, they can nevertheless "serve as a plausibility argument for natural selection of human behavior by assuming that natural selection has operated on different genes in . . . two species but has produced convergent responses as independent adaptations to similar environments." Allegedly, however, such a plausibility argument "is not even worth considering unless the similarity is so precise that identical functions cannot be reasonably denied, as in the classic case of evolutionary convergence—the eyes of vertebrates and octopuses." I understand the caution. But the caveat is far too rigid. Basically it comes down to a case of close-mindedness, otherwise known in the present context as a charge of "false analogies." The charge is improper, for false analogies are roughly as real as unicorns. Analogies cannot really be false. Rather, they vary along any number of dimensions, for example, informativeness, detail of comparisons, and heuristic value. The latter criterion is especially noteworthy. It corresponds to the crucial scientific question, "how much can we learn about an unknown phenomenon from the perspective of a known one, however gross the similarities between the two may be?" In short, the SSGSP disinclination is rigid because it implicitly denies legitimacy to the main fuel of scientific discovery, namely, the audacity to suspect systematic relationships even in unlikely areas. I strongly suspect that the future of evolutionary science has in store significant proof in favor of the hypothesis that at least some human–subhuman behavioral similarities are homologous. For the time being, however, we need not worry too much about the problem of interspecific homologies. Establishing plausibility in favor of the operation of natural selection in sociocultural behavior is achievement enough, for that helps to confirm the hypothesis that human universals, if properly isolated and constructed, are founded on a homologous basis. After all, despite genetic differences not only between individuals but also between populations, *Homo sapiens* does constitute a single species, and intergroup genetic differences are simply not so large (e.g. Dobzhansky, 1967; Dobzhansky *et al.*, 1977) as to negate a priori the hypothesis that major cross-cultural patterns are founded on such tight genetic kindredness.

The problem is very complex and its solution, if possible at all, rests in direct genetic evidence of homologously evolved cross-cultural uniformities. Still, we need not be too eager to demonstrate homology. It would be nearly as useful to be agreed beyond a doubt on a given number of analogies or convergent patterns of similarity. For then we could reasonably well estimate the improbability of coincidental or convergent similarities in a broader program of comparisons, and thus establish firmer grounds for the hypothesis of homologous developments. Ironic as it may seem, for any given comparison, the likelihood of random similarity is proportional to the number of independent traits of similarity. For n such traits, the estimated improbability is equal to 2^{n-1}. As Lorenz (1974) points out in a deliberately farfetched example, if we find the property known as "streamline" in a swallow, a dolphin, a shark, and in the man-made airplane or torpedo, we can safely assume that the common streamline is the result of the same need to reduce friction and obtain fast motion.

Does such a common "need" have a common genetic basis, thus pointing to a homology, or does it refer to something else? I think the answer is homology, but for the problem at hand, and very possibly for countless others of like interest, an answer to the question is really not essential. Our explanation is acceptable on logical grounds alone—grounds suggested by the nature of modern science and the role in it of organizing general principles like the maximization principle. When we say that self-sacrifice for others in one's own kin group is universal because it is in principle adaptive in the sense that it can enhance fitness, we may make a statement of cross-cultural uniformity that technically may hint at a homology in action. But from my understanding of science, what matters more is not whether the homology is true or false. What is more important is that we organize into one class of facts a large number of facts that may reasonably be said to be related from the viewpoint of a general theoretical principle—

the maximization principle—that, given present knowledge, does "make sense."

Likewise, I have no interest in establishing homology between, say, human jealousy and jealousy among mallard ducks, which is real and obvious. Indeed, my endeavors would be destined to fail, were I foolish enough to pursue such homology, for humans and ducks are too far removed on the phylogenetic scale. But it is not at all foolish to argue, by resort to the general principle, that in both species jealousy is an adaptation, that is, it is rooted in similar survival, or fitness-enhancing, value. It tends to guard our genetic investment. To conclude, analogies, especially when aided by general principles, are very useful tools of research and theorizing, and their results may often substitute fairly effectively for the confidence engendered by homologies.

2.5 Nature and Nurture

Explanation in social science, of which social psychology, cultural anthropology, and sociology are the prototypes, is typically predicated on two basically environmental assumptions. One postulates the extreme malleability of biology or "nature"; the other, correlatively, emphasizes as legitimate concern the extreme malleability or variability of sociocultural behavior. The two together predispose heavily in favor of the "nurture" hypothesis—we allegedly behave as we do because we are taught to behave that way. But a radical change has for some time been afoot along with the growing rapprochement between biology and psychology, the behavioral science most concerned with learning.

Whatever one's inclinations, the issue at hand is complex. Any attempt to solve it requires that we theoretically bridge the gap between the logic of phylogeny and that of ontogeny—between phylogenetic processes and life cycle ones. Beyond that, there is the lingering problem of certain rather peculiar psychological assumptions made in earlier decades about the nature of learning. Although their influence is diminishing, they have nevertheless had great impact in social science. To an extent, they are summarized in the following statement by Thorndike: "Formally the crab, fish, turtle, dog, cat, monkey and baby have very similar intellects and characters. All are systems of connections subject to change by the laws of exercise and effect" (cited in Bitterman, 1965).

In principle, this emphasis on interspecific similarities is attractive to the student of biocultural science. Its appeal, however, disappears the moment we realize that it implies a *tabula rasa* assumption and the hypothesis that behavior is a result of learning under the influence of the effects—for example, reward or punishment—that react on behavior. Moreover, despite Thorndike's enumeration of different sorts of animals, the statement was based almost exclusively on observation of a single animal, the rat. More recently, the "general-process learning theory," as psychologists sometimes term it, or "radical behaviorism," as it is called in the philosophy of the mind (e.g. Fodor, 1975, 1981; Block, 1980), has given way to richer sorts of psychology. In part, this change has taken place in view of findings by behavioral biologists who have had a much richer repertory of animal observations at their disposal. Their studies support the following statement by ethologist Niko Tinbergen (1951): "The student of innate behavior, accustomed to studying a number of different species and the entire behavior pattern, is repeatedly confronted with the fact that an animal may learn some things much more readily than others. . . . Different species are [furthermore] predisposed to learn different parts of the pattern. So far as we know, these differences between species have adaptive significance." It is unlikely, therefore, that different species have very similar intellects. Differential learning has been fashioned by natural selection under the influence of varying environmental pressures. That means that learning is directed, or *preferential*; that is, there are certain things that we can learn and certain others that we cannot.

This position, in turn, would seem to imply a number of interesting possibilities, not least of which is the likelihood that behavioral predispositions are really "teaching devices," as Lorenz (1966) has noted, which unfold through the life cycle and guide behavior more or less efficiently according to the maximization principle. They teach to choose environments and situations, to define reinforcers positively or negatively, to perceptually isolate stimuli, to react to various types of contingencies, and so forth (e.g. Tinbergen, 1951; Lorenz, 1966; Barash, 1977; Lumsden and Wilson, 1981). The point is perhaps most evident when we observe vestigial types of behavior (or leftovers from the past). Despite the commercialization of innocuous types of mushrooms, for example, many human beings are still wary of eating this foodstuff. Likewise, we have a strong aversion to snakes. Christianity teaches about the latter that it is our natural enemy for having induced us into original sin. That is a splendid example of a sophistry. The fact is that both those organisms were dangerous to us in the wild and we have not yet "forgotten" the fact. The aversion is genetically pre-programmed. Conversely, it is hardly surprising that we are typically nonchalant about many perfectly lethal things around the house—for example, electric outlets—until we happen to experience their danger. They were unknown in the wild, and our biology has not yet adapted to them (Seligman, 1971; Ruse, 1979: 185). It is possible indeed that we shall never adjust genetically to this sort of problem in the sense that learning and culture will greatly reduce the danger. It is interesting to note, however, that for the time being the danger is there and, in the absence of "pre-programmed" behavior, learning seems to be slow and ineffective, at least at early ages.

Today there is a growing recognition by psychologists that learning is a function of the type of problem confronted and of the sort of animal involved as well. In a pioneering experiment, John Garcia and his associates showed that rats become averse to tastes that are followed by radiation sickness, but they develop no aversion to audio-visual stimuli that are linked to the same sickness. The findings are analogous to the comments just above about the human reaction to snakes and electrical outlets; they suggest that rats are genetically constituted to associate sickness with certain foods but not with certain audio-visual stimuli (Garcia et al., 1966; Garcia and Koelling, 1966; see also Tinbergen, 1951; Bandura, 1977; Ruse, 1979; Pulliam and Dunford, 1980).

Inherent in the two environmental assumptions mentioned above is "that old lemon," as McBride (1971) has defined the nature–nurture debate. Social scientists (and not a few biologists as well) have shown a strong inclination to dichotomize behavior into the learned and the innate. In fact, however, a rigid line cannot be drawn between the two dimensions. Predispositions instigate behavior, but the plasticity of the latter varies with the variety of environmental conditions to which human beings can respond. The innate and the learned are not mutually distinct; analysis of one is otiose without analysis of the other. The dichotomy is sometimes justified, perhaps only at a methodological level, when considering the *origin* of a causal influence. Sometimes it is useful to know where "the first move" takes place. But the dichotomy makes no sense in the product, "which must be viewed as the result of accumulated interactions between inheritance, environment, and preceding phenotype" (Brown, 1975: 609). Indeed, ethologists and psychologists have recently concluded that the contrasting of the innate and the learned as totally distinct facts is a fallacy. The behavioral predispositions of a species constitute a potential which is differentially exploited by learning processes as the species and its members come to face changing environmental conditions (e.g. Lorenz, 1965; Hinde, 1970, 1974; Lehrman, 1970; Bandura, 1977; Lumsden and Wilson, 1981).

It follows that predispositions are not exact templates for specific topographies of behavior. Biological adaptation determines not the precise forms that behavior takes but merely certain limits they cannot overpass. From another perspective, behavioral predispositions are blueprints, codes, or instructions that "prescribe the *capacity* to develop a certain array of traits. In some categories of behavior, the array is limited and the outcome can be altered only by strenuous training—if ever. In others, the array is vast and the outcome easily influenced" (Wilson, 1978: 56–7; see also, Tinbergen, 1951; Brown, 1975; Bandura, 1977; Irons, 1979b; Pulliam and Dunford, 1980; Lumsden and Wilson, 1981).

Given the deep concerns of social science, it bears stressing that the crucial point about behavioral forces is that they predispose us to learn and manifest behavior along certain channels rather than others. Typically, the latter, however, have numerous outlets. This circumstance constitutes the leeway that frees us from strict biological determinism; it also helps to give the impression that we have totally free will. In a sense, then, the old question of free will can be reduced to a methodological issue. If we stress the fact that behavior develops along predetermined channels, for example, ritual consumption, we may conclude that we have little if any free will. The scientific mentality is quite at ease in this posture. Conversely, if one focuses on the outlets that radiate from each channel, the conclusion is justified that we have free will to a high degree. A case in point is the constant movement of converts across the various Christian churches, which entails certain definite changes in ritual and belief: for example, the form that communion takes. To a large extent, such changes are indeed a matter of personal decision.* But the fact of being religious is less freely willed.

The entire complex of sociocultural behavior is determined by both biology and environment acting in a state of mutual dependence. Under close examination, it thus transpires that what, even for convenience sake, is termed innate is really a compound in which the experientially unmodified element predominates. What, likewise, is labeled learned is not likely to be entirely devoid of a predispositional component. It follows that, to the extent that we can attribute a causal component to behavioral predispositions, they may be viewed tentatively as evidence that, in the course of evolutionary history, at least partially directed behavior has been more frequently adaptive than totally random behavior. In a sense, then, predispositions may be thought of as a priori hypotheses that make useful bets about the relative appropriateness and efficiency of certain classes of behavioral responses to environmental conditions.

The road to whatever useful grasp we may have today of the relationship between nature and nurture has had to pass through the conceptual labyrinth known as "instinct." There is still no full agreement about the precise meaning of this concept (see, e.g., Tinbergen, 1951; Dobzhansky, 1962; Schneirla, 1972; Richards, 1974). Nevertheless, while modern literature bestows a certain flexibility of interpretation on the term, two major meanings have rather wide diffusion. One merely sensitizes us to the presence of a genetic component in behavior without prejudging its magnitude or permissiveness. The other emphasizes the idea of an unlearned behavior pattern that is not susceptible to modification through

---

* A similar point is made by E. O. Wilson (1978: 77) in the form of a "paradox of determinism and free will." The broader the categories of behavior, the more determined our actions appear. Conversely, the narrower they are, the more justifiable it is to speak of free will. There are other perspectives on the problem, of course. William James (1890) argued that the life of hesitation and choice that conveys the sense of free will may be a result of the great variety of instincts which "block each other's path." Free will, therefore, would seem to be an illusion, the by-product of genetic noise.

experience. The latter is the sense in which most social scientists understand instinct, and no end of confusion arises from that circumstance, for modern biologists reject this outdated meaning.

Behavioral predispositions and behavior together constitute a set of devices for tracking changes in the environment. Responses to environmental change vary, however, because the actions of environment are not entirely predictable. Insofar as the problem of learning and instinct is concerned, there are various types or degrees of adaptation associated with this circumstance. Thus, the above twofold conception of instinct may be elaborated on to distinguish between various types of tracking systems. One, for example, belongs to "the complete instinct-reflex machine," which depends largely or wholly on stimuli to guide it. In the "directed learner," tracking employs learning that is typically related to a narrow range of stimuli. The tracking system of the "generalized learner," on the other hand, reveals an organism who undergoes a prolonged and complex process of socialization, is capable of insight learning, and indeed has a perception of history. That is, in the latter case:

> The organism's knowledge is not limited to particular individuals and places with attractive or aversive associations. It also remembers relationships and incidents through time, and it can engineer improvements in its social status by relatively sophisticated choices of threat, conciliation, and formations of alliance. It seems to be able to project mentally into the future, and in a few, extreme cases deliberate deception is practiced (Wilson, 1975b: 152).

This third type is allegedly characteristic of chimpanzees, baboons, macaques, and some Old World primates and social Canidae as well as ourselves. But it is likely that as observation of animal behavior accumulates, the class of generalized learners will expand.

Thorpe (1956, 1974), in fact, has provided a more extensive treatment of tracking mechanisms and offered analytical distinctions that avoid the tendency to associate directed learning with species low on the phylogenetic scale, and generalized learning with the "higher" species. He cites numerous studies demonstrating that the two major learning processes characteristic of the generalized learner (insight learning and latent learning) are, at close examination, widespread among the "lower" species.

Unfortunately, due to the eagerness to establish the uniqueness of human behavior, social scientists have largely neglected the terrific amount of learned behavior that is observable in the animal kingdom. In this, our attitudes are similar in all but style to those attitudes of a time not so remote which relegated "primitives" to the status of "heathens" and "animals endowed with a language." Consider white-crowned sparrows, a species phylogenetically far removed from us. Research shows that in addition to the species-specific skeleton of their song, these birds are capable of learning many regional dialects. In fact, even the skeleton requires some elements of learning, in the sense that it does not develop entirely without imitation and guidance (Marler and Tamura, 1964; Konishi, 1965). Again, Menzel and Erber (1978) have shown that bees learn quickly and remember for long periods the color and the fragrance of flowers that contain nectar or pollen. Analogous evidence is truly voluminous (see Thorpe, 1963, 1974, for useful bibliography).

The learning theorist Albert Bandura has argued that higher animals at least are highly imitative organisms. Thus, merely by watching they can acquire complicated sequences of responses even though they do not perform them until some time after the original observations (Bandura, 1977). Chimps reared in human families, for example, imitate typing, the use of mirrors, the application of lipstick to the face, and the opening of cans with sharp implements, among other activities (see, e.g., Hayes and Hayes, 1952).

Such behavior is expressed without tutoring. Under the pressure of tutoring, animal learning can reach almost sensational stages. A case in point is provided by the work of Gardner and Gardner (1971) who have shown that chimpanzees can be trained to feature the rudiments of syntactical construction. Using American Sign Language (ASL), these scholars were able to teach a young chimp, Washoe, about 300 two-sign combinations, an achievement comparable to the earliest two-word combinations of children. Premack (1970) allegedly had even better results with his chimp, Sarah, by teaching language with plastic words. As a result, he concluded that chimpanzees understand some symmetrical and hierarchical sentence structure and, therefore, are "competent to some degree in the sentence function of language." The implications of this research are summarized by Thorpe (1974: 301) as follows:

> although no animal appears to have a language which is propositional, fully syntactic, and at the same time clearly expressive of intention, all these features can be found separately (to at least some degree) in the animal kingdom. Consequently, bearing in mind the work on chimps . . . we can say that the distinction between men and animals, on the ground that only the former possess "true language," seems far less defensible than heretofore. . . . the difference between the mind of animals and men seems to be one of degree—often the degree of abstraction that can be achieved—rather than one of kind.

As might be expected from the novelty and uniqueness of the problem, these claims have not gone unchallenged. Basically, the criticisms concern the question of the very meaning of learning and language. Do chimps and related primates merely mimic their human interlocutors? Do they really have any comprehension of grammar and syntax? Are they capable of spontaneous conversation? Indeed, is their alleged linguistic ability merely a case of the Clever Hans effect, namely, the ability to pick up human cues as to when, how, and what to "speak" (e.g. Terrace, 1979; Sebeok and Umiker-Sebeok, 1980)? These are sobering questions, but common sense and the responses of the earlier researchers to these criticisms suggest that while the latter may have claimed too much to begin with, their critics may be demanding even more by accepting as language acquisition among apes only what is purely human in nature.

The built-in capacity for learning may be specified further (Dawkins, 1976). Genes cannot, strictly speaking, control behavior in a direct sense. Their direct action consists of protein synthesis. Ultimately, this is a powerful way of manipulating behavior, but it is an awfully slow technique. It cannot cope directly with the fact that successful behavior often requires responses to events that take place in split-second periods. There is nothing in human genetics, for example, that counsels us specifically against people pointing a gun at us. But there is probably room in it for an instruction that in effect says: beware especially of people who aim what may be a missile at you. And the nervous system springs to action in milliseconds. The environment, however, is excessively rich. It contains unpredictable stimuli. It follows that the more varied the life conditions of a species, the lower the specificity of genetic programs and the more general their strategies. Human beings can make a go of it in every corner of the earth. That suggests highly general genetic predispositions toward behavior. It may safely be said, furthermore, that such genetic program is both cause of worldwide adaptation, in the sense that it made it possible in the first place, and effect of it, because the multiplicity of challenges associated with the universality of our ecological niche has probably reinforced the flexibility of our genetic program.

In sum, instinct and learning, nature and nurture, the innate and the learned—in short, biology and culture—are inextricably interwoven together. As a minimum, it is necessary to consider that learning is itself a biological fact—that a capacity for learning is part of the logic of evolution. In this sense, learning is now the companion, now the handmaid of instinct, never its alternative or opponent. Learning, that is, is a mechanism evolved by the species as a tool for dealing with a complex, changing, and often unpredictable environment.

The adaptive value of learning naturally varies as we go from one species to another. That is not to say that the greater the capacity for learning, the greater the chances of survival, either at the organism or at the population level. In some species, it apparently pays to be "stupid." Thus, the "perennials" (for example, the oyster and the shark) do not seem to be particularly rich in learning ability. In other species, learning may be more adaptive. Those individuals who are more adept at deriving lessons from their environment have a better chance of survival, and thus of at least partial genetic self-perpetuation, than those less proficient in the art. Likewise, those individuals who can draw inferences, predict, and simulate may achieve greater fitness than those who learn by a simple and often wasteful trial-and-error process.

In any case, those scholars who would put all their theoretical eggs in the learning basket must come to terms with a simple fact. Human learning is certainly not random. We do not learn just anything. We do not, for example, learn to woo our mates in the manner characteristic of ducks. Indeed, we would be unable to behave like the average chimpanzee, our closest relative, even if we put all our science, genius, and stratagems to the task. The truth is that our learning is to an extent structured. Like systems that pursue preferred goal states, for example, guided missiles, we are programmed to move along certain channels and not others. Tiger and Fox (1971: 20) may be stretching the point only slightly when they state: "we behave culturally because it is in our nature to behave culturally, because natural selection has produced an animal that has to behave culturally, that has to invent rules, make myths, speak languages, and form men's [and women's] clubs, in the same way that the hamadryas baboon has to form harems, adopt infants, and bite its wives on the neck."

It bears stressing that by our own accounting, the extreme weight placed on life cycle processes has led to a sterile posture. Consider the manner in which the old Hobbesian "problem of order," as it has been termed (e.g. Parsons, 1937), is solved in cultural science. How is social order possible? In a very influential paper, Dennis Wrong (1961) isolates two major answers given by contemporary sociologists to the query. One states, in a nutshell, that the individual becomes tractable to the existing social order through the search for social approval. At least implicitly, this answer postulates a fundamental, possibly biological, need associated with social behavior. To that extent, it may take us a little closer to a more nearly complete theory of social order, a phenomenon that in its substance, if not in its details, is far older than we have imagined. The answer, however, is startling in its excessive simplicity, given a discipline that claims, rightly, to be one of the most complex ever devised by the human mind. The focus on social approval recurs with devilish regularity, and within limits it is put to very good use—as, for example, in the exchange theories of George Homans (1961, 1974) and Peter Blau (1964). It will be useful to us, too, especially in Chapter 6.

The second, and more frequent, answer typically given to the Hobbesian question concerns socialization and amounts to saying that human beings grow to internalize the norms of their society. The answer is disturbingly sterile. It merely informs that the human animal is not born socialized but is capable of being socialized. It says nothing about the content and the origin of the norms themselves. It is also surprising because it reveals a fickleness and a confusion that at least two centuries of sociological theorizing should have effaced. When years ago George Homans (1961, 1964) dared suggest that the "general propositions" of sociology are fundamentally psychological in nature, a loud cry of indignation arose in sociology that was intended to obliterate all of Homans's temerarious "psychological reductionism." When later he was inducted into the National Academy of Sciencies, a venerable member of the discipline saw fit to protest in one of our popular organs that it was certainly not as a sociologist that Homans had earned his accolade. Yet, what if not psychological is the socialization solution to what is presumably the fundamental sociological problem? As suggested earlier, it is clearly based on process-learning theory.

The theoretical problem at hand requires an answer that takes into account the why, the when, and the what as well as the how of our learning. The socialization answer gives the impression of addressing the why, but in fact it is at most directed at the how, because the why cannot be understood without consideration of the what and the when. What we learn and when we learn it in the life cycle concern crucial theoretical problems, and give powerful clues for answering the why as well.

The when of the problem is particularly instructive, and from this perspective it may be said that, ironically, it is precisely the concept of socialization that, when dynamically understood, underscores the importance of the biological substratum. At bottom, socialization refers to the process of developing and putting in ordered action certain behavioral predispositions that grant the organism the power of receiving, interpreting, digesting, and putting to use, sometimes through modification, pre-existing symbolic systems whose existence is an historical as well as an ontogenetic fact.

Few perhaps have clarified the process of socialization as well as Jean Piaget, a scholar who considered himself neither a psychologist nor an educator but a genetic epistemologist and a "man in the middle" on the nature–nurture debate. Piaget showed that the mind follows predictable developmental stages in which the individual creatively articulates the interaction of genetic heredity with the environment to elaborate concepts of meaning and causality, quantity, quality, space, time, class, play, and so forth. During the first two years of life, or the first stage of mental growth, the child's learning activity is primarily focused on physical objects. During the next stage, lasting four or five years, the symbolic world comes into focus through such activities as speech, dreams, and fantasy. In the third stage, up to about age 12, the child masters numbers and other abstractions. Finally, in the next three years, the budding adult learns to tackle purely logical thought and becomes adept at introspection in a fashion reminiscent of G. H. Mead's (1934) concepts of "me" and "generalized other"; that is, the individual behaves at least in part according to an attitude that has been shaped by significant others. Now he can also grasp symbols in highly compound form, as in irony, metaphors, and aphorisms (e.g. Piaget, 1970, 1971, 1976).

One thing seems to be clear from these findings. Learning does not take place in a vacuum or on a blank slate. The organism learns at a pace and in a sequence that are prenatally programmed. In short, as Burton (1972) has noted, socialization works only when the organism is receptive to, or ready for, environmental stimuli. All this is to say that leaving the explanation of sociocultural behavior at the indefinite level of "socialization" or "internalization of norms" is a bit like saying, for example, that we have governing classes because certain individuals learn to be political leaders. It says too little, and it is mostly false in the first place. We have governing classes for a variety of ancient reasons that have little to do with learning. They include the tendency to exploit one another's resources.

Likewise, it is not exactly true that we become tractable to social order through socialization. The proof lies in the terrific amount of conflict existing at all times between children and their archetypical socializers, the parents. Contrary to widespread, if tacit, assumption, the interests of parents and children do not always coincide. The point has been shown with extreme cogency by R. L. Trivers (1974) within a sociobiological context. For example, food scarcity and an excessive number of offspring (among many other factors, of course) sometimes lead a parent to favor the survival of one child over another. Parental manipulation is a fact, unpleasant as it may be to even mention it. The response of the underprivileged child, to the extent that it has one, can hardly be thought to be purely a result of socialization: few people would rather die than live. Again, according to Mendel's First Law, I share 50 percent of my genes with my mother, and on the average half that many with her siblings. She shares 50 percent of her genes with each of us. Therefore, according to the logic of natural selection, my being "nice to Auntie" is normally more beneficial to my mother than it is to me. Trivers logically points out that parents' behavior tends to be at odds with the "behavioral tendencies of the offspring insofar as these tendencies affect related individuals."

We do not, of course, have to take evolutionary biologists' word for it. There is copious evidence of intergenerational conflict in literature: a writer like Ivan Turgenev is in plentiful and illustrious company indeed. Within our own ranks, Lewis Feuer (1969) has convincingly shown that intergenerational conflict at the aggregate, extra-familial level, has been throughout history at least as important a factor of social upheaval and revolution as the class conflict immortalized by Karl Marx.

In short, what we become has doubtless much to do with what is taught us. But there is something for a remainder that no socialization theory can account for. To an extent we internalize norms. To another extent, we struggle over them and bend them to our diverse interests and the needs with which we enter the world. In conclusion, an exclusive focus on learning provides an inadequate and distorted view of sociocultural behavior. Taking behavioral predispositions into account broadens the theoretical vista and renders it empirically more fruitful.

3 The Interplay of Biology and Culture

Far from constituting an insurmountable interdisciplinary barrier, the debate over nature and nurture points compellingly to the fact of coevolution, namely, the interplay or mutual dependence existing between biology and culture. The dichotomy of behavior as either learned or innate has a shrinking number of supporters as knowledge of behavioral processes increases (Bandura, 1977). Heredity and culture interact both in individual human behavior and in the evolution of human societies (Anderson, 1967; Eckland, 1967). The present chapter seeks to defend this hypothesis of coevolution, in part through a critical examination of certain theoretical classics of social science that on the surface appear to be especially intractable to a biocultural treatment.

There are two main sorts of relationships between biology (or genetics) and culture that I wish to consider here. In one, culture may be viewed as being in a state of mutual dependence with natural selection itself. In the other, we observe a more immediate type of interplay: sociocultural behavior and behavioral predispositions act and react upon each other in such a fashion as to heighten or to lessen their respective salience. The strategy, therefore, is to focus on two major properties of biological and sociocultural factors. They will be termed *Distribution* and *Intensity* (or *Variability*), respectively. Distribution will help to examine the mutual influence of culture and natural selection. Intensity will guide the study of the modifiability of behavioral predispositions, universals, and variants when these are viewed synchronically or ontogenetically. For simplicity's sake, no distinction will be made between cultural universal and cultural variant in this type of analysis; the two levels will be treated as one, representing together the input of culture in biology.

For similar purposes of simplification no distinction will be made between the natural selection of inelaborate hereditary material, for example, simple genes, and the natural selection of more complex genetic material, for example, those genetic compounds that are the bases of the behavioral predispositions. The study of this latter problem, while strictly speaking required by my focus on behavioral predispositions, is beyond the present means of evolutionary science. It is not at all unjustifiable, however, to hypothesize that natural selection operates at all levels of complexity in genetic structures and processes, although undoubtedly it acts more slowly on behavioral predispositions than on simpler genetic bases. We may continue by turning to the problem of intensity and set out with a few introductory comments of a critical nature.

3.1 Intensity of Causation

"The genetic and cultural evolutions of mankind are not independent but interdependent. They are tied together in a system of feedback relationships" (Dobzhansky et al., 1977: 459). Such statements by biologists promise well for a scientific rapprochement and for the future of a partially *culturalized* Modern Synthesis. For the time being, however, there is, with few exceptions (e.g. Lumsden and Wilson, 1981), a certain hollow ring to them that is as unfortunate as the lingering resistance in the cultural sciences to concede the possibility of a biological input in sociocultural behavior.

To the extent that evolutionary biologists have considered the influence of culture on biology, the tendency has been to focus on life cycle factors that produce fluctuations in the intensity of genetic forces without affecting the latter's distribution in populations. The concession may be illustrated by E. O. Wilson's (1975b: 19–21) discussion of "behavioral scaling," which is defined as "variation in the magnitude or in the qualitative state of a behavior which is correlated with stages of the life cycle, population density, or certain parameters of the environment." Surprisingly, even this notion, which prescribes a quantitative or variable view of behavior and the biological forces underlying it, is apparently fairly recent in biological studies. Thus, again according to Wilson (1975b: 19), the early studies of vertebrate sociobiology tended to consider traits as invariant: they either existed or they did not. As in much of social science, there was a lack of sensitivity to the fact that a given behavioral force may have a range of degrees of manifestations, and thus may be plotted in time along a fluctuating line.

Wilson provides several enlightening illustrations of behavioral scaling. Consider, as a typical example, two studies of vervet monkeys carried out, respectively, by T.T. Struhsaker at the Amboseli Masai Game Reserve in Kenya, and by J. S. Gartlan in Uganda (reported by Wilson, 1975b: 19–20). In the first setting, the vervets were observed to be regularly engaged in fights to maintain rigid dominance hierarchies. According to the second study, by contrast, these animals had no visible hierarchy at the time of observation; fighting was rare; and males moved rather freely between troops.

Known behavioral variations of this sort are now quite common, and involve various sorts of animals. They are sometimes mistakenly viewed by social scientists as inconsistencies in the biologist's emphasis on innate forces of behavior. The variability is apparently associated with a variety of factors. In the case in question, Wilson (1975b: 20–1) judiciously begins by considering the possibility that it reflects genetic differences between populations of a same species that have had diverse environmental histories—the possibility, in short, that environmental factors and natural selection acted differently, to some degree, on the relevant behavioral predispositions of the two vervet troops.

There are a number of environmental factors, however, that seem to relate strictly to the intensity variable, at least in the short run. They include group size, availability and quality of food, the manner in which the food is distributed in the environment, and population density, among others. The latter factor is probably the most frequently reported parameter governing behavioral scaling. We may represent the regularity with the statement that, with few exceptions, the greater the population density, the greater the tendency toward aggressive behavior and activities to establish dominance. Shifting to human society, sociobiology would, then, correctly predict the well-known sociological finding that, *ceteris paribus*, the crime rate tends to increase with an increase in population density. Again, human sociobiology would have little trouble predicting the urban recurrence of corner-boy gangs whose activities include the disputation of "turf," *inter alia* (e.g. Whyte, 1943).

Human sociobiology is still very new, however, and it is little surprise, therefore, that it has been faulted for insufficient sensitivity to the two-way influence existing between biology and culture (e.g. Piaget, 1976). The strongest reaction has come from those whose specialty is the study of culture, namely, cultural anthropologists and sociologists of culture. Perhaps as powerful antidote, the critics have sometimes left no room for compromise. Marshall Sahlins (1976: 13), for example, has exhibited extreme eagerness to defend "the autonomy of culture" and reached the point of absurdity in declaring that causality does not flow from biology but "the other way around." According to this scholar, "a meaningful system," that is, culture, engages human "natural dispositions" "as the instruments of a symbolic project."

A related viewpoint has been put forth by Clifford Geertz (1973: 81–2), according to whom "not only ideas, but emotions too, are cultural artifacts in man." The rationale transpires in several ways, some fairly satisfactory. "In order to make up our minds," for example, "we must know about things; and to know how we feel about things we need the public images of sentiment that only ritual, myth, and art can provide." There is excess here in the form of a heavy rationalistic emphasis on knowing and making up our minds. There is more to human behavior than this. Still, the underlying point conveys a compelling message, and the unfolding of a culturalized Modern Synthesis will have to come to terms with it to some extent. It states, basically, that the biological substratum is not in itself sufficient to evoke, and thus explain, identifiable types of behavior. This position will be qualified by our later discussion of what will be termed "the imperative to act." Still, it is symbols—for example, ritual, myth, language, indeed a painting or a sculpture—that typically call forth and modify the action of innate predispositions, even though it is also true that without the latter no "public images" would be possible in the first place. Earl Count (1958: 1081) is, thus, at least partly right in stating: "Cultures are the idioms for eliciting and expressing . . . innate nature."

While Count's position does not grasp the other function of culture, which is to react on natural selection, it expresses clearly the ontogenetic type of cultural effects on biology, which I have associated with the property of intensity. Consider the statue of the Pietà, a sublime expression of cultural activity. There is a real sense in which it may be said that its "meaning" has roots deeper than cultural underpinnings. It engages our "knowing" and our understanding, but it does not leave our recondite pathos unprobed. It is a splendid piece of art that elicits, or gives dramatic "public image" of, a complex of feeling states that are a trademark of our species: the sense of the sacred, the need for redemption, the pathos inherent in self-sacrifice, the profound sense of tragedy in a mother's incomparable grief, the sense of the absolute evil inherent in the destruction of the sacred, personification in myth, and so forth.

Note at the same time that the feeling states elicited or "released" (Hinde, 1974; Eibl-Eibesfeldt, 1975) by this exquisite configuration of symbols are not likely to be the same throughout human society. One may reasonably doubt, for example, whether such a symbolic structure will call the above feeling states into public action equally among, say, Christians, Animists, and Buddhists. That is to say that a given behavioral predisposition has multiple symbolic manifestations (variants) that may be properly viewed as functional alternatives, insofar as their etiological import vis-à-vis the predisposition is concerned, and therefore may to a large extent be classified in the same cultural taxon (universal), despite their variability in form.

There appears to be little or no cause, therefore, to question the position taken above by Sahlins and Geertz, at least as restated by Count. Culture not only manifests biological forces; it also elicits their action. But to the extent that biological forces are "selective," as it were, of their responsiveness to cultural traits, they also benefit from a certain etiological primacy, at least from the viewpoint of the intensity variable. In short, it is precisely the great variability of symbolic systems that ensures the methodological propriety of looking first from the narrow end of the funnel rather than from its fan side.

We are led back to the question of the relative etiological status of biological and cultural forces. The latter vary much more than their biological mates both in time and place. In a sense, then, they constitute causal repertories of *special* kinds of behavioral science. The more constant behavioral predispositions are better suited for the construction of a more general theory of human behavior.

For all that, however, one must hasten to add that a focus on behavioral predispositions has its own limitations. Such forces are likely to suggest general laws accounting for uniformities of fact that constitute but small segments of sociocultural "reality." Thinking of this type of limitation is analogous to thinking of a law of falling bodies without awareness of the theoretical tools necessary to bridge the gap between an idealized vacuum state and the real states offered by multiple complicating factors. In this sense, then, the proximate causation represented by cultural phenomena offers a richer, and perhaps more challenging, yield, even though it is necessarily fickle in the degree to which it helps to explain and predict events. In the last analysis, we are down to this: proximate causes show facts unfolding in their immediate and rich complexity without explaining them except in an ephemeral and highly particular way. Ultimate causes, conversely, harness the tendency to deny the notion of ordered events universal in time and place, but in so doing they sacrifice much information that is of immediate significance to the human experience. This deficiency, however, is largely remedied by the fact that ultimate causation tends to bring into theoretical action a general paradigmatic principle, for example, natural selection, whose fundamental function is to guide complex levels of proximate explanations toward an overriding organizer of knowledge: for instance, the conception of the bottom-line utility of behavior for the perpetuation and amenities of life. That is an important aspect of the scientific enterprise. Without some such principle, there are constant disorientation, incoherent bickering, and an accumulation of propositions and facts among which it is impossible to separate the significant from the trivial. Above all, without it, explanation proves at close examination to wear a false label. Thus, most social theory is vacuous, highly transitory, and shallow, or at best offers explanations that beg the question.

3.1.1 SOCIAL STRUCTURE AND ANOMIE: THE CASES OF E. DURKHEIM AND R. K. MERTON

In his famous essay on "social structure and anomie," Robert K. Merton (1968: ch. 6) sets out to disprove, through a presumably "structural" (and functional) analysis, the biological hypothesis, presumably held by some scholars several decades ago, that nonconformity in society is "anchored in original nature." His "primary aim is to discover *how some social structures exert a definite pressure upon certain persons in the society to engage in nonconformist rather than conformist conduct*" (Merton, 1968: 186—emphasis in the original). This orientation is termed the "sociological perspective." A careful reading of the essay, however, reveals a complex problem that, willy-nilly, calls into action biological and cultural factors as well as social or structural ones.

The essay is to a degree an application and intended extension of Durkheim's (1897: 241–76) theory of anomic suicide, a fact that transpires explicitly from Merton's (1968: 189) own exposition. It is advantageous, therefore, to proceed by turning first to Durkheim's work. The logical structure of Durkheim's theory runs approximately as follows (Lopreato and Chafetz, 1979). The human "passions" or "needs" are in principle "inextinguishable." While they are left unspecified, they apparently include the innate desire for power, wealth, and other resources that are clearly correlated with Darwinian fitness. They are potential mischief makers, for "No living being can be happy or even exist unless his needs are sufficiently proportioned to his means." To avoid their mischief, "the passions must be limited."

Durkheim finds the "regulative force" in "society," that famous entity of very indefinite meaning already remarked upon. Still, poor conceptualization notwithstanding, one may fairly surmise in the present context that societies represent regulative forces in the sense that they have stratification systems and normative systems or ideologies associated with them that tend to align needs with relative availabilities or means. In short, "at every moment of history there is a dim perception, in the moral consciousness of societies, of the respective value of different social services, the relative reward due to each, and the consequent degree of comfort appropriate on the average to workers in each occupation."

The probability of suicide, among other pathologies, rises "when society is disturbed by some painful crisis or by beneficent but abrupt transitions" and is no longer capable of performing its regulative function. The passions now have relatively free rein. Old rules break down; new ones take time to crystallize and provide "a new scale" for "the reclassification of men and things." We are in a state of relative normlessness or "anomie." In the sphere of trade and industry, due to the continuing Industrial Revolution, anomie is chronic. As anomie rises, the rate of suicide also rises.

Now, what sort of etiology is at work here, whatever the awareness and language of Durkheim's analysis? The scoundrels, clearly, are those needs, or passions, that constantly threaten to break through; thus, they are the cause *in ultima instantia* of (anomic) suicide. Their force, however, is modified (facilitated or attenuated) by: (*a*) the differential access to the means whereby needs may be satisfied, along with the tendency to accept the ideology that supports the differential distribution of means; (*b*) the varying efficiency with which existing normative rules check the needs. In this sense, these cultural and social factors are *intervening* variables whose function is analogous to the function of a dimmer electrical switch. If the regulative force (the switch) works, the rate of anomic suicide may be low, in principle null. If it does not work, or works inefficiently, the suicide rate increases.

Plainly, validity of theory aside, logically there is an intimate interconnection between biological and sociocultural factors in the manifestation of a given suicide rate. Regulator and perpetrator are but parts of a systemic whole. Hence, the rate of anomic suicide is a function of the mutual adjustment between organismic forces (the passions) and environmental ones (the regulators). It would be a mistake to attribute total causality to the regulative subsystem, for it has a function to perform only in the presence of the other subsystem that, in the form of passions, constantly threatens to break through. If the latter is absent, the regulative apparatus is superfluous. Similarly, the electrical switch is useful if there is a flow of electrical energy along the wire attached to it; otherwise, it is at best an ornament on the wall. Indeed, one can light a bulb without a switch but not without energy, the analog of need. In this sense, one may go further and state that the switch is of secondary importance in the inclusive etiological system in question—that the system is overdetermined by the energy subsystem. By analogy, the same may be said of the needs component in the needs-regulators system.

Merton's theory has, at bottom, the same logical structure, with the major difference that Merton, unlike Durkheim, denies the theoretical relevance of some such entity as human nature or innate needs. He prefers to focus on two major "elements of social and cultural structures." One refers to "culturally defined goals, purposes and interests," some of which "are directly related to the biological drives of man" but "not determined by them." The other element "defines, regulates and controls the acceptable modes of reaching out for these goals" (Merton, 1968: 186–7). Merton uses various terms to refer to this latter concept, among which are "regulatory norms," "procedures," "expedients," and "institutionalized means." The former concept is at the same time referred to as "culture goals."

Societies and cultures differ in time and place in the degree to which they balance the relative emphases on the goals to be pursued and the means needed to achieve those goals. Some sociocultural systems, for example, place a heavy emphasis on providing the means for achievement without a corresponding stress on the virtue of achieving. Other systems may emphasize inordinately the propriety of striving, and striving high, without a parallel concern with providing the means necessary to reach the lofty goals.

If we now introduce the problem of social stratification into the argument, we reach what may be termed Merton's principal *structural* focus. Specifically, the various social classes and strata of a society have differential access to the goals that, allegedly, are culturally prescribed. Moreover, suppose we have a society like the United States of America where an ideology like the "American Dream" presumably reigns supreme and dictates that "low aim" is "a crime," at the same time that some individuals and groups find insurmountable obstacles on their approach to success through legitimate means. Then we have a situation in which a cultural component operates to "tease"—excite, as physicists might say— what we may term the need for achievement. But given the differential access to the coveted goals, some people are destined to failure *if they execute their pursuit with the help of institutionalized, or legally and morally acceptable, means.* Therefore, not everyone is likely to pursue the American Dream through recourse to unimpeachable means of attainment.

In fact, there are several major "modes of adaptation" in a population that features this disjunction between cultural goals and institutionalized means of achievement. Merton terms them *conformity, innovation, ritualism, retreatism,* and *rebellion.* The first reaction is the "most common and widely diffused"; that means, *inter alia,* that many (the majority, most— we are not told which) must engage in a difficult juggling act in which the unbalanced goals and means are somehow brought into mutual alignment. Another way of saying the same thing is that for a large portion of the population the American Dream becomes a mere abstraction to be taken very lightly, for the lofty prizes dangled by the Dream are after all precious few.

Surprisingly, Merton does not explain this modal adaptation. His emphasis falls on innovation, a response wherein the lofty goals are taken only too seriously and, when the institutionalized means prove inadequate, the goals are pursued with whatever means at one's disposal. This strategy sometimes runs afoul of the law: "deviant" behavior has been perpetrated and recorded. The theoretical use of the present mode of adaptation is especially intended to explain such phenomena as the rise of the American Robber Barons and the deviant behavior of deprived minorities. The famous gangster Al Capone, for example, "represents the triumph of amoral intelligence over morally prescribed 'failure' " (Merton, 1968: 200).

Very appealing, of course. But why so much conformity? Why do some individuals, very likely the vast majority in every group, apply no amoral intelligence to triumph over morally prescribed failure, while others, the minority, living in the same social-structural situation, pursue triumph at any cost? Indeed, why do other individuals "retreat"—typically relinquish both lofty goals and success means of any type once they have tasted the moral conflict that is caused by the disjunction between prescribed goals and morally appropriate means? They include vagrants, outcasts, tramps, chronic addicts, and so forth (Merton, 1968: 207–9).

Still others become "ritualists"—they "play it safe" and end by making a fetish of the existing rules while rejecting, except in passive hope, the goals of lofty achievement. A case in point is "the zealously conformist bureaucrat [the paper and pencil pusher] in the teller's cage of the private banking enterprise or in the front office of the public works enterprise" (Merton, 1968: 203–7). Finally, there are the "rebels"—these find something radically wrong with the entire system of cultural goals and legal means of achievement and seek to transform it all, as the Spaniards say, *de pie*, from its roots (Merton, 1968: 209–11).

Merton enriches his analysis with the usual recourse to socialization processes. Thus, different social classes and strata are socialized to respond differently to moral-ideological systems, such as the American Dream. Ritualism, for example, is most typical of lower-middle class Americans, for "it is in the lower middle class that parents typically exert continuous pressure upon children to abide by the moral mandates of the society, and where the social climb upward is less likely to meet with success than among the upper middle class" (Merton, 1968: 205).

Still—and the point seems crucial to me—not all lower-middle class workers become ritualists, even if given the proper employment opportunity. Nor are all lower-strata workers innovators. Indeed, as noted, the most common and widely diffused reaction in all social strata is conformity, an "adaptation" that merits all of two paragraphs in Merton's analysis. The topic is disposed of with the vacuous explanation that "Were this [tendency to conform] not so, the stability and continuity of the society could not be maintained" (Merton, 1968: 195). And so it develops that the "sociological perspective" is a perspective that seeks to explain the exceptions but not the rule. This theoretical position is excessively modest as well as inordinately restrictive.

First preference should go to the explanation of conformist behavior; that is where the uniformity of fact, the focus of a scientific law, resides. Furthermore, while conformity is no doubt a result of socialization processes, it is evident that if we were not so constituted as to be molded—indeed to seek to be molded—by rules of behavioral conformity, socialization processes would be tilling on infertile soil. As we have seen, modern social learning theory has shown that behavior is a result of the mutual influence of biological predispositions and learning processes and techniques. Accordingly, Chapters 5 and 6 will discuss a behavioral predisposition of conformity and its role in various aspects of the sociocultural life. Before that, in Chapter 4, I shall discuss predispositions of self-enhancement.

Here, I shall conclude the discussion of Merton's still very important work by suggesting, broadly, that predispositions of self-enhancement and conformity are not equally intense in a population, with the result that some individuals are in principle more inclined toward conformity than others. The two sorts of predispositions, further, may be viewed as being at least in part in a relation of dialectical tension with each other, for the more intense is the drive to achieve, the lower the propensity to obey rules of conformity that condemn us to failure. However, whether it is one predisposition or another that achieves behavioral predominance has to do, as often implied already, to a large extent with the manner and the degree in which environmental factors, among which are the socialization processes and techniques, are brought to bear on the predispositions in question.

Finally, to say, for example, that conformists are those in whom forces of sociality, of which conformity is one, weaken the maximization drive under the pressure of cultural injunction is to make a biocultural statement that is anchored in a general scientific principle. Likewise, to say that nonconformists are energized by intense self-enhancement forces under the pressure of the maximization drive (and despite in-principle mitigating cultural injunctions) is to state an evolutionary proposition that is anchored in a general principle. Conversely, to state that society would not be possible if there were not a great many conformists around is to state the obvious and to explain nothing; it is one of those tautologies that not a few social scientists have long decried (e.g. Davis, 1959; Homans, 1964, 1967).

3.1.2 THE PROTESTANT ETHIC AND THE SPIRIT OF CAPITALISM: THE CASE OF MAX WEBER

The case of Max Weber is equally instructive. This scholar is without a doubt one of the most renowned figures in the history and practice of social science, and is credited with a large number of contributions, including a (useful but very limited) theory of social stratification that many unconvincingly consider superior to Karl Marx's own monumental, even if in part misguided, endeavor. One of Weber's most characteristic features is an emphasis on meaning. Indeed, at least in one context, he (1949: 81) defines sociology as the study of "all human behavior when and insofar as the acting individual attaches a subjective meaning to it." A striking application of this definition may be found in his (1904–5) volume on "the Protestant Ethic and the Spirit of Capitalism," which is perhaps the most famous book in sociology, despite the widespread confusion surrounding the nature of the causality contained therein, or indeed, the formal impossibility of extirpating the confusion (e.g. Zetterberg, 1963).

It is not my intention to engage in a systematic critique of this work or to give a full account of its theoretical contents (but see, e.g., Tawney, 1926; Samuelson, 1957; Bendix, 1960; Braudel, 1979; Cohen, 1980; Houghton and Lopreato, 1981). My aim is rather to examine Weber's thesis to the extent necessary to show that the extreme stress intended by Weber, and celebrated by many of his readers, on meaning and ideas as causal factors is quite misplaced. Indeed, it makes no sense at all apart from sensitivity to the interplay of ideas and their biological substratum.

Writing at a time when economic determinism reigned imperious in some scholarly quarters, Weber's (1904–5: 90) express aim was to offer "a contribution to the understanding of the manner in which ideas become effective forces in history." He turned to the relationship between religion and capitalism for an implementation of his goal. More specifically, therefore, he (1904–5: 91—emphasis provided) wished "to ascertain whether and to what extent religious forces have taken part in the *qualitative formation* and *quantitative expansion* of [the] spirit of capitalism over the world."

I have underscored two crucial expressions. Clearly, Weber intended to allow for the operation of multiple forces in the qualitative development of the spirit of capitalism, one set of which was allegedly religious in nature. Evidently, too, Weber intended to advance religious ideas as forces that played an accelerative role in the maturation of modern capitalism, namely, an economic system rationally controlled by the owners of capital or their agents for the purpose of pecuniary profit, and resting on the organization of legally free wage-earners. At one time, presumably, capitalism was philosophically influential enough to set its stamp on the entire sociocultural system: it had a "spirit"—a spirit with a number of "fundamental elements." Ascetic Protestantism, according to Weber, contributed some of those elements, including the one that most parsimoniously described the spirit of capitalism and at the same time most effectively accelerated the development of capitalism. Weber (1904–5: 180) put it this way: "One of the fundamental elements of the spirit of capitalism, and not only of that but of all modern culture: rational conduct on the basis of the idea of the calling, was born—that is what this discussion has sought to demonstrate—from the spirit of Christian asceticism."

In fact, Weber did not demonstrate that the birth took place in that fashion. At most he pointed to certain analogies existing between the spirit of ascetic Protestantism and the spirit of capitalism. The fact that he showed only analogy, rather than determinism, between *some aspects* of the spirit of capitalism and *some elements* of ascetic Protestantism was documented by Weber (1904–5: 180) himself in the continuation of the above-quoted passage.

"One has only to re-read the passage from [Benjamin] Franklin, quoted at the beginning of this essay, in order to see that the essential elements of the attitude which was there called the spirit of capitalism are the same as what we have just shown to be the content of the Puritan worldly asceticism [a sort of ideal type of Protestant asceticism], only without the religious bias, which by Franklin's time had died away." A cruder analogy or equation of one institutional phase with another is hard to find in behavioral science.

What, then, is the connection drawn by Weber between the spirit of ascetic Protestantism and the spirit of capitalism? Examining Calvinism, the type case of ascetic Protestantism, by way of the Westminster Confession of 1647, Weber stressed the etiological importance of certain tenets in it, and especially "the doctrine of predestination [which] was considered its most characteristic dogma" (Weber, 1904–5: 98). To people steeped in the Christian principle of salvation, this doctrine was allegedly the cause of a profound emotional stress. This is a crucial point. "The question, Am I one of the elect? must sooner or later have arisen for every believer and have forced all other interests into the background" (Weber, 1904–5: 110).

Fortunately, Protestant doctrine contained a number of other helpful tenets. Some enjoined the believer to work assiduously for the Glory of God—as a calling—and, by condemning the "lusts and temptations of the world," to practice systematic self-control or asceticism. Not leisure, luxury, and enjoyment, but only intense bodily and mental labor serves to increase the glory of God. The result of such intense attachment to labor as an end in itself, coupled with an ascetic approach to its fruits, was the accumulation of wealth. But because wealth is the result of obeying God's injunctions, and because a just and prescient God would not have been likely to condemn to eternal death those who would in their worldly life obey His injunctions, the Calvinist gradually developed the conviction, allegedly, that wealth and the good works were in fact the signs of grace. This is another crucial point, for this conviction, and the acquisitiveness going with it, then allegedly combined with limited consumption to produce capital accumulation. As Weber (1904–5: 172) put it, "When the limitation of consumption is combined with this release of acquisitive activity, the inevitable practical result is obvious: accumulation of capital through ascetic compulsion to save."

There remains to say, in this necessarily brief exposition, that Weber buttressed his argument about the connection between the spirit of Protestantism and the spirit of capitalism in a variety of ways. We need not go into them; they are as good as the theoretical nucleus around which they revolve. There are many errors in Weber's analysis. A fundamental and surprisingly elementary one lies in his failure to appreciate the logical requirements of the system concept in scientific analysis. There is a variety of unfortunate expressions and consequences of this deficiency. They may be epitomized by reference to his (1904–5: 90) pivotal statement of intention: "we are merely attempting to clarify the part which religious forces have played in forming the developing web of our specifically worldly modern culture, in the complex interaction of innumerable different historical factors."

In recognizing the complex interaction of multiple historical factors, Weber in principle acknowledged the systemic nature of the elements at work in the formation of modern capitalism. The logic of this recognition is simple and straightforward. We have a phenomenon, X, that is constituted by the interdependent operation of forces $a, b, c, d \ldots n$. So far, so good. But if this be the case, we simply cannot "clarify the part" which religious forces, a, have played in the formation of X unless at the same time we determine also parts $b, c, d \ldots n$. Formally speaking, a system analysis requires the solution of simultaneous equations. That is too much to ask for, then as now. But even a metaphorical recognition of this fact would have helped. If it be true that these constituents of X are interdependent, it is also true that: (1) X is a resultant of the multiple interdependences of $a, b, c, d \ldots n$; (2) each element in this set of forces is at one and the same time partial determinant and result of the other elements. Thus, Weber's narrow and exclusive focus was simply not adequate to "ascertain whether and to what extent religious forces have taken part in the qualitative formation and the quantitative expansion of the spirit of capitalism over the world" (Weber, 1904–5: 91).

What is even more ironic is that Weber, the presumed historical sociologist *par excellence*, attempted to confront a supremely historical phenomenon without a grasp of history, either in the narrower or in the broader sense of the term. He divorced, for example, the development of modern capitalism from the rise of the bourgeoisie, whose historical vicissitudes had already spawned more than half a millennium, from the Renaissance and the rise of the modern arts and sciences, and even from the discovery of the New World. He showed little or no awareness of the fact that spirit of capitalism and Protestant ethic were both by-products and at once ingredients of the emerging industrial order. In substituting, *in practice* though not in idle intention (Weber 1904–5: 183), Marx's economico-technological analysis of the new age with a "spiritualistic" argument, he proved himself to be not an historian but perhaps the most insidious of the well-intentioned Social Darwinists.

He also failed to descry the unfolding of history *in magno* when he could not make theoretical capital out of his insight that most social action is "governed by impulse or habit" rather than by reason (Weber, 1947: 112). Ultimately, the kernel of the Protestant ethic is located in the emotional stress allegedly produced by the doctrine of predestination in view of the Protestants' need to know whether they had been destined for "eternal life" or for "eternal death." Weber's (1904–5: 110) point bears repeating: "The question, Am I one of the elect? must sooner or later have arisen for every believer and have forced all other interests into the background." It was this predominant interest, combined with an intense consideration of the implication of the ascetic doctrine of the calling, that led to the fulfilment of the need to know one's eternal fate.

Why should the doctrine of predestination have produced such tremendous stress? The magnitude of the deception that it led to—namely, the conviction that wealth was a sign of grace, *despite* the Confession's denial that God's verdict was knowable—is a good clue to the answer we are seeking. The need to know one's eternal fate must be, or must have been, extremely intense, overwhelming. What was its nature? Its furious intensity suggests a noncultural fact.

There is some fairly serious evidence of this. The belief in an eternal soul is universal in human society (Murdock, 1945), though it, too, is subject to historical fluctuation, and hence it is not universally, or always, entirely evident. What is perhaps even more remarkable is that, as we shall see in Chapter 6, all known religions have featured an entire system of beliefs and practices intended to "save" or facilitate the proper survival of the soul (see, e.g., Brandon, 1962, for a brilliant review and discussion of this phenomenon).

The moral of the story is that ideas may indeed have been powerful forces in the maturation of modern capitalism, but only because they were "public images" of a biological substratum. Without the latter, they would have fallen on deaf ears. The doctrine of predestination was in a sense an attempt to stultify a profound biological need; namely, what in Chapter 8 will be termed the denial of death. The attempt could not but fail, although it may very well have modified predispositions and behavior enough to have contributed a major thrust in the then developing modern capitalism. Little wonder that seventeenth-century Protestants found a clue to their eternal fate after all. It is even less surprising that Protestant religion has since evolved far afield from the doctrine of predestination. A Gallup poll taken in the mid-1970s found that 34 percent of the people in the United States of America described themselves as "born again," that is, as having had a turning point in life of total commitment to Jesus Christ (Gallup, 1977: 50). What is interesting about the evangelical movement is that salvation of the soul may be obtained by *faith alone*—by merely "surrendering" to Jesus Christ and doing one's best to obey his laws.

In conclusion of this section, I wish to stress that it has not been my intention to downgrade sociocultural explanations of behavior. Rather, my aim has been to suggest that such explanations, if divorced from evolutionary conceptions of a continuing interplay of biology and culture, are necessarily limited in scope and sometimes logically vacuous. Fundamentally, they are ahistorical even if they present themselves in historical garb. They thus beg the question. They fail to regress the why, to engage in the favorite sport of the scientific enterprise; and they consequently fail to see the rich interconnections existing between cultural and biological forces.

3.2 The Mutual Influence of Culture and Natural Selection

But that is not to say that they are worthless. A renaissance of social theory depends not merely on an opening to evolutionary behavioral science in our future endeavors but also on taking inventory of our past accomplishments from this wider perspective. The above critique notwithstanding, Weber's thesis, for example, makes good sense when viewed from a biocultural perspective. Indeed, it implies a provocative lesson for biocultural science. Particularly striking is the manner in which the stress created by the tenet of predestination is alleged to have been resolved. Unbeknown to the great master, the resolution came forth with more than a dash of Darwinian logic. As the theory of altruism in Chapter 6 will note, a widespread feature of religion is the association of the idea of eternal salvation with the moral injunction to do good unto others in one's own group. Religion, thus, tends to mitigate the self-serving behavior that is inherent in natural selection. The alleged affirmation by seventeenth-century Protestants that wealth was a sign of grace was, conversely, a sort of Darwinian vengeance against the encroachment normally perpetrated by moral rules on the search for genetic fitness. The point is that the moral and spiritual quandary ostensibly created by the tenet of predestination was expunged not with the aid of moral agents but with the help of one of the classical material factors working at the service of genetic fitness.

That such a development, if real, should have taken place at a time when human society was undergoing a radical transformation of its technological basis is especially significant from our perspective. At such historical junctures, the competition for technologies, resources, and space—that is, the struggle for existence—is singularly acute. Some come upon enormous opportunities for self-enhancement and aggrandizement; others encounter nothing but dangers and endless causes for failure. Weber's emphasis on economic achievement was, therefore, extremely well placed.

From a broader perspective, there is a complex concatenation of actions and reactions between behavioral predispositions and sociocultural factors. Not all people have an equal capacity to take advantage of economic opportunities, just as not everyone has a special aptitude for mathematics, sports, the monastic life, or any number of other activities. Likewise, at the level of whole groups or societies, the capacity for inventiveness is not uniformly intense. As a result, the different societies proceed at varying speeds toward the discovery of new resources and the invention of new technologies. Similarly, different societies have unequal access to pivotal natural resources, and consequently, even if we hold constant the capacity for innovation, they vary in the rate at which they invent, borrow, and apply new technologies. Thus, in the struggle for space, resources, and supremacy, the swifter and the more efficient is the use of technology and other types of cultural information, *ceteris paribus*, the greater the Darwinian success.

This problem of "group selection," or differential success between societies, will be discussed in more detail in Chapter 6. Here we may note that, while sociobiologists allow for this type of selection, the models so far developed do not appear to be particularly efficient to establish the mechanism as a common factor of evolutionary history. This circumstance is due to the fact that, as we shall see, the problem of group selection tends to be fused with the problem of altruism, and, finding the latter difficult to conceptualize, sociobiologists have difficulty conceiving of group selection as well.

Later I shall argue against this conjunction. For the moment, one conclusion is inescapable. Human societies have often competed to the death, and those endowed with technological and other cultural advantages have also been at a selective advantage (Darwin, 1871; Sumner, 1911; Davie, 1929; Alland, 1968; Otterbein, 1970; Alexander, 1971; Wilson, 1975b; Durham, 1976; Lenski and Lenski, 1978). In particular, the horticultural revolution, the agrarian revolution, the industrial revolution, and now the nuclear revolution may be recorded among the bloodiest in the history of the species. A survey gives evidence that the vast majority of recent horticultural societies endowed with the use of metal tools were in a state of either perpetual or common warfare (Leavitt, 1977). Again, according to a sampling by Pitirim Sorokin (1957) involving eleven countries and many hundreds of years, European history has been marked by military action nearly 50 percent of the time. But one of the most obvious examples of group competition and achievement of differential fitness at that level is afforded by events on the post-Columbian American continent where large numbers of peoples have been either entirely wiped out or reduced to genetic and cultural vestiges of their former selves.

Indeed, the history of human society is to an exceeding degree the history of hunting-gathering peoples equipped with bows and arrows and other weapons devastating those who lacked such implements; of horticultural societies endowed with metallurgy raining death upon those who made no use of metal implements; of agrarian societies whose discovery of iron ore and whose invention of the plow and related tools yielded an economic surplus of sorts and stimulated the imperial and destructive inclination; of industrial societies whose highly developed mechanical means of transportation, communication, and warfare incited them toward a global colonization and often wholesale destruction of less endowed societies and cultures. What would the human genotype be like today without such differential developments and applications? One is in any case tempted to hypothesize that, at the intersocietal level, technological innovation has played a role analogous to that of differential reproduction at the individual level.

Such innovation is at least in part promoted by what in Chapter 7 will be termed predispositions of variation or combination. Intense predispositions of this type may lead a given society on a rich path of cultural development. The latter change, in turn, may lower its mortality rate or lead it to conquer and plunder the resources—human, material, and cultural—of its neighbors. The added resources may further act positively on the genetic fitness of the conquerors.

The interplay of natural selection and culture is equally evident at the individual level. The ancient man who was best constructed to throw a spear had, on the whole, the best chance of coming away from the battlefield alive. His advantage was magnified when superior innate endowment was combined with a superior weapon and exceptional training. Again, the chances of survival to the age of reproduction are not equally distributed in a society. Nor are the chances that survival to this stage will translate into actual reproductive activity. Hereditary factors may explain these facts in part. But, as will be seen in Chapter 4, we must also consider such sociocultural factors as education, occupation, and life style in general.

3.3 The Modifiability of Predispositions

We still know very little of a specific nature about the influence of biology on sociocultural behavior. We are probably in a slightly better shape insofar as the reverse relationship is concerned. So, we may well inquire here, as we approach the theory of the biogram: what sociocultural factors influence behavioral forces, whether along the intensity or the distribution property? I shall focus on a few obvious ones.

3.3.1 SYMBOLS

One has received considerable treatment already. Broadly speaking it refers to culture in general; more specifically, it takes the form of language, rituals, artifacts, ideas, and so forth. For convenience sake, we may name the class collectively with the label of *symbols*. "Never have so few sacrificed so much for so many." "Ask not what your country can do for you; ask what you can do for your country." Famous pronouncements such as those have frequently in history "fired up" endless numbers of people, stimulating primeval forces of self-sacrifice, patriotism, reverence, or for that matter intense hatred and hostility toward someone or other (e.g. Whorf, 1956; Chomsky, 1973, 1975).

From the perspective of another category of symbols, we note that the mere sight of a cross, the unfolding of the flag even in a football stadium, the beating of drums in a parade, such events and numberless others like them sometimes make our spine tingle and our hair stand on end. At another level still, the sight of a gun may frighten us "to death" or, conversely, "make us see red." The examples are legion.

While it could be shown that symbols have an influence on natural selection itself, for example, through the effect of technology and weaponry on population phenomena, their more obvious effect is on the intensity of behavioral predispositions. Sophistries, for example, can be powerful tools of persuasion by harping on predispositions that incline toward the desired behavior. Often, their cogency lies in the fact that they express clearly ideas that people already hold in a confused sort of way. Hence, sophisms discharge also a clarifactory function: they are organizers of ideas and information; in short, like culture in general, they are implements of learning.

The relationship between symbols and predispositions helps to explain the great importance of silence in the social life. Politicians understand the fact, as when, for example, they deliberately ignore each other's chatter. Silence is frequently an effective instrument for sapping the strength of a given statement by preventing a clarification and possible flowering of the predisposition(s) that it is rooted in. Conversely, argument, however brilliant from a logical standpoint, may excite the predispositions which are linked to that statement and end by granting it wide acceptance.

The opposites of silence, reiteration and harping, are powerful tools in the hands of preachers, lawyers, and advertising experts, among others. The technique is to hammer constantly on a theme, often with simpleminded formulas. It helps to reinforce the simplistic formulas with visual aids or dramatic imageries of the message being conveyed by the sophistries.

One form of repetition has implications for an exceedingly widespread phenomenon in human society: persecution. Basically, persecution has two major expressions. One would impose on a people practices, means, and formulas that are disharmonious with their traditions and the behavioral forces that these are anchored in. The other aims at preventing a people from giving free expression to their behavioral inclinations and the beliefs and practices associated with them. In both instances, the persecuted are bombarded with stimuli that sharpen up sentiments and often end by rendering them altogether sacred. Persecution, thus, tends to produce the very opposite of the behavior desired. Christianity might never have broken through the theological labyrinth of Roman paganism without the horrendous persecution of its early converts. Likewise, Judaism today might be more theological and less political if another expert manipulator of symbols had not sacrificed 6 million Jews to that primeval demon whose first commandment is natural selection: the Holocaust wiped out a tragic portion of an entire ethnic people.

3.3.2 GREAT EVENTS

One important set of factors involved both in natural selection and in the varying intensity of behavioral predispositions is represented by such epoch-making phenomena as war, revolution, pestilence, major earthquakes, and profound economic crises. "Events that deeply impress a population modify sentiments in the individuals who have witnessed them very considerably" (Pareto, 1916: 1839). A people who has experienced a great economic depression, for example, is likely to develop and sustain frugal habits. Thus, one would guess that those Americans who experienced the ravages of the Great Depression are today less susceptible to the devastations of economic cycles than those who were born since World War II. It is interesting to note, for example, that they have one of the lowest suicide rates in the United States of America (US National Center for Health Statistics, 1980).

Likewise, people of the Hebrew faith who remember the unspeakable Holocaust may be expected to be under the continuing influence of certain powerful sentiments that were stirred by the Event: for example, an intense attachment to family, a strong sense of peoplehood, a profound reverence for tradition, and a heightened suspicion of the "outgroups," among others. The deep attachment of Diaspora Jews to the state of Israel is to a considerable degree a reflection of the gravity of such predispositions. Similarly, what some term the intransigence of Israel in relation to "the Palestinian problem" and the status of the post-1967 territories is to some extent a product of those same experiences and the elements of human nature in which they are rooted. It is hard to be "fair," "reasonable," "flexible," and "confident" of a future described by "them" rather than by "us," when the past shows a poor record of reason, fairness, and flexibility. By the same token, Jews are probably on firm ground in fearing the continuing hostility of Arabs in general and Palestinians in particular, for the latter have experienced grievous losses that either make restitution irrelevant or are not easily erased by restitution.

Great events, however, are not equally influential on those who know of them only by hearsay or tradition. In the absence of personal experience, their force tends to wither and eventually disappear. Thus, bypassing the hypothesis of youth idealism (e.g. Feuer, 1969), it is probably safe to predict that younger Jews are more flexible toward the Palestinian problem than their elders. Likewise, the further we move away from 1945, the horrendous year of Hiroshima and Nagasaki, the less likely are the Japanese to hold a special attitude against nuclear arsenals.

3.3.3 SOCIOECONOMIC AND POLITICAL CHANGES

Another set of factors responsible for the modification of behavioral predispositions both in their intensity and in their distribution refers to major technological and economic transformations, such as the Industrial Revolution. I have already touched on this topic, particularly from the viewpoint of group selection. Viewed more broadly, major technological changes are associated with radical socioeconomic and political modifications, and these have an effect both on the distribution and the intensity of behavioral predispositions.

The effect on natural selection takes place because the changes in question entail dislocations, transfers of wealth, and an alteration in the rhythm with which socioeconomic conditions act differentially on the life chances of different individuals and groups.

The effect on the intensity of behavioral forces is due basically to the fact that socioeconomic changes involve changes in interpersonal and interclass relationships, and these, in turn, require readjustments in the tolerance for relations of superordination and subordination, *inter alia*. For instance, the economic expansion taking place in recent decades in many agrarian societies has undermined both the economic and the social grip that landowners and the gentry in general held on the peasantry for many a century (e.g. Lopreato, 1967; Singelmann, 1981). In having to adjust to these changes, the old privileged class has also had to experience and accept the reduction of what may be termed a need for dominance in the social hierarchy.

The story of the Industrial Revolution is too complex to go into here. But consider, as a major index of its enormous effect on natural selection, that in preindustrial times the world population doubled roughly every 700 years; at today's rate of increase it may be expected to double in less than forty years (United Nations, 1974; Dumond, 1975; Lenski and Lenski, 1978). Ironically, this enormous change has been brought about not by an increase in the birth rate but rather by a decrease of the birth rate accompanied by an even greater decrease in the death rate. There has thus been a major shift from quantity of reproduction to quality accompanied by improved living conditions in general—better food, better health and hygiene, and, among many other things, better knowledge of the relationship existing between the human organism and its habitat, including other types of organisms that prey on us.

But the Industrial Revolution has also entailed the political awakening of the masses of people throughout the globe. It has stimulated the people's need for recognition and their social climbing propensities, among other behavior predispositions that will be treated in the next chapter. That may be readily seen in the rise and the great strength of trade unions, in the organization of anti-establishment political parties everywhere, and in the widespread demand for what T. H. Marshall (1950) termed "citizenship rights."

In short, the Industrial Revolution came along with the sociopolitical revolution. It is a remarkable fact that virtually each and every one of the sociopolitical upheavals that social scientists label revolutions—whether it be Cromwell's revolution, the French Revolution, the Soviet Revolution, or any other—has taken place in a society that has seen the dawn of industrial development. People rebel not when economic conditions are poor but when they are relatively good and the expectations of further improvements are not realized rapidly enough (e.g. Marx and Engels, 1955, I: 94; Pareto, 1916; Brinton, 1938; Davies, 1962). They revolt, in short, when they have achieved a strong enough sense of security and self-confidence to afford taking a potentially lethal risk. This fact is reminiscent of the phenomenon of "bimaturism" in many animal species whereby the males mature sexually later than the females. This circumstance is adaptive for the males, for they thus defer challenging the dominant males for the sexual services of the females until such time as they have a good chance of avoiding being seriously hurt and of succeeding in their challenge (e.g. Barash, 1977).

3.3.4 MIGRATION

One of the major effects on genetic material is provided by the movement of people from one society to another or even from one region to another within the same society. Insofar as natural selection is concerned, the effect on it derives from the fact that migration shuffles different people about who represent also different genotypes. The effect is twofold, corresponding to the concepts of emigration and immigration. Specifically, migration has a genetic influence, large or slight, on a given population by subtracting a given number of individuals from it (emigration). It also has a genetic influence by adding new individuals to another population through immigration.

The effect of geographic movement on the intensity of behavior forces is both more complex and easier to conceptualize. It frequently involves the introduction of variations in language, attitudes toward politics, education, obedience to the law, religion, indeed aspects of the entire institutional framework. This change is a potential source of conflict that tends to intensify different behavioral predispositions among the different people in often unwelcome contact. The old-timers, for example, may experience a strong need for recognition or superiority; the newcomers may desire social approval, the humbler version of recognition. The potential for conflict is exacerbated by the competition for jobs, space, and, not infrequently, for members of the opposite sex. The history of the United States of America is to a large extent a history of cultural conflict due precisely to the continuous inflow of new people from places abroad and from the hinterland to the urban centers (e.g. Park and Miller, 1921; Davie, 1936; Wittke, 1939; Whyte, 1943; Park, 1950; Handlin, 1951; Glazer and Moynihan, 1963; Lopreato, 1970; Greeley, 1971; McLemore, 1980). To a lesser extent, the same applies to many another country, for example, Australia, Israel, and Great Britain. I shall return to this topic in Chapter 9 in connection with a discussion of ethnicity.

3.3.5 SOCIAL MOBILITY

Related, to an extent, to the action of geographic mobility is the effect of social mobility. Both types of movement recall in part what geneticists term "gene flow." It is very likely that "the genetic selective processes will operate differently depending upon the degree of social mobility a given society permits and upon how much equality of opportunity its citizens enjoy" (Dobzhansky, 1962: 245).

A given behavioral force is not at a level of uniform intensity within a population. To a degree, the different social classes and strata, indeed the different occupations, are characterized by different intensities in behavioral predispositions. That is due to the fact that different social categories are to a degree endowed with different interests and cultural propensities. Thus, social mobility transports individuals from one social class or social stratum to another, and in the process it produces perturbations in predispositions by the simple mechanisms of addition and subtraction.

The consequences of social mobility for behavioral predispositions are so numerous and complex that they would deserve a full-length treatment in their own right. Here I can only touch on a few points that have special salience for the purposes of this book. In general, it may be said that as the rate or probability of social mobility increases, the stratification system (dominance order, or whatever one may choose to term the phenomenon of structured social inequality) tends to become less rigid. To avoid the apparent tautology of such a statement, it might be better to say that the permeability, flexibility, or openness of stratification systems is enhanced in direct proportion to the amount of mobility that at any given time they allow for. The greater the movement of personnel between one social stratum and another, the more blurred are the lines that separate the one from the other. To this extent, there is a tendency toward a levelling of intensity in behavioral predispositions.

At the same time—and there is a certain paradox in this—an increase in the rate of social mobility tends to stimulate and energize the drive toward self-enhancement. The rationale for this proposition may be found in the fact that people tend to expend energy toward self-enhancement in direct proportion to the real and perceived possibilities of success (e.g. Keller and Zavalloni, 1964). This is probably the main principle behind the common finding that lower-class people are presumably less ambitious or achievement-oriented than upper-class people (e.g. McClelland, 1961). The alleged paradox lies in the fact, then, that the struggle for existence is greatest there where the freedom of competition is at its maximum.

A second major effect of social mobility is that its rate is positively related to the perceived legitimacy of existing social structures. Consequently, to the extent that we may be justified in hypothesizing a predisposition toward conformity, as Chapter 5 will propose, the tendency toward conformist behavior intensifies with an increase in the rate of mobility. Thus, while mobility undermines stratification systems that are rigid, it also tends to support them when they feature a high degree of openness. Concerted action, what Karl Marx termed "class consciousness," tends to give way to an individualistic type of action. In a sense, whatever degree of group selection may in principle be operating at the level of social class is taken over by individual selection. As Dahrendorf (1959: 60) put it:

> Where mobility within and between generations is a regular occurrence, and therefore a legitimate expectation of many people, conflict groups are not likely to have either the permanence or the dead seriousness of caste-like classes composed of hopelessly alienated men. And as the instability of classes grows, the intensity of class conflict is bound to diminish. Instead of advancing their claims as members of homogeneous groups, people are most likely to compete with each other as individuals for a place in the sun.

Finally, social mobility is a factor in the historical viability of societies. In general, the more open a stratification system, the greater the evolutionary success of the society possessing it. Suppose, for simplification's sake, that we divide a given society into a governing class and a governed class. Under the most historically viable circumstances, we observe a governing class consisting of those individuals best qualified to occupy positions of political administration. The tendency, however, is imperfect: certain factors interfere with the free circulation of talent and skill. The relatives, friends, and connections of governing individuals have easier access to governmental positions than do individuals lacking in any sponsorship. Then again, people of great wealth often succeed in purchasing a position in the governing class, regardless of their qualifications. As a result, as Pareto noted (1916: 2026–39), there is at all times a certain disjunction between title, or label, and the capacity requisite for inclusion in the governing class.

The more closed the governing class, and the greater its incompetence, the greater the probability of revolution, for the thrust toward dominance from below is overwhelming. Thus, whether through revolution or through a more normal type of circulation, social mobility is an enduring mechanism in the evolution of societies. The basic reason, of course, is the pressure of self-enhancement forces. But mobility also performs a vital evolutionary function. Specifically, it provides the personnel along with the behavioral forces and cultural traits necessary for the maintenance of internal social order and for the organization of defense against competing sociocultural systems. Hence, the most evolutionarily viable system is one that, *ceteris paribus*, is characterized by free social mobility.

There is a variety of other modifiers of behavioral predispositions. My intention, however, has been to illustrate not to be exhaustive. Furthermore, many other types can probably be subsumed under those already briefly discussed. For example, social planning, which immediately comes to mind as a major candidate for inclusion in the present section, can in all likelihood be encompassed under the headings of symbols and socioeconomic and political changes.

3.4 The Form of Sociocultural Change

I conclude the chapter by turning briefly to a problem that is inherent in the phenomenon of intensity in the special sense in which this term has been understood in this volume. To say that behavioral predispositions are subject to modification in their intensity is to say that there is a rhythmic movement or unending fluctuation in all sociocultural phenomena that are in any way connected to them. That introduces a large number of problems into behavioral science. For example, over a period of time, the frequency of a given phenomenon may, on the average, be constant, expanding, or declining. In view, however, of the fact that the three hypothetical cases necessarily manifest themselves along fluctuating lines, there is ever the danger that a basically constant tendency is mistaken for a rising or decreasing one, a tendency that is at bottom on the increase is judged to be stable or declining, and so forth. The problem is analogous to the situation encountered in navigating a river. It seems to flow in all directions of the compass; in fact, however, in view of its length as a whole, it flows in only one direction.

Suppose we represent the changing rate of murders committed in a given country within a century with an ascending line marked by depressions *a*, *c*, *e*, and by peaks *b*, *d*, *f*, as in Figure 3.1. If a social scientist were to compare points *b* and *c* or *d* and *e* in order to get at the general trend, the conclusion would be that the murder rate was decreasing. Such a conclusion would be false. Nevertheless, miscalculations of this sort abound in social science, and not a few endless debates and controversies in our journals have no other basis than the failure to take into account the fluctuating nature of the social movement. Thus, as we go from one year to another, from one decade to another, we first claim, and then polemically disclaim, or vice versa, that "alienation" is associated with a given variable, that "status inconsistency" has a liberal political orientation as an effect, that the sense of political efficacy is high, and so on *ad nauseam*.

If these and other facts move along a fluctuating line, it is important: (1) to recognize the fact and avoid ill-founded squabbles; (2) to specify clearly the period during which a given phenomenon is said to present itself in a

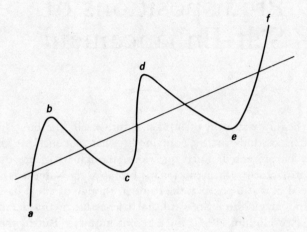

Figure 3.1 *A hypothetical curve of murders committed in a given country within a century.*

given way; (3) preferably to incorporate as large a span of time as possible in effecting one's measurements; and (4) to specify, if possible, the factors that are responsible for the minor fluctuations and those that account for the overall tendency itself.

4 Predispositions of Self-Enhancement

We now begin discussion of human nature itself (the biogram). How many predispositions shall it contain? In what sequence can they most efficiently be presented? Such questions, though basic, are difficult to answer satisfactorily. In evolutionary biology, the subject is relatively new; indeed, it is still posed at the frontier, though of late it has certainly sparked the imagination. Sociocultural science has a considerably richer tradition (see Hullum, 1980, for a recent inquiry). But in the absence of an explicit link to natural selection, or some other general principle of uniform causation, they have also tended to be theoretically chaotic.

On the whole, then, there is at present little basis for hope of extensive treatment of the problem at hand. What follows is offered as a rough approximation of what will some day be a more fully developed theory. The presentation will conveniently be divided along the lines of two fundamental problem areas of behavioral science. One refers to the question of sociocultural transformation, and is the object of the last three chapters of the volume. The other concerns roughly the goals that human beings pursue and the quality of interpersonal relations that their behavior both engenders and must adjust itself to. The present chapter and the next two are devoted to this second aspect of the problem. Together, the two topics call into action four classes of behavioral predispositions. Table 4.1 gives a bird's eye view of them.

We begin, in this chapter, with the predispositions of *self-enhancement* because, while all predispositions are at least in part the work of natural selection, we may think of the present class as most directly reflecting the basic notion of competition that is inherent in natural selection, a crucial concept in this book. Predispositions of self-enhancement are thus conceptualized as forces most closely associated with the search for individual advantage. There are various types of advantage, and my job is, to some extent at least, to view them as related from the vantage point of the maximization principle.

Before joining our task, several introductory comments are in order. First, like chemical elements, behavior predispositions do not appear in their pure form. What we observe in behavior is the working out not of

Table 4.1 Classification of Behavioral Predispositions

Self-Enhancement

The Climbing Maneuver
Territoriality
The Urge to Victimize
The Need for Vengeance
The Need for Recognition

Sociality

Reciprocation
Domination and Deference
The Need for Conformity
The Need for Social Approval
The Need for Self-Purification
Asceticism

Variation

The Explanatory Urge
Symbolization
The Imperative to Act
Homologous Affiliation
Heterologous Affiliation
Incest Avoidance
Heterologous Contraposition

Selection

Relational Self-Identification
The Denial of Death
Reification
The Susceptibility to Charisma
The Need for Ritual Consumption

simple predispositions but of mixtures or compounds thereof. Or to say it with Maslow (1954), behavior is multimotivated. That necessarily creates a problem of classification, and may invite the charge, occasionally levelled at theories of human nature (e.g. Mulkay, 1971), that some items of the classification overlap insofar as the sociocultural manifestations attributable to them are concerned.

Classifications invariably create some difficulties, even as they solve others. As far as the problem of overlap is concerned, the solution lies in recognizing that, in a given compound, a particular predisposition may be dominant over others; for convenience sake, the compound as a whole is thereby given the label of that behavioral force. The solution, however, is imperfect. We are really confronted by a problem of systemic interdependence. Ideally, therefore, we would wish to say something about the quantitative relationships existing in the compound. Moreover, in view of the fluctuating nature of phenomena, predispositions appearing in compound form do not recur with invariable causal asymmetry, with the result that, again ideally, we should have a dynamic view of those relationships.

Many other considerations would be possible, but they may be overridden by the necessary admission that at the present level of development, imperfection in classification is unavoidable. One exception, requiring a brief comment here, concerns the scope of the biogram. Some sociobiologists hold that human nature is basically mammalian if not altogether vertebrate in nature. Until recently, Wilson (1978; but cf. Lumsden and

Wilson, 1981), for example, held that "behavioral development is channeled in the direction of the most generally mammalian traits." That may very well be true, although ultimately the usefulness of this statement depends on what sorts of behavior we choose to focus upon and on the weight that we place on the differences that gradually surface between human and mammalian behaviors. Nevertheless, because I am dealing with a highly cultural species, and because, further, I have some interest in underscoring the influence of culture on biology, as well as the opposite relationship, I am bound to come up with a biogram that to a considerable extent is uniquely human. We shall have to deal with such phenomena as ritual, magic, self-deception, belief in the beyond, abstractions, personifications, ascetic behavior, and so forth, which appear to have no counterparts in other animals, and therefore to be anchored in uniquely human predispositions. On the other hand, much of human nature is probably generalizable to other mammals, although in very broad, perhaps all too sweeping, terms.

4.1 Introduction: the Problem of Inclusive Fitness

An introductory comment of a different sort takes us back to the maximization principle, which henceforth will assume a major theoretical role in the discussion. Most sociobiologists hold that this principle is implied by the principle of natural selection. Terming it the "Central Theorem" of the New Synthesis, Barash (1977: 63) states it as follows: "When any behavior under study reflects some component of genotype, animals should behave so as to maximize their inclusive fitness." In fact, the principle is considered problematic by some scholars, at least as presently stated (e.g. Slobodkin and Rapoport, 1974; Sahlins, 1976; Stearns, 1976; Schulman, 1978; Lopreato, 1981), despite some very impressive evidence in its favor. Chapter 6 will argue that the evolution of culture has imposed considerable constraints on the maximization tendency without denying the relevance of a genetic component in sociocultural behavior. Still, the notion of a quest for genetic fitness inheres in the behavioral core of natural selection, namely, the idea of competition for genetic survival. Under the circumstances, the maximization principle is within limits theoretically compelling. To the extent, further, that sociocultural behavior may be linked to evolution by natural selection, it is hardly possible to escape the argument that the evolution of behavior has not departed too far from a strategy of genetic maximization.

Certainly, looking at culture retrospectively we have no choice but to view its rise as an adaptation conferring on both individuals and groups an enhanced capacity to survive, though of course in varying degrees. As Durham (1978: 429) puts it, "the developing culture characterizing a group of people, whatever else it was, must have been adaptive for them in terms of survival and reproduction." Unless, further, matters have radically changed in this respect—and the evidence does not suggest this possibility—culture continues to be used by human beings to enhance their survival and reproduction. That would mean that sociocultural selection and retention operate at least in part in terms of the criterion of reproductive success. Accordingly, Chapter 7 will argue that the more reproductively advantageous a cultural innovation is, the more likely it is to be incorporated into existing tradition. This proposition follows from what may be considered one of the fundamental propositions of this volume as well as the central tenet of biocultural science: *Fitness-enhancing sociocultural behaviors are favorably selected, and then react on natural selection by influencing the distribution of genetic material and the distribution of sociocultural behaviors that are associated with it* (see also Lumsden and Wilson, 1981: 99).

The connection noted in Chapter 3 between the Industrial Revolution and the population explosion is proof enough, in a general key, of the intimate connection existing between culture and genetic fitness. Studies are also available, however, that explicitly test the hypothesis of the fitness-enhancing nature of culture. In a 1973–4 study, William Irons (1979a), for example, found that much of the daily activity of the Turkmen of Persia was devoted to economic production according to a strategy that approximated the maximization of wealth (see also Irons, 1975). The common technique was capital accumulation in money, jewelry, consumable goods, agricultural land, and livestock. Wealth thus accumulated was used in many fitness-enhancing ways. In general, it made possible a high standard of living, which included good food and medical care. Perhaps more importantly, it facilitated payment of the expensive bride price, not uncommonly consisting of 100 sheep or more. Indeed, wealth served a variety of fitness-enhancing functions. For example, the wealthier the family, the lower the rate of involvement in strenuous and high-risk labor. The wealthier the family, the younger the age at which a bride could be purchased, the quicker the tendency to remarry after the death of a wife, and the greater the probability of polygynous, or multi-wife, marriages. The latter point is especially important because the bride price necessary to acquire a second or third wife was three times what one had to pay for a first wife (Irons, 1979a: 267–8).

As a result, wealthier families could attain greater genetic representation in future generations than poorer families. The proof is compelling. Dividing his sample into a wealthier half of the population and a poorer half, Irons ascertained that 100 male individuals of the former were represented by 813 offspring in the next generation, as contrasted to 464 for the latter. Likewise, 100 wealthier females in generation one could count 561 offspring in generation two, while the total second-generation representation for the corresponding poorer females consisted of 499 individuals.

Note the clearcut case of biocultural interplay. Among the wealthier families, the higher fitness of males reflects the interaction of economic means and the polygynous tendency that, as we shall see in Chapter 9, is associated with a high reproductive capacity among males. Men exceed women by far in the production of sex cells, and this genetic overabundance tends to stimulate intra-gender competition, which in evolution has resulted in a polygynous tendency. This circumstance also produces an artificial scarcity of women in the wealthier families, for the polygynous competition in them leaves the less successful males without peer mates. It follows that those who lose the competition among their socioeconomic peers tend to redirect their superior resources toward a successful competition for mates vis-à-vis their male counterparts in the poorer families.

Such a displacement or nuptial mobility is facilitated not only by the superior material resources of high-status males but also by a certain hypergamous tendency (upward mobility) among low-status women. Under the influence of the maximization drive, the latter tend to be attracted to the resource-rich men of superior status. As a result, among the poorer families it is the females who achieve a higher genetic fitness, showing in effect that poor males are the most disadvantaged in the struggle for reproductive success (see also Burchinal, 1964). This scenario, we might add, is to an extent indicative of what happens in the animal world in general (e.g. Trivers and Willard, 1973; Dawkins, 1976; Barash, 1977; Wilson, 1978). Culture has accompanied, instead of cancelling, a biological phenomenon.

At close examination, the idea that culture tends to enhance survival and reproductive fitness is central, though typically implicit, in much social science—for example, in the literature on penology, social stratification, demography, poverty, and epidemiology. Parsons (1951), a major social theorist, saw the individual operating simultaneously in various systems, including the biological and the cultural, all of which feature a degree of mutual support. Another scholar, among many others, argued that there must be a degree of harmony between behavioral predispositions and our modes of living; else society would break down and cease to exist (Pareto, 1916: 1932).

Increasingly now there are explicit attempts to establish the causal fit between sociocultural behavior and the maximization principle (e.g. Blumberg and Hesser, 1976; Dawkins, 1976; Barash, 1977; Wilson, 1978; Chagnon and Irons, 1979; Freedman, 1979; Lumsden and Wilson, 1981). One scholar, as I have implied, has been especially active in this area. William Durham (forthcoming) has cogently shown the validity of the coevolutionary hypothesis in a study of an annual festival held in many areas of tropical West Africa to celebrate the conclusion of a rigid but temporary prohibition against the consumption of the new yam crop. New yams are rich in cyanogenic glycosides, which, through enzymatic conversion in the blood stream, tend to reduce the genetic resistance to *falciparum* malaria conferred by sickling hemoglobin. Various factors point to the coevolution of yam-related culture and the genetics of sickling hemoglobin. For example, the prohibition against the consumption of new yams is carefully timed to coincide with the rainy season and to end with the reduced density of *Anopheles* mosquitoes in the midsummer dry season. Furthermore, the prohibition is conspicuously absent among both people who live outside the spatial range of the midsummer dry period and people who inhabit unusually dry areas within the range. Finally, native beliefs appear to associate fairly clearly the timing of the new yam harvest with the reduced danger of malaria infestation (for a review of other, more proximate but complementary, explanations of the same cultural prohibition, see Coursey and Coursey, 1971).

In still another study, Durham (1976) shows equally compelling evidence that cultures evolve by the selective retention of cultural variations that enhance the survival and reproductive success of individuals in their particular environment. The heart of the study is a model that accounts for the presence or absence of intergroup competition for resources, and predicts conditions under which intergroup aggression yields a net fitness benefit for individuals despite the inherent costs of aggression. Data about war among small, stateless societies uphold the general hypothesis that individuals maximize their chances of survival and reproduction by participating in collective aggression or war when access to scarce but valued resources is at stake.

In conclusion of this section, it bears reiterating a point emphasized in Chapter 1 in connection with the debate over the concept of adaptation in sociobiology. Culture may be said to be adaptive; but precisely because cultural evolution appears to be tied to natural selection, culture can be adaptive only in a *selective* sense. That is, it is more adaptive for some individuals and groups than for others. Similarly, different degrees of culture—say, of education—have different adaptive effects. Otherwise, culture would interfere with the sorting process of natural selection. "Adaptiveness of culture," therefore, is a phrase whose meaning must always be grasped in a relative sense.

An example will help to firm up this point and at the same time give a maladaptive glimpse of sociocultural behavior. Cochrane (1979), among others, has argued that education has negative effects on fertility. The higher the level of education, the more likely is the regulation of fertility, even in marriage. More specifically, as education increases, the probability of postponing marriage and parenthood also increases. This circumstance, in turn, increases the chances of one-child parenthood or childlessness altogether. This relationship is in part a special case of the fact that age at marriage has consistently negative effects on fertility (Rindfus and Sweet, 1977: 150). In recent years women in the United States of America and other highly industrialized societies have postponed marriage or motherhood at an unusually high rate; and an exceptionally large percentage of them have avoided marriage altogether. Such behavior has had its effect on fertility. Thus, in 1965 childless marriages represented 12 percent of those women who married when they were 25 to 29 years of age. By 1976, such marriages accounted for 22 percent of the women who waited that late to attain a marital status.

In concomitance with these developments, the rate of first-child motherhood at a late age has also become a widespread phenomenon, and education correlates highly with this postponement. Thus, in 1979 almost half of the first-time mothers aged 30–34 years had completed at least four years of college, compared with just 28 percent in 1970 (Ventura, 1982). Such mothers tend to have healthier babies since they are more likely than younger mothers to seek early prenatal care (Ventura, 1982). But this type of advantage is not likely to balance off, in terms of genetic fitness, the fact that the older mothers are more likely than younger mothers to end their reproductive cycle with a single child. Consequently, it seems safe to conclude that education, an aspect of culture that is valuable in so many respects, also tends to contribute to the reduced fitness of certain people—perhaps those who from the viewpoint of the maximization principle may have too much of it.

4.2 The Climbing Maneuver

We may now enter discussion of the behavioral predispositions that together constitute what has been termed human nature or the biogram. One of the most interesting and most rich of sociocultural repercussions has often been referred to with varying degrees of explicitness in the theorizing of political sociologists, among other experts of sociocultural science. For convenience sake, I term it the *Climbing Maneuver*, but this label should be understood as a convenient shortcut. It really stands for the expression "the inclination to maneuver into a social climb," and refers to a tendency whereby an individual, perhaps a social class, or any type of aggregate, maximizes its chances of moving up from a lower social position to a higher one. The effort at movement is cloaked by all sorts of sophistries that emphasize the interests of a collectivity, such as a whole class, instead of personal interests. The deception is not necessarily consciously executed. Indeed, very often the climber has no awareness of the difference existing between apparent and underlying goal, self-deception frequently preceding deception of others (Mosca, 1896; Michels, 1914; Pareto, 1916).

At close examination, however, the real goal is not the collective good (for example, equality and justice for all) but something issuing from a behavioral force with which egalitarianism must in fact contend hopelessly, at least in the foreseeable future. The predisposition suggests that people do not really wish equality for all. Rather, they are inclined toward achieving a superior position for themselves, often by replacing those who are already more fortunate. It is, therefore, analogous to the force that, for example, impels a lion seal without a harem to engage in combat a more fortunate one with a view to gaining control over the harem.

"Whatever name we give it," Alfred Adler (1932) argued, "we shall always find in human beings this great line of activity—the struggle to rise from an inferior position to a superior position, from defeat to victory, from *below* to *above*. It begins in our earliest childhood; it continues to the end of our lives." Other scholars have taken notice of this trait. Auguste Comte (1875–7, 1896), for example, termed it the "instinct to dominate." Later, Giddings (1896), in a milder form, labeled it the "desire to be as well off as others."

The struggle for achievement, as we might otherwise term the climbing maneuver, is a fundamental feature of the social existence in more species than our own (for some general comments, see van den Berghe, 1974; Wilson, 1975b; Barash, 1977). Striking, however, among the many peculiarities of its manifestation in human society is our tendency to conceal its true nature with a veneer of verbiage—the sophistries—that tends to present it as a force of egalitarianism. This work of camouflage normally takes the form of appeals either to the individual interest of others or to a collective interest of some type (Pareto, 1916: 1477–97, 1498–1500; see also Pareto, 1906: ch. II, sections 104–6). In the former instance, A would falsely convince B that a given action is beneficial to B. Sharing the responsibility of management with the workers, for example, is alleged to be advantageous to management because workers would have a more direct interest in the enterprise and, thus, would be more hard-working and productive. The sophistry lies in the fact that those most interested in co-management are not the workers themselves but those who call themselves such, namely, union officials who have a keen eye to the advantage of greater gain and power. The sophistry is used with equal success by the person who already enjoys a privilege. For example, it was once said that docility was beneficial to the slave because it afforded him the protection and goodwill of the master.

The other sophism, accord with collective interest, is generally used by people who want something for themselves but ask for it in the name of a group or a society as a whole. Politicians and business people are masters of this variant. All ruling classes and aspirants thereto have at all times identified their own interests with the interests of the country, and much of their behavior has been directed toward misleading the masses into serving their interests. Thus, American politicians often ask for the public vote in order to preserve the two-party system and the "American way of life." Likewise, capitalists only too often forget about the principle of free enterprise, allegedly so dear to their hearts, and ask for governmental subsidies, guaranteed loans, and the like in order to save a tottering industry "for the benefit of the working classes" if not altogether for the national prosperity. So, the American automobile corporate leadership and its union counterpart ask for quotas on Japanese imports in the name of the American worker. What at bottom is true in all this is that the two-party system is dearest to those who are out of power, and the benefit of the working people is closest at heart when it is essential to the maintenance of the businessman's and the union leader's position of power. The climbing maneuver has tenacious defense of the mountain top as its corollary.

Class consciousness, or class struggle, long a pivotal focus of sociocultural science, would seem to be real enough. But upon close examination it is really an *individualistic* struggle. It is a means of self-improvement for a few fortunate individuals. That is why I have termed the predisposition in question the climbing maneuver. I shall return presently to this question of class consciousness. In the meantime, it bears stressing that the predisposition in question is really a sentiment of privilege. It provides the impetus to scale the walls of inequality in order to improve one's own condition. During this stage, the individual is rich of talk about equality, justice, fairness, classlessness, and the like. Once success has been achieved, the need for a society protected by hindrances against the less fortunate—what will be termed the need for domination in Chapter 5—then asserts itself with a vengeance.

Examining carefully the climbing maneuver, one is struck by an interesting contradiction between two tendencies in human behavior, namely, what is sometimes termed the need for democracy and the need for aristocracy. On the one hand, there is a tendency to include the largest possible number of persons in the advantages that we ask for ourselves. On the other, there is a tendency to restrict as far as possible the number of those who actually share in our benefits. The paradox vanishes the moment we consider that the latent strategy is to admit to the advantages all those whose cooperation helps us toward obtaining or increasing them, so that their inclusion yields more than it costs. Conversely, we exclude others when their participation costs more than it yields. A warlord wants a large army for the fighting, but he prefers to have as few soldiers as possible with whom to divide the booty. As sociologist Lenski (1966: 44) puts it in a related key, "men will share the product of their labors to the extent required to insure the survival and continued productivity of those others whose actions are necessary or beneficial to themselves."

The predisposition to climb is normally vigorous precisely because, instead of being a sentiment of equality, it expresses the direct interests of individuals who are bent on setting up systems of inequalities that maximize their own benefits. And yet this canny behaviour is not without redeeming grace—or without evolutionary wisdom, for that matter. Without it, given distributions of power and influence might undergo little if any change. Social mobility—so crucial an associate of behavioral predispositions, as we have seen—would be greatly reduced. Ruling classes would deteriorate to levels that would invite group predation and possibly extinction.

But we must not lose sight of the selfish undercurrent. It was Aristotle, I think, who noted that conscience lies deep inside the granary. It follows that hypocrisy, the loathsome label for fake altruists, is an especially pronounced property of privileged position. For, the rich, the high, and the mighty not only have more to defend; they also have, by virtue of their power and possessions, greater clarity of vision about valuable goals and the means whereby to achieve them. This greater vision expresses itself partly in the manipulative behavior that hypocrisy usually entails.

Human societies offer two chief means of acquiring useful resources. One is to produce them directly through one's own work; the other is to appropriate the resources produced or possessed by others. Direct production is often very arduous, whereas appropriation is at times extremely easy. Behavior develops along the line of least resistance, and sophistries cunningly employed to appeal to popular predispositions and goals not infrequently dupe fellow human beings into facilitating such movement. Again, governing classes are masters of such techniques, often deploying the maintenance of law, order, and security in the service of their self-serving operations. Exploitation tends to receive the sanction of established religion and the law, for those institutions are intrinsically oriented toward ordered and stable relations, and thus usually favor the rulers.

In general, the masses stand toward their governing classes much as bees stand to apiarists who make periodical extraction of the honey produced by those busy insects. The techniques of appropriation are varied. They include plunder, individual assaults, assaults on inheritance, religious and political favors, various sorts of clever fraud, special tax laws, issues of bonds and certificates of public debt that are then repudiated in whole or in part, currency devaluations, monopolies and protective privileges, and so on.

> The simplest form is the direct spoliation by violence of certain numbers of savers, often selected by chance and with a view only to their wealth. That corresponds in a certain way to the hunting of wild animals. Forces that are progressively more complicated, more and more ingenious in character, and more and more general in bearing keep appearing as we come down in history; and they correspond to the rearing of domestic animals. The analogy holds even for the consequences. The first sort of chase destroys incomparably more wealth, and occasions far greater disturbances than the second. (Pareto, 1916: 2316)

Nothing I have said, of course, can gainsay the fact that the collective interest is not necessarily a mere metaphysical fiction. If nothing else, a selfish pursuit can have a collective good as an indirect consequence. Moreover, hypocritical sophisms may be of great importance to social order, though again more from the viewpoint of the more fortunate than from that of the less fortunate. They tend to intensify predispositions of sociality and may even facilitate genuine, non-self-serving, altruism. To a degree, therefore, they obviate deleterious conflicts between purely selfish predispositions and predispositions that may serve to enhance broader sets of interests.

Hypocrisy, like beauty, is in the eyes of the beholder. It tends to engender disgust and contempt. Disgust for hypocrites is a good sign of the viability of sociality forces. It may even reveal that residual force, that pinch of asceticism, that in some cases leads to accept the principle that to give is more sublime than to receive. Indeed, it may represent the anguished cry of a tormented conscience; it may betray the deep emotions of guilt and shame for the failure to achieve the nobility of self-abnegating behavior and to emulate the better measures of virtue. One is reminded of Mr Pip, the immortal social climber created by Charles Dickens in *Great Expectations*. Thinking back to the time when he, now in a lofty social position, received the embarrassing news that Joe, his blacksmith brother-in-law and childhood friend and protector, was about to pay him a visit, Pip reveals the dramatic, if delayed, human capacity to condemn one's own evil, often perpetrated at the service of the evil of others: "I had little objection to his being seen by Herbert or his father, for both of whom I had respect; but I had the sharpest sensitiveness as to his being seen by Drummle, whom I held in contempt. So, throughout life, our worst weaknesses and meannesses are usually committed for the sake of people whom we most despise."

The climbing maneuver is typically executed in a situation that features that eminently social property known as power. Max Weber's (1922) classical definition of power refers to "the probability that one actor within a social relationship will be in a position to carry out his own will despite resistance, regardless of the basis on which this probability rests." Closer to my perspective, Richard Adams (1975: 9–10) has written of power as being "derived from the relative control by each actor or unit over elements of the environment of concern to the participants." The power concept has received a great deal of reflection in social science, and given rise to some interesting reformulations and applications (e.g. Bierstedt, 1950; Blau, 1964; Gamson, 1968; Homans, 1974; Schacht, 1978). Basically, power is insignificant apart from the conceptual polarity of domination and submission. Power, thus, is a property of social stratification or, more broadly, of dominance orders.

Within this context, power has a causal significance analogous to the joint significance of gravitational force and friction in physics. Specifically, it may be viewed as the capacity of individuals or groups to place impediments in the path of other agents' pursuits or to reduce these in one's own favor (Lopreato, 1977). Hence, power is a measure of the differential access to the scarce goods and services that are distributed through a dominance order. The easier the access to, and the greater one's share of, those goods and services, the greater the power. By the same token, the more successful the climbing maneuver, the greater the power, for the achievement of power—the reduction of friction to one's own advantage—is precisely the basic goal of the climbing maneuver. Equally true is the proposition that the greater the power, the easier the climb, but within limits, for power is greatest at the top, which constitutes the ceiling of the power structure. The limitation, however, disappears if we visualize shifts from one power structure to another. On the whole, for example, it is easier for a senior executive than for a junior executive to become a member of a national cabinet.

Ultimately, what we call stratification systems are the results of self-enhancement predispositions, for it is these that generate power. But as the climbing maneuver attests, once a power structure has been established it also facilitates the expression of those very predispositions. Thus, the climbing maneuver ends by defending the existence of the power structure. But it began by making it work to one's own advantage. The result is the maintenance of inequality of opportunities (or power) and the concomitant reshuffling of the distribution of those same opportunities.

This manner of conceptualizing the problem introduces a sort of theoretical paradox. On the surface of it, it would seem that the greater the rate of social mobility in a society, the more egalitarian the society is. The paradox lies in the fact that for any given individual the climbing maneuver introduces fluidity in social systems—it combats social friction—while simultaneously it constitutes impediment of movement (friction) for others in direct proportion to the success of the maneuver.

We may illustrate the paradox by returning to the problem of class consciousness or struggle. In *The Poverty of Philosophy*, Karl Marx (1847: 173), the great revolutionary who added his own peculiar brand of ideology to the evolutionary theory he so much admired, wrote as follows of the transformation of the working class in industrializing society:

> Economic conditions had first transformed the mass of the people of the country into workers. The combination of capital has created for this mass a common situation, common interests. This mass is thus already a class as against capital, but not yet for itself. In the struggle, this mass becomes united, and constitutes itself as a class for itself. The interests it defends become class interests.

This passage epitomizes what is meant by class consciousness in classical studies of social inequality: namely, the gradual unfolding of a corporate feeling that inclines to the aggressive pursuit of the interests of a group as a whole against those of another group. On this basis, Marx looked forward to the proletarian revolution and envisioned the following (classless or communist) end-result:

> If the proletariat during its contest with the bourgeoisie is compelled, by the force of circumstances, to organize itself as a class, if, by means of a revolution, it makes itself the ruling class and, as such, sweeps away by force the old conditions of production, then it will, along with these conditions, have swept away the conditions for the existence of class antagonisms and of classes generally, and will thereby have abolished its own supremacy as a class (Marx and Engels, 1848: 29).

Gone will be the struggle for dominance and privilege. The wealth of a nation will become "social power"—in Marx's terminology, the equitable and collective ownership by the citizenry. The principle that will presumably regulate and define interpersonal relations on the marketplace will be: "From each according to his ability, to each according to his needs" (Marx, 1875).

In fact, much human misery has been wrought under the pretext of such worthy principles. The "Marxist" revolution has come in a considerable number of societies, preeminently in Russia where, after a lifetime of presumed efforts to bring about the classless society, conditions still invite theoretical speculation about the real "motives" that underlay the revolution. Existing facts in Soviet Russia suggest the conclusion that, if we account for the economic improvements that in varying degrees attend the process of industrialization everywhere, what was allegedly meant to be the revolutionary movement toward a classless, unexploitative, society has in fact gradually restored, in substance though not in form, the political *status quo ante*. Political sociology and now biocultural science could have predicted the outcome. According to Robert Michels's (1914) "Iron Law of Oligarchy," it is always a small minority that governs, and its primary goal is to maintain itself in the saddle at any cost. The power of the Soviet dominant class in relation to the masses is probably no less today than it was in the period prior to the revolution. This is the message about socialist societies in general conveyed by Milovan Djilas (1957), the Yugoslav scholar who, as Marshall Tito's colleague, saw communist conditions from within and later spent much of his life in jail for daring to unmask the cruel sham and afflictions perpetrated against the proletariat in the name of the proletariat (for some "Marxist experiences in destratification," see also Lenski, 1978; Jones, 1978; Ulč, 1978; Volgyes, 1978).

"Everything happened differently in the USSR and other Communist countries from what the leaders . . . anticipated." The state did not wither away, it grew stronger. Classes and class exploitation did not vanish, collectivization and destruction of capitalist ownership notwithstanding. Rather, a "new class" of owners and despots evolved. Djilas refers to it as the "political bureaucracy." With the Communist Party as its base, it consists of those who have special privileges and economic preference because of the political monopoly they hold. Their dominance over the masses is much greater than the dominance of ruling classes in capitalist societies, for the new class is absolutely accountable to no one but itself. As it matures, the political bureaucracy becomes increasingly impermeable at the highest reaches. "Open at the bottom, the new class becomes increasingly and relentlessly narrower at the top" (see also Jones, 1978). Theoretical equality gives way to an arrangement whereby even the promising children of the humble are with relatively few exceptions channeled into secondary training, whereas the offspring and other relatives of high party functionaries attend the University of Moscow where they are trained to rule tomorrow.

Facts such as these and the logic of the climbing maneuver require that a distinction be made between class struggle as *action in process* and class struggle as *action in retrospect*. The paradox encountered earlier inheres in this distinction. As action in process, class struggle may indeed be viewed as a collective phenomenon, and the individuals who engage in aggregate action may be thoroughly convinced that they are pursuing a collective good. However, at a certain point in the unfolding of the collective good, action becomes clearly egoistic, at least for some. The impulsion toward individual self-enhancement is triggered off in direct proportion to the benefits derived from collective action. The closer one gets to the reins of power, the more likely one is to forget the collective goal and to pursue one's own interest. In this sense, however, it is correct to conclude that what appeared to be collective action was in fact only a mere *summation of individual actions* motivated by forces that would by nature impede free social mobility and the emergence of true collective results. In retrospect, thus, "class action" was selfish, *individual* action.

The class struggle destroys one source of power, or friction, only to introduce another. The fundamental force behind such a turn of events is natural selection as manifested in part by the behavioral predisposition now under consideration. In part, and at a higher level of abstraction, the climbing maneuver may be considered an element of a large adaptation relevant to group survival. Specifically, it may be viewed as part of a mechanism whereby certain positions of high executive importance are emptied so as to be refilled with other individuals who, for a while at least, feature a higher level of administrative competence. That, at the aggregate level, is the fundamental significance of socio-political revolutions. They replace inept and incompetent rulers with elements who, for the time being at least, are better qualified to mobilize and guide group action according to the basic functions attributable to viable governing classes: organization of defense against other societies, maintenance of internal social order, and stimulation of technological development.

These considerations naturally cast serious doubt on certain aspects of class analysis normally practiced in social science, especially in sociology and political science (for a conceptual analysis and comparative discussion of class consciousness, see Lopreato and Hazelrigg, 1972: pt II). According to a national study of class behavior in Italy, a country whose intense political life would suggest a highly developed class-consciousness, the most common type of interclass relations perceived by the respondents is one of hostility and suspicion; the response accounts for over 40 percent of the sample. The inclination is to interpret such a finding as evidence of a class type of behavior (Lopreato and Hazelrigg, 1972: 187–8). In fact, however, it may represent only an individualistic form of social conflict that is merely cloaked in collectivist garb. Much evidence points in that direction. For example, a considerable percentage of people spontaneously (without a forced-choice question) identify themselves as "workers," as "powerless," and so on, and yet claim occupational categories situated at the upper reaches of the occupational structure as members of their own self-assigned class (Lopreato and Hazelrigg, 1972: 208). Likewise, in Italy 40 percent of those who have skidded from above into positions of skilled blue-collar workers continue to think of themselves as members of the middle class or altogether of "the bourgeoisie" (Lopreato and Hazelrigg, 1972: 210). Ironically, however, that does not prevent skidders in general from espousing left-wing politics and complaining almost in unison that the ruling elites are those who profit most from national prosperity. Hence, the "revolutionary politics" of skidders really represent "an individualized, pragmatic, and self-seeking calculation" (Lopreato and Chafetz, 1970: 448).

Little wonder that in some societies, where the political means are relatively weak for such a work of behavioral camouflage, the bases of class self-identification bear little relation to socioeconomic status. Thus, while Americans in the USA are quite capable of identifying with a given social class on the basis of a forced-choice question, their identification is only weakly related to such conventional indices of social class as occupation, income, and capital holdings (Hodge and Treiman, 1968). These authors conclude that "patterns of acquaintance and kinship between various status groups, as well as their residential heterogeneity, are no less important than the socioeconomic position of individuals in the formation of class identities." In later chapters, we shall learn to appreciate this emphasis on kinship. The only likely problem with these authors' conclusion lies in the assumption of *class* identification in the first place. If the appeal to class is a convenient means whereby to hide a strictly self-interested orientation, identification with some class is greatly facilitated when we must choose a label out of a set preconceived for us.

Max Weber (1946: ch. 7) rightly argued that it is easier to develop a consciousness of kind on the basis of kinship and shared social status, or honor, than on the basis of shared economic status, or class. That makes very good sense. The status dimension is where we encounter friends and relatives with whom we readily exchange favors and forge alliances that tend to defend or increase our inclusive fitness. The economic or class dimension is where the battle line is drawn, where our competitive spirit is at its maximum, where every other is a potential enemy.

I should stress, in conclusion, that if egalitarian ideology has failed to negate the climbing maneuver, and thus to help bring about the egalitarian society, the ideology has not been without consequence for the climbing maneuver. The nature of the effect has already been alluded to, and recalls the discussion, in Chapter 3, of the factors that tend to modify behavioral forces. The primary modification relevant here has, ironically, taken the form of a great intensification of the climbing maneuver.

This powerful environmental influence may be due to the fact that the egalitarian ideology is a synthesis of most, if not all, the modifiers discussed in Chapter 3. Certainly it is a complex historical phenomenon composed of intricate actions and reactions. Fundamentally, the egalitarian ideology is a compound of direct and indirect results of a very great evolutionary event, namely, the transformation of agrarian society by the Industrial Revolution. Especially worthy of mention perhaps is the fact that the bourgeoisie, the social class that had its historical ascendancy during that Revolution, prevailed over the *ancien régime* and the feudal aristocracy in part by virtue of its democratic appeals to the lowly masses. The legitimate parent of socialism is capitalism. The "commonwealth," "the people," "the common interest," "the general will," and such other political formulas, as Gaetano Mosca (1896) called them, were all creations of the rising bourgeoisie which, in its historical battle for political supremacy, needed the material support of the popular masses. It is those fictions, basically, that by the eighteenth century had blossomed into a popular egalitarian ideology that a little later had its sharpest expression in the great synthesis of Karl Marx.

The rise of the trade unions and the multitude of progressive parties was the major organizational result of the ideology. Organization, in turn, bestowed on countless people the power to challenge established privilege and numerous opportunities to climb the proverbial ladder of success. This stimulation of the climbing maneuver was assisted, of course, by the enormous economic expansion of industrializing society and by the common observation that upward mobility, the climb onto a spot in the sun, more or less central, was no mere illusion but a reality of everyday occurrence. The subject is complex. We may recapitulate, roughly, by stating that great historical opportunities for achievement stimulate the mind toward spawning great ideologies of success, and these, along with the opportunities themselves, energize the climbing predisposition. But as the predisposition intensifies, it reacts, in turn, on the ideology and stimulates it further.

4.3 Territoriality

Some animals, "including nearly all vertebrates and a large number of the behaviorally most advanced invertebrates, conduct their lives according to precise rules of land tenure, spacing and dominance. These rules mediate the struggle for competitive superiority. They are enabling devices that raise personal or inclusive genetic fitness" (Wilson, 1975b: 256; for some of the controversy surrounding the territoriality concept, see Noble, 1939; Carpenter, 1958; and especially Peterson, 1975, for a very useful, brief review). *Territoriality* is the name commonly given to this behavior syndrome, and its wide diffusion in the animal kingdom as well as its apparent universality in human society suggest that it is rooted in a behavioral predisposition. The territoriality concept is so widespread in evolutionary behavioral science that it may be helpful to use it in the present volume to refer to both the cultural universal and the predisposition possibly underlying it.

Ours, however, may not be the most felicitous choice of terms. For the human species, the term "territoriality" is justifiable when we consider that until very recent times in our history the vital resources were direct properties of the terrain. But to some extent, the direct salience of terrain has been reduced in modern society; thus, labels such as "acquisitive force or impulse" and "sense of property," which are widespread in social thought, might usefully serve as substitutes for territoriality.

Still, the subject of our inquiry directs attention to behavior that is frequently observable within a given space or territory, and maintaining a link to the latter concept will have its usefulness. In evolutionary ecology a territory is understood to be "an area occupied more or less exclusively by an animal or group of animals by means of repulsion through overt defense or advertisement" (Wilson, 1975b: 256; see also Brown, 1964, 1975; Wilson, 1971a). Accordingly, territoriality as a predisposition may be defined as a genetically based tendency to acquire and defend space containing resources that directly or indirectly enhance fitness.

The use of the term "defend" in this definition may at first blush appear to direct attention to a passive form of behavior. In fact, however, defense is irrelevant without the correlative concept of attack or aggression, and thus the term conveniently conveys the relational aspect of the concept in question. Indeed territoriality is closely interwoven with aggressive or agonistic behavior. Wilson (1975b: 249) notes: "The strongest evoker of aggressive response in animals is the sight of a stranger, especially a territorial intruder." Likewise, the quest for territory is frequently pursued as an end worthy of any means.

The conflict over territory may take place between individuals or groups of the same species, and it may obtain also between species. Indeed, interspecific competition over territory is one of the chief forces of evolution, for, although two or more competing species may coexist stably, more frequently it happens that one will push the other(s) out of the overlapping area. Even when the competitors are territorially compatible, they are bound to reduce each other's niche space and biomass (Wilson, 1975b: 276; also Murray, 1971), for the basic tendency is to establish a territory that is small enough to be easily defensible but at the same time large enough to suffice for sustenance. Thus, sharing may reduce a space unit that approximates minimal needs, and accordingly tends to reduce the fitness-enhancing property of the unit. My interest here is exclusively in intra-specific territorial behavior.

As the quoted passage that opened the present section suggested, territorial behavior is widespread in nature, although it probably reaches its widest diffusion among birds (e.g. Stokes, 1974). It does not, however, appear to be universal, or universally very salient. Indeed, some of our closest relatives—chimpanzees, gorillas, baboons—have haunts that are neither exclusive nor typically defended to any great extent (DeVore, 1965). On the other hand, there are animals that, while genetically much further removed from us, nevertheless have feeding habits that are remarkably similar to those characteristic of our hunting and gathering ancestors, and they maintain *group* territories. I am referring to such social carnivores as lions, hyenas, wolves, and African wild dogs. These species regularly patrol a territory that is roughly the size of that patrolled by hunting and gathering peoples. They hunt in packs, and like human hunting parties, they manage thereby to capture prey that is much larger in size than themselves. The similarities in these respects are so striking that such species have been termed the ecological analogs of the human species (Wilson, 1978).

To say, however, that territorial *behavior* is not universal is not necessarily to say that the territorial predisposition is not universal. Recall that behavioral predispositions are, *qua* behavioral forces, subject to scaling or varying intensity. Accordingly, even if we had absolute knowledge that all living animals are predisposed toward territorial behavior, we would be in no position to predict such behavior in the absence of knowledge about particular environmental factors that impede or facilitate the expression of territoriality. Thus, the question whether *Homo sapiens* is motivated by an acquisitive impulse is a question that, while subject to lively debate (e.g. Reynolds, 1966; Montagu, 1968; Esser, 1971; Crook, 1973; Cohen, 1976;

Durham, 1976), can be answered only with much circumspection and not a few qualifications.

Understanding territory in a general sense, as space, social scientists have with few exceptions been keen to its theoretical importance. Robert Park (1967: 63), for example, once stated:

> Since so much that students of society are ordinarily interested in seems to be intimately related to position, distribution, and movements in space, it is not impossible that all we ordinarily conceive as social may eventually be construed and described in terms of space and the changes of position of the individuals within the limits of a natural area; that is to say, within the limits of an area of competitive cooperation.

Another scholar explicitly postulated "relations with places" as a basic human behavioral tendency (Pareto, 1916: 1041–2). More recently, a sociologist has argued that territoriality is a biological force and, together with hierarchy, mediates the aggression that is engendered by competition for resources within a group. Resource competition is aggravated by population pressure and, in recent millennia at least, by culturally created needs. The greater the demographic pressure and the more intense the culturally induced needs, the greater the competition for resources and, therefore, the greater the level of aggressive behavior. Territoriality mediates aggression in the sense that it "defines monopolistic rights of resource exploitation within a given space," while hierarchy, a subject to be taken up in the next chapter, establishes "a rank-order of access to, and/or a scale of distribution of, both material and social resources." More specifically, territoriality and hierarchy mediate aggression because, while on the one hand they suppress or regulate it to the extent that the distribution of existing resources is acceptable to the underprivileged, on the other they stimulate aggression in the degree to which they predispose to a challenge of the status quo (van den Berghe, 1974).

This scholar rather convincingly demonstrates territoriality in human nature. "Man is not only highly territorial, but he is territorial at practically every level of social organization: the family has its home, the juvenile gang its 'turf,' the ethnic group its ghetto, the state its national territory. Even monastic orders and 'counter-culture' communes dedicated to ideals of equality and propertylessness post no-trespassing signs against outsiders, and establish pecking orders between their members."

Considerable evidence of human territoriality is available from ethnography where, until recently at least, interest typically focused on pre-agricultural societies, whose conditions of life were much less removed from the zoological realm than is the case with agricultural and industrial societies (see King, 1976, for a review). Tylor (1871), Durkheim (1912), and Radcliffe-Brown (1930), among other great masters, saw ample evidence in their rich sources to argue that the territorial band was the typical, and thus the optimal, pattern of spatial organization for hunter-gatherer peoples (see also Service, 1962; B. J. Williams, 1974). This position has been convincingly reaffirmed more recently by several studies. Nicolas Peterson (1975) has written that Australian hunter-gatherer peoples have displayed territoriality in several ways, for example, through the notion of a "personal space halo," the concept of landed property, a willingness to defend areas of ground, and fairly specifiable limits to the range of daily, yearly, and lifetime wanderings. Territorial behavior, moreover, has had fundamental consequences in the evolution of Australian societies: for example, through intergroup warfare (see also Durham, 1976). Chances, then, are that territoriality is indeed an innate tendency. But whether or not this proposition is based on fact, "spacing is basic to Aboriginal society survival and therefore of biological significance."

Many scholars have noted a flexible form of spatial arrangement (e.g. Lee and DeVore, 1968; Damas, 1969; Martin, 1974; Dyson-Hudson and Smith, 1978), and some have thereby tended to dismiss the territoriality concept as unimportant. Conceptual confusion aside, however, the latter position is often based also on a lack of sensitivity to the intensity property—to the fact that behavioral predispositions have varying degrees of expression, depending on various environmental contingencies.

In a recent paper, Dyson-Hudson and Smith (1978) reinterpret the heterogeneity and complexity of human spatial organization and argue that it parallels fairly well that discovered by sociobiologists in animal territorial behavior. These scholars choose to avoid the question whether humans are innately territorial, but, specifically using the sociobiological concept of territory quoted above, they apply a well-reasoned cost-benefit model that focuses on "economic defendability" (Brown, 1964). The model predicts territorial behavior when the benefits of exclusive use and defense of a space outweigh the costs incurred in such behavior. Further, the model assumes that such a strategy tends to maximize fitness, for the energy extracted from the defended territory is superior to that expended in its defense.

In general, according to the model elaborated by Dyson-Hudson and Smith, territorial behavior is a function of economic defensibility, and the latter, in turn, increases with a rise in the *predictability* of resources in their spatiotemporal distribution, up to a point, and with an increase in the average *density of resources*. Predictability of resources makes possible economy of effort and risk on the part of the occupant(s). Density of resources reduces the space that must be defended, and thus also the defense costs.

The model is applied to three human cases: (1) three ethnolinguistic groups comprising the Basin-Plateau Indians of North America (Western

Shoshoni, Southern Paiute, and Northern Shoshoni); (2) the Northern Ojibwa of the Northeast; and (3) the Karimojong of northeastern Uganda. The first affords a chance to demonstrate intergroup variation in a single region. The second is examined historically, and thus affords a study of variation in time. The third case features intra-group variation at one point in time.

The Western Shoshoni of the Basin Plateau lived in an area of low and patchy rainfall. For their sustenance they had to rely primarily on grass seeds and pine nuts. The seeds, furthermore, ripened sequentially from the lowlands to the highlands as the seasons changed. Due to the nature of rainfall, they were both scarce and unpredictable. Pine nuts, on the other hand, could be superabundant in the winter, but they were unpredictable and ephemeral. As a result, "the Western Shoshoni exhibited changes in degress of nomadism and dispersed in general accordance with our model. . . . territoriality was an economically undefendable option."

By contrast, the Southern Paiute, the second of the Basin-Plateau Indian groups for which information is adequate, lived in a smaller and much more productive area featuring a diversity of resources, including both edibles and an abundance of water. The sufficiency of water facilitated irrigation and enhanced both density of resources and their predictability. As a result, "groups of villages were organized into well defined bands which delimited and defended territories." It is interesting to note, however, that spatial organization with respect to hunting, which involved a high degree of movement, was much more flexible; while some hunts were limited to village members, others, especially the deer hunts, involved the participation of several bands (see also Steward, 1938, for the Indians of the Great Basin region).

The Northern Ojibwa lived originally in Canada east of Lake Superior, and they subsisted by gathering a wide variety of vegetable produce and by hunting moose and caribou, among other prey (Bishop, 1970, 1974). As long as they depended on such highly mobile and unpredictable game for a large portion of their livelihood, the Northern Ojibwa did not defend territories. But by 1820 large game was depleted, competition for fur-bearing animals became fierce in response to the fur trade, and the Ojibwa came to rely on small game, primarily hare and fish, for their subsistence. Accordingly they developed family hunting territories. According to Bishop (1970: 13), "the loss of large game, caribou and moose, and the forced reliance on hare and fish constituted the crucial factor in the development of family hunting territories."

Finally, the Karimojong have a great variety of resources. The major sources of food energy, however, are cultivation of plants, especially the sorghum grain, and the husbanding of cattle. The grain is predictable both in time and in space, and although yield per acre varies widely, it is generally also a dense resource. It becomes denser still when, after the harvest, it is stored in granaries. Under these conditions, the Karimojong may be said to be a territorial people engaged in the protection of a particular resource and area.

Their behavior, however, undergoes an interesting transformation in the sphere of pastoral activity. Cattle, too, is predictable and may even be considered abundant. Hence, it is protected as a valuable resource. However, defense of cattle does not entail the delimitation and defense of particular territories. Cattle is highly mobile, following the availability of grazing land, which is unpredictable due to drought and a relatively low density of plant resources. "Because of the patchy and unpredictable nature of resources and the individualized pattern of herd movements, territorial ownership of fixed grazing areas is not a viable strategy.... Only in times of extreme scarcity do Karimojong herdsmen defend grazing areas, and the defended areas are those which have relatively abundant grazing at that particular point in time" (Dyson-Hudson and Smith, 1978). They also defend their tribal grazing lands against non-Karimojong, killing trespassers with spears and taking their cattle.*

It appears, then, that the territorial impulse is expressed in many forms and in different degrees of intensity, varying in time, in place, and depending on an assortment of environmental factors, including predictability of resources, defensibility of resources, density of resources, requirements of movement in space, and who your neighbor is. It bears stressing the point again that whether we are justified in the hypothesis that territorial behavior is rooted in a biological predisposition cannot be determined with any degree of certainty without direct genetic evidence. But, the fact that certain environmental parameters regularly trigger off territorial behavior strongly suggests that a predisposition for such behavior does exist and is, furthermore, in a relationship of mutual dependence with those environmental factors.

The territoriality of the human species can perhaps be best surmised from aspects of the international relations pursued by modern peoples. Consider European colonialism. As the scientific revolution engendered a technological revolution in navigation, transportation, communication, and commerce, European nations—some more, others less—hastened to chart new pathways across the oceans, discover and map new lands, and with their superior weapons conquer endless peoples spread out across the continents. While bringing "Christianity and Civilization" to the vanquished, the conquerors plundered and pillaged, divesting the losers of their resources and devastating their populations and their traditional ways of life.

It is important to note, however, that rarely did conquest fail to meet

---

* A cursory examination of the Human Relations Area Files reveals ready evidence in support of the Dyson-Hudson and Smith perspicacious analysis (see, e.g., Gusinde, 1937; Takakura, 1960; Schram, 1961; Watanabe, 1964).

with resistance. In the Americas, in Africa, Asia, and Europe itself, conquest and colonization engendered nationalism, guerilla warfare, and wars of national liberation. As the colonial powers became exhausted by the hardships of defending conquered real estate and by endless wars among one another, many of the vanquished reclaimed their lands and expelled the conquerors. In something like fifty years of colonial turmoil, the United Kingdom, for example, slipped from the position of the greatest power on earth to the status of a prominent European country.

Recall, further, that a behavioral predisposition can flower into hypertrophic expressions that have little or nothing in common with its original adaptive function. So today, like much of time past, peoples may set out to conquer not merely, or not at all for that matter, in order to appropriate the wealth of others, but rather in order to impose on others their particular designs for life, their ideologies, their legal-political systems. Thus, in the course of a few centuries, and all the more so in recent decades, Russia has become the largest empire in the world by conquering both in Asia and in Europe vast numbers of different religious and ethnolinguistic peoples that may constitute an economic liability rather than an advantage.

Territorial clashes are no one's monopoly. Looking around the globe, one of the starkest features of spatial organization is the interstitial area that pits virtually every existing nation against at least one other. The British Isles are living a seemingly endless tragedy fashioned by territorial disputes. France and Germany have long bred a more or less dormant animus with respect to Alsace and Lorraine, one of their traditional bones of contention. Poland, the Sudetenland, the Tyrol, the Basques' region, Cyprus, Corsica, England, Argentina provide but a few of the vivid reminders of recent, if not current, territorial disputes between the nations of Europe or former Europeans.

In the Middle East, an area of ancient and complex societies and cultures, the territorial question has long been, and is especially today, *the* question that engenders destruction and threatens a conflagration. Further to the East, one encounters Iran and Iraq struggling for a piece of space; India has its territorial disputes with China and Pakistan; China with the Soviet Union, among others; and the countries of Indochina with one another. The litany is far too long to recite in full.

Field studies of animal behavior have revealed various phenomena associated with territorial behavior that are important enough to be considered special properties of territory, and useful for our own purposes (Wilson, 1975b: 270–4). For example, territorial size typically varies with population density. At the individual level, the territory contracts as population density increases, so that the individual is forced to make room for the population surplus. "But there is a limit beyond which the animal cannot be pushed. It then stands and fights, or else the entire territorial system begins to disintegrate" (Wilson, 1975b: 270).

The circumstance introduces one of the more intriguing properties of territory, the "invincible center." The center is usually the location of either the animal's shelter or the densest concentration of resources. Unless the occupant of such an area is ill or grossly overmatched, he may be expected to prevail physically over territorial intruders of the same species. One of the classic studies of this phenomenon was executed by Tinbergen (1953) in an experiment involving a fish tank with two male sticklebacks who had established nests at opposite ends of the tank. Tinbergen found that when the two individuals were brought close to each other, they became highly aggressive. But he found something else. In the various combinations of approaches between the two territorial defenders, the closer one was to one's own nest, the more likely he was to assume an attacking posture; conversely, the further removed from it, the more likely he was to retreat. This type of behavior is very frequently observed in nature in various guises. Among humans, for example, a man's home is his "castle," and there he may be expected to be most violent against those who would invade his lair without permission or proper warrant. It bears recalling, moreover, that criminal law is most lenient precisely in relation to those who kill intruders of their domicile, and sometimes of their property in general.

Dawkins (1976: 85) offers an ingenious explanation of invincible-center behavior. If resident individuals did not have an advantage over intruders, or, worse still, if they were at a disadvantage, they would always be striving to avoid being caught as residents. That, of course, would be an absurd strategy: costly of time and energy, and devoid of the advantage of knowing, among other things, the habits of predators and prey. In short, to hold your ground, as folk wisdom teaches, bestows a survival advantage; it both expresses and reinforces the sense of property that aids the quest for useful resources.

There is, however, another aspect of this special property of territorial behavior that is of particular interest to the social scientist. To defend invincibly the irreducible patch of one's territory is to create the conditions whereby an increase in demographic density leads to an expansion rather than a contraction of the territory. The reversal normally comes about through migration, which is typical of human as well as much animal behavior. Thus, a country like the United States of America, indeed all the countries of the American continent, have been populated largely by migrating contingents of many other societies. Particularly numerous among the immigrants have been certain nationality groups whose home societies experienced famine and/or intense population pressure, for example, the Irish and the Italians.

It follows that territoriality, through the operation of such mechanisms as invincible behavior and migration, may be viewed as a factor of population control, *inter alia* (see, e.g., Wynne-Edwards, 1962; Brown, 1969;

Wilson, 1975b). Likewise, it follows that the intersocietal territorial disputes referred to above may be the proximate results of migration from one society to another. Sometimes the immigrants are confronted by powerful injunctions, on the part of the indigenous population, to shed quickly the old ways and resocialize to the new before they may expect to partake fully of the citizenship rights and the cultural life of the host group. Such has been the case, for example, in countries like England, Australia, Canada, and the United States of America, where at one time or another there have been vigorous attempts at "pressure-cooking" acculturation. Other times, migration is directed to seemingly empty "niches," that is, to foreign areas that for one reason or another are poorly defended. In either case, sooner or later the result is intergroup conflict: when it is not conflict over territorial rights, it is conflict inspired by the clash between old cultural traits that are slow to disappear and the new traits that demand obliteration of the old.

Sometimes defense and aggression are only latently associated with territoriality. To the extent that territoriality is an evolutionary adaptation, we may hypothesize that in time it has surrounded itself with mechanisms that keep it vigorous even without individual or population clashes. The idea may be better conveyed through an analogy: for example, the quest for inclusive fitness is aided, in the human species at least, by the phenomenon of being in love. The territorial counterpart of this mechanism is *emotional attachment* to a place. It expresses itself in various ways. One is the tendency of individuals to return to ancestral places for a variety of "reasons": for example, worship, reproduction, and feeding. Other cases include the class reunions and the frequently associated return to the favorite "watering holes" of college days; the visits of the immigrants to the "home country"; and, of course, the return to the ancestral lands that has often been observed among American Indians, and today is fashionable among the descendants of the immigrants and the "uprooted," as in the search for African roots among black Americans.

Still other examples include such phenomena as long-distance devotion to the homeland and nostalgia for the hometown. The tendency derives its strength in part from emotions of patriotism and kinship. Such sentiments may be seen at work in innumerable settings. For example, TV show hosts, MCs, and the like invariably captivate an audience's enthusiasm by recognizing and appealing to the region, the state, the city, or even the quarter of town that various segments of the spectators originate from.

Dramatic evidence of territoriality as emotional attachment, something like imprinting in animals, is afforded by the behavior of people whose families and properties have been devastated by floods and earthquakes. There are cities and towns throughout the world that are under constant seismic siege. Their inhabitants remain undaunted generation after generation. For instance, on 6 May 1976 an earthquake measuring 6.9 on the Richter scale shook the Friuli, in northeast Italy, for fifty-five interminable seconds, killing 984 people, injuring nearly 3,000, and leaving about 100,000 homeless. The earthquake came again on 11 September of that same year, and still again, twice, four days later. Many survivors left. "But all of those who left will return" was the consensus of four women in Venzone, one of the affected towns. Why did they and others not leave?, representatives of the public media wished to know. One of the residents put it this way: "There's the land, the mountains, the valleys, the people, the families, the history of the people from generation to generation—you can't leave."

Four and a half years later, the earthquake struck further south in that same country. One day, buses showed up under military escort in one of the ravaged villages. The captain sought to convince the homeless to board the buses and temporarily move to the thousands of hotel rooms vacated for their convenience in nearby towns. The townfolk listened politely to the captain's arguments. Then an old man spoke the collective reaction: "You are a good and capable man, but don't come here again. It would be better for your sake. This is where we live, and this is where we want to die" (*Time*, 1980). Likewise in his novel on *The Immigrants*, Howard Fast gives a dramatic account of the horrendous 1906 earthquake in San Francisco and the bustling activity of reconstruction whereby "perhaps never before in history—or since for that matter—did a new city arise from the ashes of the old as quickly, as hopefully, as vitally as San Francisco." "Go elsewhere?," asks Fast rhetorically. "Live in another place? Be damned if they would!"

4.4 The Urge to Victimize

To seek control of the necessities of life in a situation of scarcity is in principle an indirect way of competing for genetic representation in future generations. Those who succeed in the competition are not necessarily disposed to perpetrate harm against those who fail. There is, however, one way to prevail over others that is much more direct in character; and that suggests an additional, though closely related, behavioral predisposition. Specifically, individuals and groups may be observed enhancing, or attempting to enhance, their life chances, and thus their reproductive fitness, by sacrificing the wealth, the freedom, or altogether the life of others. I hypothesize that such behavior flows from a predisposition that may be termed the *Urge* (or drive, impulse, etc.) *to Victimize*—or, more conveniently, merely *Victimization*. It is probably one of the most effective forces of self-enhancement.

The phenomenon that in social science is known by the label of exploitation can be placed under the present rubric, and it is interesting to note that at least one sociologist has defined exploitation as "increasing one's own fitness at the expense of others" (van den Berghe, 1978b: 143–4). More broadly, we may think of exploitation as the gratuitous and sustained appropriation of the resources of another for one's own benefit. The action tends thereby to increase the exploiter's genetic fitness at the expense of the exploited.

Exploitation takes numerous forms, most of which are interrelated. Consider education. It is a fairly well-known fact that the secondary schools, colleges, and universities of various societies vary widely in the quality of education they can impart in view of their resources, which sometimes originate in public funds. Moreover, schools do not provide egalitarian access to their facilities. The two factors combine with others that might be mentioned to yield socioeconomic as well as educational and other advantages and privileges to certain individuals and groups as against others.

There are all sorts of constraints against equality of educational opportunities, which in today's world are the key to many important forms of success in life (e.g. Porter, 1965; Blau and Duncan, 1967). There is evidence aplenty, for example, that teachers, being mostly of middle-class origin or orientation, are not immune to the implications of the old saying that birds of a feather flock together. In a classic study, many times confirmed before and since, American sociologist August B. Hollingshead (1949) showed that the youth of more "respectable" families were much more welcome in the classroom than those reared among less respectable people. As if that were not sufficient work of impediment and victimization, scholarships were also likely to go to those who needed them least and did not always deserve them most.

Findings of this kind are abundant, and many scholars have been led to argue that the principal culprit responsible for the relative scholastic failure of the socioeconomically underprivileged is not poverty itself but the expectations that it engenders in teachers. William Ryan (1976: 54–5), for example, maintains that in the United States of America "the primary effect of poverty, race, and family background is not on the children, but on the teacher, who is led to *expect* poorer performance from black and poor children. . . . the expectations of the teacher are a major determinant of the children's performance." Likewise, a study of a child-centered school in England revealed that, official policy notwithstanding, teachers tended to emphasize genetic and psychiatric explanations of good or poor performance, often linked to teacher expectations, that were disadvantageous to lower-status youngsters (Sharp and Green, 1975).

Some observers of the USA public school system have convincingly shown that school is in various ways "rigged for failure" as far as the economically weak are concerned. One institutional means of failure is the so-called track system. Originally this innovation was presumably enacted in order to make schooling more relevant to all classes of students, to single out and promote the divers special aptitudes, and to reduce the absenteeism of the underprivileged. But good intentions, even when they exist, often go astray under the burden of more primitive forces. As a result, the track system has become a formal basis for translating class-based biases into academic criteria which separate students into "those who will drop out; those whose diplomas will not admit them to college; those who will be able to enter only two-year or junior colleges; and the lucky few in the honors classes who will go on to elite institutions and to graduate or professional schools" (Lauter and Howe, 1970).

Lest this development be considered a pattern typical of capitalist society, as some students of this topic are wont to declare, we should recall that an analogous situation obtains also in so-called revolutionarily egalitarian societies. Thus, Inkeles (1950), among others, reported the emergence of a "labor draft" in the Soviet Union whereby, by the 1940s, when a post-revolution stratification system had fairly well crystalized, youngsters not attending schools and not needed on the farms—that is, youths from families low in the social class hierarchy—were put to work under conditions that stunted their occupational as well as their educational careers. Inkeles also noted the preference given to high-status students for attendance at the elite schools, such as the military cadet schools, which were, in turn, among the main recruiting grounds for the managerial and other important positions of Soviet society.

"Regardless of any special capacities for academic achievement," according to two authors critical of "capitalist" education in the United States of America, "white, upper-class men are provided the best educational opportunities" in American colleges. Furthermore, schools feature "vertical differentiation," "with students destined for managerial positions being encouraged to develop leadership and decision-making skills, while students channeled to subordinate positions are socialized to comply with authority" (Bowles and Gintis, 1976—for a good study of curriculum-tracking and educational stratification, including a useful review of literature, see Alexander et al., 1978).

However much privilege-dominated schools may victimize the poor, it may also be true that family environments do not encourage equally a positive evaluation of education; furthermore, the factors accountable for such a difference are not all traceable to exploitation. As Porter (1965: 173) hazards to note in his study of the stratification system in Canada, it is "reasonable to assume that in any given human population there is a wide range of general ability depending for its development on the appropriate social environment." On the whole, the greater the ability of parents and/or children, the more positive the evaluation of education, and the less formidable the obstacles, artificial or otherwise, to be surmounted. Thus, if we could assume that there is less ability in the lower reaches of society than in the higher ones, we could also hold that sociocultural arrangements and behavioral predispositions are mutually supportive, the two together accounting for the low achievements of the poor and the underprivileged.

Porter (1965: 168–72) discusses several sociocultural barriers to equality of educational opportunities in Canada, a society that features what is perhaps an average degree of differentiation along ethno-religious as well as class lines. One barrier refers to the influence of religion on educational politics. In Quebec, for example, Catholic boys abandoned their studies much earlier than Protestant boys. So, at 16 years, only one-quarter of the former were still in school, compared to one-half for the latter. While such variations may in principle have reflected differences between the two ethno-religious groups in socioeconomic status and family size, Porter convincingly reasons that the tax burden on education was relatively lenient in Quebec, so that "the resources made available for education are a reflection of the dominant [French-Catholic] values." He further points to various indications that education in Quebec has been considered much more a function of the church than one of the state: the unique absence, for example, of a minister of education in that province. Accordingly, education is inevitably affected by religious belief, and both its quality and its quantity may place the Catholic student at a disadvantage in the secular competition for achievement.

The second social barrier to equality of opportunities, according to Porter, refers to family size. The relatively lower socioeconomic status of Canadian Catholics, together perhaps with their Catholicity per se, accounts for a relatively large family size. As the size of the family increases in such economic circumstances, children become an economic liability unless they, or at least some of them, are put to work in order to help meet the family needs. It is, therefore, worth noting that a large family size is no guarantee of high inclusive fitness. If it interferes with socioeconomic success, it may in the long run reduce inclusive fitness by depressing the life chances of the future generations.

As the above considerations imply, the most obvious social barrier to education is the existing inequality in wealth. That applies not only to Canada but to every other society as well. Education is costly not only in terms of financial expenses but also because of the economic gains that must be forgone during the training years. The poor are, therefore, the least able to afford degrees of formal education that normally tend to improve one's life conditions. This relation of mutual dependence between poor means and poor access to education is sometimes caught in a vicious circle.

In a careful analysis of the American occupational structure, Blau and Duncan (1967: 401–10) found that socioeconomic background (or social origins) exerted a great deal of influence on one's occupational career. However, one's own training and early work experience were even more important for future success. In the typical case, the strength of one's education and first job was such that in a reasonable span of time relative disadvantage in social origins was cancelled out. That was the case, for example, for sons of immigrants and for southern whites moving to northern areas. But there was the major exception of blacks. Their handicaps tended to create "a vicious circle" through which poverty was perpetuated from generation to generation. Education simply did not produce the same career advantages for blacks that it produced for whites. Remarkably, for example, the occupational careers of well-educated blacks lagged further behind the careers of comparable whites than did the careers of poorly educated blacks when compared to their white counterparts. The vicious circle lies in the fact that, since education did less to advance the occupational career of blacks, these were more likely to drop out of school early—a fact that had depressing effects on their career and socioeconomic status.

It bears noting that school systems and schoolteachers, while morally blameworthy for their favoritism toward the privileged classes, belong to a large class of culprits. Some years ago, for example, sociologist Feagin (1975) reported the results of a public opinion study of the United States of America which included an inquiry into the "reasons some people give to explain why there are poor people in this country." The author divided respondents' choices into *individualistic explanations*, which attributed the cause(s) of poverty primarily to the poor themselves, *structural explanations*, which blamed socioeconomic arrangements, and *fatalistic explanations*, such as bad luck and illness. A majority of the sample blamed the poor for their poverty. What is more remarkable is that differences in this respect were not large when Feagin compared responses across such control units as socio-religious groups, racial groups, age categories, education, and family income. Consider the latter. Fifty-one percent of those in the lowest income bracket were high on the individualistic attribution. Among those in the highest income bracket, the corresponding percentage was 50. Findings almost identical to these have also been reported from Italy for the middle 1960s (Lopreato and Hazelrigg, 1972).

The exploitation of the economically weak sometimes takes astonishingly subtle forms. In a classical study of social class and mental illness, Hollingshead and Redlich (1958) found, for example, that people of lower socioeconomic status were more likely than individuals in the higher reaches of the stratification system to suffer from some form of psychopathology. Furthermore, the less privileged were apparently more susceptible than the more privileged to the more severe forms of illness, the psychoses. Yet, even when the capacity to pay was controlled for among those who sought therapy, the researchers discovered that people of lower status were likely to be treated by interns and other less experienced personnel, despite the fact that their more severe illness required the more expert attention. They received a great deal of drugs and electric shock. Conversely, the experienced psychiatrists dispensed their services to the socioeconomically more privileged, who needed expert help the least. There seems to be an unbridgeable gulf between the middle-class culture and mentality of the psychotherapist and the mentality and culture of the low-status patient. Where the best medical treatment requires communication almost as an art, the low-status individuals will have to do without that treatment.

The temptation in the literature is to conclude that the lack of communication between therapist and patient is purely a matter of class-related cultural factors. And, of course, it would be foolhardy to deny the influence of culture altogether. But it is equally erroneous, as we shall better see in the next chapter, to understand class or class-like phenomena in purely cultural terms. Social differences reminiscent of human class differences are widespread in the animal kingdom: the lowly get exploited in more ways than one in many more species than one (for convenient brief reviews, see Wilson, 1975b: Barash, 1977).

What do the millions of mentally ill poor do? Most of them never even see a therapist. They live with their neuroses and psychoses, and die with them or of them. Their illness is rendered the more crippling by vagrancy, alcoholism, addiction, a high tendency to suicide, delinquency, murder, and jail (Coles, 1966). Many of the afflicted do not have the freedom or the opportunity to reproduce. Others who do reproduce, tend to transmit to their children their own disadvantages—social, cultural, and clinical—with the result that subsequent generations tend to thin out in the throes of great suffering and humiliation (for an attempt to systematically assess the status of mental illness and its study in the United States of America, see Dohrenwend *et al.*, 1980).

Another way to victimize the poor—and again with untoward consequences not only for the quality of their life but also for their reproductive fitness—is to come down hard on them with the legal system. The point, note, is not the degree of leniency of a legal system; it is rather the question of differential leniency. By and large, however severe or lenient a legal apparatus is, it somehow manages to have keen awareness of a pool of "undesirables" in the population under its control.

In a study of federal district judges and race relations cases in the US American South during the racially awakening period of 1954–62, Kenneth Vines (1964) found that, in spite of their reputation for finding justice in the federal courts, blacks were successful with their rightful grievances only in a little over 50 percent of the cases. The areas of success varied greatly. In cases connected with education, they achieved success nearly 61 percent of the time. In cases involving voting, transportation, and desegregation of government facilities, they won nearly one-half of the cases. However, in areas less well marked by precedent or government intervention, such as employment, defense of black organizations, and other civil rights, success was achieved in less than one out of every five cases.

There are various proximate explanations for this type of discrimination. For example, federal district judges typically come from the district which they serve; they are graduates of the state law schools; and they have held office in the state. Consequently, their ties with the state judicial district are deep and characterized by cronyism. But such explanations and others which they imply give also powerful clues to the possible operation of deeper causes at work. The fact remains, for example, that sentiments of family, friendship, and neighborhood often overrode the supreme law of the land. There thus appears to have been an auxiliary undercurrent of parochialism at work, or a behavioral force, later to be termed homologous affiliation, that is also widespread among non-cultural species and is associated with such phenomena as nepotistic favoritism and phenotypical favoritism (see Chapters 6 and 9 below). Note the biocultural interplay: discrimination is rooted in biological forces, but the intensity of its practice, in view of various sociocultural factors, can in turn reinforce or mitigate the underlying predispositions.

In a study of juvenile offenders in four municipalities of the United States of America, Nathan Goldman (1970) found "a differential disposition of Negro offenders." For example, their arrests were twice more likely to reach the court. Much more consequential is the fact that the death penalty has been meted out to blacks in the United States of America in disproportionate numbers when compared to whites found guilty of the same offenses (Bowers, 1974).

Years ago Short and Nye (1957) noted that five out of six boys in training schools for delinquents were drawn from the bottom half of the US stratification system. However, an examination of *self-reports* of behavior showed that there were no significant differences in the distribution of delinquent acts among different socioeconomic groups. The poor were simply more likely to be caught and punished for deviant behavior that knew no class lines. It is hardly surprising, therefore, that official crime data grossly underestimate the real incidence of crime. Thus, a recent study in the United States of America estimated, on the basis of survey data and methods, that unreported crimes were roughly eleven times more numerous than those recorded in police statistics (Sparks *et al.*, 1977).

Findings of this sort are legion, all pointing to the tipped scales of justice. The Bill of Rights and several amendments to the US Constitution purport to guarantee the equal protection of the laws in a variety of respects. Yet, as Stuart Nagel (1970: 115) has noted, the least equal in America are generally those the Fourteenth Amendment was apparently designed specifically to protect—the black, the poor, and the uninformed. Using a large nationwide sample with data taken at all stages of criminal procedure, both federal and state, and with concentration on grand larceny and felonious assault, Nagel found that the law frequently failed unconstitutionally to release on bail, to hold a preliminary hearing, to provide counsel, to try by jury, and so forth. More important for our purposes, the indigent were those to suffer most from the failure of proper criminal procedure. They were also the most likely to be found guilty and the least likely to be recommended for probation, even in cases of no prior record.

Again, we must take care not to take anything away from cultural factors in explaining such findings. For example, to the extent that the indigent benefit from the right to counsel, their attorneys are not likely to be among the most effective advocates before judges or juries, whether because they are court-appointed or ill paid by private sources (Carlin *et al.*, 1966). Likewise, however, we must again pause and wonder why judges and juries—individuals selected for presumed objectivity and probity—are so selective in the cultural aspects that they readily honor and those that with equal facility they reject. Could it be that such cultural discontinuities are reflections of the continuing struggle for existence whereby the more powerful, instinctively or otherwise, interfere with the enjoyment and genetic continuity of the life of the less powerful? Certainly, the hypothesis cannot be dismissed out of hand.

The hypothesis receives some solid support from the data on crimes typical of upper-middle class people, the perpetrators *par excellence* of white-collar crime (embezzlement, price-fixing, and so on). Many studies have shown that, with negligible exceptions, such crime is either undetected or goes largely unpunished by being dealt with through non-criminal regulative procedures (see, e.g., Sutherland, 1949; Ryan, 1976; Geis and Meier, 1977). The favorable legal treatment of those high in the social hierarchy is not unique to complex societies. In many less technologically advanced societies, justice is meted out according to social rank. The point was made in a typical key by Firth (1959: 348) who stated about Maori litigation that the "rank and the social circumstances of both parties [in dispute], as well as the nature of the theft, determined the course to be adopted."

A most severe expression of victimization bears the label of slavery. The historical records of this human phenomenon are scant, and there is a strong inclination to associate it with the advent of the horticultural revolution some 10,000 years ago, and perhaps more especially with the invention of metallurgy. Thus, according to Murdock's *Ethnographic Atlas* (1967), as horticultural societies moved from the pre-metal to the metal stage, the percentage of societies practicing slavery increased from 14 to 83. Chances are that human slavery is as old as human warfare. No doubt, however, with the advent of more efficient weapons and the associated emergence of property accumulation due to the cultivation of plants and animals, the victors had powerful reason to spare the vanquished and put them to forced labor instead. So, slavery at a grand scale is probably of rather recent vintage, going back perhaps to the dawn of the empire-building era. By contrast, among certain species of ants, slavery or something resembling it goes back many millions of years (Wilson, 1971b).

One of the most recent and glaring cases of slavery was practiced until little more than a century ago by what fancied itself the citadel of modern democracy. According to Paul Lewinson (1932), slavery was so fundamental a fact of the US American South that land and slave ownership were the joint criteria of social class, and permeated the entire stratification system. More imaginative and articulate than ants, southern whites had convinced themselves by 1830 that slavery was ordained by God, recognized in Scripture, and sanctioned by the laws of civilized nations (see also Newby, 1965). In the meantime, the slave had long been chattel, often disposable more easily than a prized artifact of a former generation. Harriet Beecher Stowe, one of the most dramatic raconteurs of "the peculiar institution," wrote in *Uncle Tom's Cabin* of the unspeakable anguish of humans who were beaten, raped, and auctioned off at will, and who therefore had no right even to a family. The slave woman was at the mercy of nearly all white men on the plantation (Douglas, 1855). Many white men considered every slave cabin as a sort of prostitution house (Blassingame, 1972).

Others have noted that, given the high economic value placed on healthy and sturdy slaves, the South practiced to a considerable degree a "stud system," whereby on some plantations a relatively small number of male slaves were used to sire many of the slave children. We may conclude, therefore, that bondage meant not only personal tragedy and immeasurable economic burden for the slaves. For the males, it often meant also systematic exclusion from reproductive activity either through the stud system or through the related practice of slave concubinage among white masters (Olmsted, 1959).

Human slavery has by no means disappeared in more recent times. It is still practiced in some West African societies, for example (Grace, 1975), and was common in China as recently as the 1940s (e.g. Lin, 1947). Besides, there are really different forms of slavery, or forms of unfreedom, including, in addition to chattel slavery, "domestic slavery" (in which the subject often works and serves in the household of the master under conditions analogous to those of an unwanted poor relative), the *apartheid* of South Africa, and even the outcastes and untouchables of the traditional caste system in India (see van den Berghe, 1978b: 150–8; also van den Berghe, 1965; Adam, 1971).

The purest type of victimization is represented by the phenomenon of human sacrifice. The natural selection implications here achieve their starkest expression. To deprive others of their life is one of the most effective means of increasing one's own fitness. This statement is especially true when the sacrificed are young males whose elimination releases for the victimizers the reproductive partnership of women who would otherwise be unavailable. At the same time, human sacrifice is often, not always, linked to religious and other cultural ceremonies that justify the sacrifice, and in this sense we may speak of a biocultural interplay.

Sacrifice takes various forms. They are not all known by our label, but they nevertheless share just about all but the name. One type, extremely widespread until recent times according to ethnographers, concerns head-hunting or war raids whereby one group sometimes succeeded in decimating another (e.g. Jones, 1939; Whiting, 1941; Chagnon, 1968). There are apparently complex cultural developments associated with such practices. It is reported, for example, that head-hunting was a frequent practice in only about 20 percent of hunting and gathering societies, but in over 90 percent of horticultural societies (Lenski and Lenski, 1978: 462, n. 80). Indeed, there is ample evidence that warfare and aggression in general were much more common among horticulturalists than among hunter-gatherers. Lenski and Lenski (1978: 166) hypothesize that combat among the relatively sedentary horticulturalists, still present on the world scene, may serve as a psychic substitute for the dwindling art of hunting. More telling perhaps is their added comment that this form of victimization is probably adaptive under the conditions of population pressure introduced by the horticultural revolution. We may, therefore, entertain the hypothesis that warfare increased *pari passu* with the technological breakthroughs of the horticultural revolution and with the elements of population increase and property-holding resulting from the domestication of plants and animals. Chagnon (1968) reports that the Yanomamö male horticulturalists of Venezuela are a quarrelsome people almost constantly engaged either in war raids with their neighbors or in endless political debate that is probably a good bromide for the boredom introduced by the sedentary life and the inordinate participation of women in productive activities. One by-product of this belligerence is the cult of the warrior, along with certain scoring techniques, for example, the number of shrunken heads preserved as trophies (Lenski and Lenski, 1978: 167).

In his reexamination of Robert F. Murphy's (1957) data on the Mundurucú of Brazil, William H. Durham (1976) has convincingly argued that the cultural tradition of head-hunting among the Mundurucú had the effect of decimating their neighbors competing for scarce protein. Accordingly, head-hunting may be viewed as a cultural accretion of the urge to victimize, whose function was to diminish competition on the hunting grounds and thus to enhance the Darwinian fitness of the perpetrators.

Frequently associated with scalping and war raids has been the ritualistic practice of cannibalism (e.g. Métraux, 1946), on which we have already touched in the discussion of ritual consumption. Whatever the cultural justification behind it, ceremonial cannibalism has been surprisingly widespread in horticultural society and has certainly served as a mechanism whereby an individual, or a group of individuals, reduced the inclusive fitness of another for one's own benefit.

Human sacrifice has lasted well into our own times, and may take extremely subtle forms. Thus, a study of the Wisconsin military personnel who died in the Vietnam War shows that the poor were about twice as likely as the more privileged to die in the war, even when efforts were made to focus on draftees alone (Zeitlin *et al.*, 1974). Other studies have shown that black inductees died in Vietnam in significantly larger numbers than was justified by their proportion in the US population. John Butler (1980: 69) notes with a certain hyperbolic flair that "blacks were called upon to expose themselves in greater numbers and with greater risks than expected of whites. Blacks were over-represented in combat units, in terms of their actual percentage in Vietnam. This meant that death came at a higher rate for blacks. . . . Of all the major combat units, blacks made up 12.6 percent. But their death rate was a staggering 14.9 percent" (see also Moskos, 1970). Badillo and Curry (1976) argue that such findings refer to class rather than race effects. For my purposes, however, the difference is not of the essence. Whether we victimize one another on the basis of class differences or racial differences, or both, does not gainsay the victimization of the underdog.

Other forms of human sacrifice are much less subtle. One refers to the wholesale destruction of entire peoples. Hamilton wrote in the *Federalist Papers* that European settlers in the United States of America arrived to build a nation based on "reflection and choice" rather than on "accident and force." In fact, however, they ended by pursuing both aims, and did a great deal of killing in the process. They embarked on a policy of systematic destruction of a large number of Indian societies and cultures. "A good Injun is a dead Injun" became throughout the period of settlement the motto whereby pious and Christian hearts were often inspired to kill Indians with furious enjoyment. They appropriated their land, took them on endless forced marches, butchered the children, raped the women, and mortified the men in numerous and horrendous ways.

It is interesting to note that from 1820 to 1852 six of the eleven major candidates for president had achieved their political prominence either as generals fighting Indians or as secretaries of war when war was waged almost exclusively against the Indians. Once a proud people consisting of hundreds of societies, some of which were endowed with rich civilizations, Indians were reduced in the course of three centuries to a handful of wretches living far out of sight on the public dole.

There are all manner of cultural theories explaining the destruction of the Indian, some exceedingly colorful and imaginative (Rogin, 1974). All probably contain an element of truth. One element, however, that is rarely if ever mentioned by social scientists concerns the simple fact that the American settlers had come here to settle. In the process of appropriating the means for the settlement, they found resistance from the Indians who were certainly not lacking in the sense of territoriality and possession. The colonists, therefore, early had a score as well as a country to settle, and their superior technology and army were as efficient as their aggressive hypothalamus was ancient and subservient to the drive toward genetic maximization.

There have been few if any societies that have escaped the impulse to sacrifice their own kind to one ideology or another. I avoid any attempt to examine the practice whereby on special occasions one individual or a small number are sacrificed for ceremonial reasons. Instances are numerous (e.g. Hubert and Mauss, 1898; Hertz, 1907–9; Monteil, 1924; Firth, 1963). Many peoples, for example, have followed the practice of sacrificing a slave or a prisoner to their gods for various and sundry reasons. I am rather interested in instances of human sacrifice at so large a scale that they nearly defy the imagination and conjure up the image of a human being who, to say it in a religious tone, has the serpent in his heart. The serpent does its work of destruction in the seeming service of one form of religion or another. In truth, religion is usually the servant, not the master.

Consider. The Rome of the Caesars sacrificed to their philosophy of *panem et circensis* countless numbers of pious Christians. The descendants of those same Christians later slaughtered one another to stamp out "heresy," "popism," Protestantism, witchery, sorcery, and so forth—all in the name of the humble Nazarene whose alleged Father had commanded "Thou shalt not kill," and who himself had advised to turn the other cheek. It is hard to imagine how culture could misfire so abysmally if it were not, at least in part, at once an instrument and an imperfect governor of ancient animal forces at work. If we in the social sciences insist on putting all eggs in the culture-learning basket, we must at least not shy away from the question as to why culture and learning produce such monstrosities, without resorting merely to name-calling and evil leader theories.

Sometimes, massive acts of human sacrifice are practiced to propitiate ideologies personified by the likes of Classless Society, Equality, Justice, Racial Purity, and so forth. It is estimated that Stalin's regime sacrificed at least 20 million Russians to a goddess demanding the advent of *Homo sovieticus*. The Nazi regime sacrificed more than 6 million Jews, gypsies, and other "undesirables" to Racial Purity. Historian Walter Laqueur (1981), one of the latest authoritative voices on the Final Solution, dislikes the word Holocaust because etymologically it refers to a burnt offering, and the Nazis did not intend to make a ritual sacrifice of this sort. On this he is doubtless right, but to dwell on words is often to miss the substance that underlies them.

Recent publications have rather convincingly suggested that the preeminent event of the twentieth century has been genocide. Between the battlefields, the refugee camps, and the death camps, in Europe, in Asia, in Africa, in the Middle East, and everywhere, an estimated 110 million have been sacrificed to some ideology or other (e.g. Lifton and Olson, 1974; Dawidowicz, 1975; Weisbord, 1975).

Among the perpetrators of massive human sacrifice, few could compare with the pre-Columbian Aztecs. These people had a great pantheon of gods who were believed to depend upon human offerings for their welfare. In addition, they considered the heart as their most precious possession and hence the most desirable gift for the gods. Hence, the Aztec priests every year offered to their gods large numbers of human hearts, usually belonging to their prisoners (see, e.g., Vaillant, 1941; Caso, 1953; Harris, 1977).

The expressions of victimization mentioned, and others that could have been included, are most assuredly varied. They represent distinctly rich clusters of facts whose roots radiate into diverse forms of religion, family, economy, and politics, among other institutions. They do, however, have one thing in common. They are mechanisms whereby some individuals and groups appropriate the vital resources of others or altogether eliminate others from the existence, increasing thereby their own life chances and inclusive fitness. This result is susceptible to cultural explanation, but ultimately it is predictable from the principle of natural selection, which sensitizes us to the fact of competition for genetic representation in future generations. Hence, it is safe to say not only that evolutionary biology is relevant to the study of sociocultural phenomena but also that at an ultimate level of analysis natural selection is an economic organizer of sociocultural data. Recourse to natural selection seems useful for purposes of singling out uniformities of sociocultural facts and, possibly, arriving at behavioral predispositions that underlie such uniformities.

4.5 The Need for Vengeance

If human beings are capable of victimizing one another for their own benefit, they, as we shall better see in Chapter 5, are also capable of utilizing each other's resources for their mutual advantage. But there is a sort of irony in this proposition: it suggests another statement, namely, that in view of the basic drive toward maximizing our own good, when an act of beneficence toward others fails to induce beneficent reciprocation, the wronged party is likely to seek reparation. An extreme form of this tendency is to seek revenge—hence the present behavioral predisposition.

In a general sense we are dealing with a force that impels humans—and many other animals as well—to return harm for harm received.

The predisposition for vengeance recalls the eye-for-an-eye principle of the Old Testament. Harm alters the feeling of integrity of affected individuals by producing a sense of discomfort, as if something in them were not quite as it should be. The sense of integrity is not recovered until the aggrieved individuals have performed certain acts that are intended to punish the offender ruinously. Typical are duels and the phenomena known as feuds, the vendetta, and the like. In many societies, even today, there are feuds between families or whole communities that sometimes endure for centuries and come to an end only when one of the parties has been completely wiped out or at least enfeebled to the point of being unable to avenge the last wrong received.

In *Romeo and Juliet* Shakespeare immortalized the pathos of feuds in the following prologue: "Two Households, both alike in dignity, in fair Verona, where we lay our scene, from ancient grudge break to new mutiny, where civil blood makes civil hands unclean. From forth the fatal loins of these two foes a pair of star-cross'd lovers take their life; whose misadventur'd piteous overthrows do with their death bury their parents' strife." In nearby Florence, Dante's *Divine Comedy* was to a large extent inspired by the divine poet's sufferings during the many years of exile and pilgrimage due to the recurring strife between the Guelphs and the Ghibellins.

Allegedly "sociologistic" Durkheim (1893: 87) wrote:

> It is an error to believe that vengeance is but useless cruelty. It is very possible that, in itself, it consists of a mechanical and aimless reaction, in an emotional and irrational movement, in an unintelligent need to destroy; but, in fact, what it tends to destroy was a menace to us. It consists, then, in a veritable act of defense, although an instinctive and unreflective one. We avenge ourselves only upon what has done us evil, and what has done us evil is always dangerous. The instinct of vengeance is, in sum, only the instinct of conservation exacerbated by peril.

The context is Durkheim's discussion of mechanical solidarity, or the type of social order that is typical of homogeneous and technologically simpler societies. The focus, moreover, is on crime viewed as an offense against the community. Hence, we are strictly at the level of group selection. But not for that is Durkheim's celebrated analysis less useful to understand the more general case. If the threat to the community is met by a reaction that corresponds to the instinct of self-preservation, ultimately it is the individual who reacts to a wrong that is individually felt because, even if the threat takes place in a social context, it is in principle a threat to any member of the group.

Every strong state of conscience is a source of life; it is an essential factor of our general vitality. Consequently, everything that tends to enfeeble it wastes and corrupts us. There results a troubled sense of illness analogous to that which we feel when an important function is suspended or lapses. It is then inevitable that we should react energetically against the cause that threatens us with such diminution, that we strain to do away with it in order to maintain the integrity of our conscience. (Durkheim, 1893: 96–7)

Toward the end of 1976, Gary Gilmore, a convicted murderer who was scheduled to be soon executed in a Utah prison, attempted suicide by ingesting a lethal quantity of drugs. The authorities rushed him to the hospital to save his life, thus seemingly engaging in a senseless act. A well-known pundit discussed the case in a newspaper column and concluded with the indignant question: "What nightmare logic is this?" He certainly deserved the sympathy of all rational persons. But he utterly misunderstood "the logic" of the case. The logic is simply that the authorities' reaction to the suicide attempt was not, and could not be expected to be, a matter of logic. On the one hand, it is customary to aid the sick. On the other, suicide would have robbed the offended social conscience of the satisfaction of revenge. It is a disturbing fact, but a fact nevertheless.

The need for vengeance often extends beyond the actual offender. When, for example, the family is the social unit, a transgression by only one of its members may be avenged upon the family as a whole. The same may be said of an entire community or society. By the same token, the offense suffered by one member of the family is felt also by the family as a whole, and often indeed by companions, dependents, and fellow citizens. Many have been the cases in history when an entire community has suffered the wrath of another for the offense of only one of its members. When the European peoples were unchallenged masters of the world, it occasionally happened that one of their number was murdered in an "uncivilized" country; the home society then reacted by destroying an entire village or by exacting a heavy indemnity. The latter may have been a reasoned act. But many people were led to approve it under the influence of the instinct of vengeance.

Acts of vengeance have varied greatly in time and space, and not uncommonly have been highly ritualized, especially in preagrarian societies. This suggests that wrongdoing in a moral order has long ceased to be a matter for the exclusive concern of either the victim or the perpetrator. In keeping with predictions suggested by the concept of inclusive fitness, crime and vengeance have typically pitted one kinship network against another. In many instances, vengeance in pre-literate societies was released not only against the perpetrator but also against particular members of his kin group. In some cases, as among the Australian Dieri, the rules of blood feud demanded that the kin of a murdered victim kill the offender's older brother rather than the offender himself (e.g. James, 1872; Howitt, 1904; Malinowski, 1926b; Hogbin, 1934; Evans-Pritchard, 1940; Gluckman, 1965; Ziegenhagen, 1977). It should be noted that among the Dieri, as well as in many other societies, wealth, authority, and even the probability of marriage tended to favor the first-born, so that the vengeance taken against the offender's older brother might wreak greater damage in terms of inclusive fitness than reprisal against the perpetrator himself. So, it would seem that vengeance tends to be dispensed according to genetic relatedness. The statement implies two specifications. On the one hand, the inclination is to be vengeful in inverse proportion to the degree of genetic relationship between the parties in conflict. On the other, the offended party, in cases of loss of life, seeks to exact the maximum loss of genetic fitness on the part of the offending party.

Acts of vengeance that go beyond the individual offender are common in technologically advanced societies, too. The building of America was accompanied by numberless family feuds. Some of them—for example, the renowned quarrel between the McCoys and the Hatfields—have achieved fame through fiction and film. Certain countries have perhaps a relatively exaggerated, but not for that entirely undeserved, reputation for the family feud. In keeping with the above statements on the relationship between vengeance and genetic relatedness, it is very likely that the feud tends to thrive best where family relations achieve their highest intensity. The correlation is predictable from the viewpoint of the maximization principle, for it is in the family that inclusive fitness is achieved and it is, therefore, between families that competition is at its highest. The following description from an area in 1943 China represents fairly closely the facts bearing on the proposition:

> Anyone entering the Lolo area cannot fail to observe the prevalent phenomena of fighting and killing between Lolo clan enemies. The extent of these feuds usually depends upon the size of the hostile groups. There are feuds between clans, clan villages, and clan branches. There is not one branch or lineage among the Lolo families in Liang Shan that is in complete amity with its neighbors or without some involvement with hostile families surrounding them. (Lin, 1947: 97)

All too frequently such feuds and vengeance in general are closely associated with territorial and other property disputes, so that the predispositions of territoriality and vengeance are in a relation of strict mutual dependence.

Closer to us, when the Vietnam War was being waged, entire villages were destroyed because of the suspicion that the residents were aiding and abetting the enemy. The best known case involved the American massacre of hundreds of civilians in My Lai. The event provoked great consternation and soul-searching at various levels of US society. It did not seem possible to many people that their own soldiers were capable of such wholesale cruelty, and any reasonable explanation was hard to come by. The reason is simply that there is little or no "reason" involved in such acts. Motivated by primeval forces, they have struck many times and in many places. Terrorists and guerrillas operating in areas of international tension have murdered many innocent victims out of a sheer need for vengeance against someone or other sometimes remotely connected.

One of the most glaring, and bizarre, cases of vengeance in recent times was perpetrated against an entire nation by Khomeini's Iran, which took prisoners, threatened to kill them, and held them hostage for over a year. It was as pure a case of instinctual behavior as one can find in the annals of human history. The United States of America had without a doubt meddled in the affairs of Iran to the extent of imposing on that nation a regime that, if it was ever fully tolerated, certainly ended by being unpopular to the highest degree (Fischer, 1980; Kamazani, 1980; Hooglund, 1982). The Shah, surrounded by a corrupt and greedy ruling class, had furthermore attempted to transform almost by fiat a profoundly agrarian society and moral order into a modern, industrialized society and political-military power. Corruption and repression aside, the attempted transformation created deep sociocultural, economic, and religious cleavages. It threatened to graft onto an old, essentially theological, order a system whose passwords were "secularism" and normative relativism—or to destroy the old order altogether. The sociocultural apparatus that related to the biological substratum of Iranian society went into a profound crisis. All this took place at the service, real or apparent in varying degrees, of a country, the United States of America, whose haughtiness was ill concealed, fanned by Christian animosity, and committed to defending the actions of a society in the region, Israel, that to many in the area seemed to make imperial sport of Islamic peoples and traditions. The United States of America, thus, came to represent to the religious masses in general and to the Persian priests of the old Islamic order in particular the symbolic counterpart of eternal biological destruction: Satan incarnate.

I disclaim, of course, any interpretation that the predisposition of vengeance and the perceived threat to the moral order that energized it bear exclusive responsibility for the Persian attack on the "American Satan." A society undergoing rapid economic transformation contains also powerful though often latent economic and political interests that compete for predominance in part through divergent international identifications. That is, it is in the throes of the climbing maneuver as well. Little wonder that many of the Iranian revolutionaries held, and have continued to hold, political views whose application would be no less destructive of the old order than American secularism was. None of this, however, can gainsay the fact that the American foreigner stimulated the ancient impulse to wash away the wrath in revenge.

Before closing this section, we should note the degree to which a behavioral force can transcend its practical operations and become the hidden partner of a purely symbolic, indeed transcendental, form of behavior. As Pareto (1916) noted, the predisposition of vengeance has two genera. One pertains to *real offenders*; and I have been writing of it exclusively heretofore. The other refers to *imaginary* or *abstract* offenders. This latter variety helps to explain such facts as the quarrels that human beings carry on with their divinities, devils, and other fictions of good and evil. It is a well-known fact that humans often relate to those abstract beings much as they relate to others in the flesh. The profanities that are common among Catholic peoples are vengeful acts against deities that are deemed unjust, unresponsive, or merely capricious. Members of pentecostal religion sometimes speak so rancorously and so vividly of the devil that it would almost seem that he is in full view of such pious people.

The phenomenon is not unique to Christians. The ancient Greeks and Romans were in constant contention with their deities. The poems of Homer colorfully represent mortals as quarreling and doing battle with their gods and goddesses. Pre-Columbian Americans often pointed vengefully to superhuman beings to explain many of their disasters and other daily inconveniences. Human vengeance spares no one. Indeed, it could not be otherwise. A species that attributes so many of its happy events and experiences to supernatural beings would have to be totally lacking in the capacity for evil, or mere spite, to avoid attributing to those beings some of its unhappy experiences as well. At the same time, the evil aimed at the immortals, whose power we both covet and dread, is by its very rashness a dramatic measure of the human need to perpetrate malevolence, in revenge or otherwise.

4.6 The Need for Recognition

Some behavioral forces are farther removed than others from the maximization drive of natural selection and more permissive of nonadaptive dimensions of culture. But rarely are they entirely shorn of adaptive value. A case in point is what has been termed the *Need for Recognition* by Thomas and Znaniecki (1918; also Thomas, 1923), the desire for reputation by Maslow (1954), and the love of praise by Herbert Spencer (1895–8) before them.

The adaptive output of the need for recognition increases in direct proportion to its association with such other predispositions as territoriality, the climbing maneuver, and the predisposition of domination to be discussed in the next chapter. Under these circumstances, the need acts as a force whereby an individual or group advertises its worth through devices that tend to secure an enviable and advantageous position. To the extent, however, that it becomes an end in itself, it is a force that in some cases at least converts adaptive pursuits into nonadaptive or neutral sublimations of them. W. I. Thomas (1923) lists self-sacrifice, saintliness, and martyrdom as activities that fetch recognition. He also distinguishes between the associated "showy motives," which he terms vanity, and "the creative activities" that he calls ambition. The latter distinction is perhaps a bit artificial; in most instances, the need for recognition entails an element of both vanity and ambition. For example, there are probably few individuals who are at once more ambitious and more proud than scholars in general and the few great minds who produce scholarly breakthroughs in particular. The history of science is laden with great rivalries and disputes concerning priority of scientific inventions (Thomas, 1923; Sorokin, 1941; Kuhn, 1962). In the proportion in which we may associate self-sacrifice and martyrdom with the need for recognition—or even scholarly ambition that may substitute immortality of name for immortality of genes—the predisposition is the origin of nonadaptive behavior.

When Beethoven reached Vienna from Bonn in 1792, he was still a young man in much need of tutelage and training. Haydn, then the leading composer in Vienna and the unmatched maestro of the oratorio, *inter alia*, consented to teach him. But the relationship was from the beginning strained by pride, envy, concealment of influence, and charges of ingratitude. Haydn was aware of the young composer's genius and sought to cope with the rising threat with paternal affection, more or less sincere, combined with demands that Beethoven concede publicly that he was Haydn's pupil, and even with requests that Beethoven place "pupil of Haydn" on the title page of his first Vienna works. Beethoven, stubborn and haughty, disdainfully refused. On at least one occasion he even had a burst of anger in which he absurdly declared to a friend that he "had never learned anything" from Haydn (Forbes, 1967; also Solomon, 1977).

Rivalry of competence aside, the struggle for recognition by the two masters was in part intensified by the competition for recognition that musical genius, in turn, occasioned in Vienna high society. The best families vied with each other for the pleasure of granting patronage to great artists, and of course for the honor of the occasional dedication and palace premiere. Patronage was lucrative, and to this extent added an adaptive dimension to the artists' need for recognition. For reasons that are not entirely clear, Beethoven obtained immediate and easy access to the highest reaches of Viennese society, while Papa Haydn was still struggling for tutelage when Beethoven arrived on the scene (Solomon, 1977).

It is not surprising, therefore, that, while Beethoven sometimes disparaged Haydn's music, Haydn, in turn, was often either unable or unwilling to acknowledge some of Beethoven's greatest achievements. Nor is it surprising that, once Beethoven had fully established himself as a great master and no longer suffered from the self-doubt that typically accompanies an intense need for recognition, he displayed magnanimity and even a touch of humility in relation to Haydn, all the more so as the latter was approaching the end of his life. In the end, Beethoven regarded Haydn as the equal of such other past masters as Handel, Bach, Gluck, and Mozart, and self-effacingly excluded himself from such company (Anderson, 1961: letter no. 376).

Beethoven never married and he was not particularly generous with his two brothers and their efforts at family building, although he later became so attached to his orphaned nephew that he drove him to several suicide attempts. Hence, while it cannot be denied that to win recognition also amounted to obtaining material resources, such as annuities, that were in principle fitness-enhancing, it is also true that the search for fame interfered with the drive toward reproductive success.

The need for recognition has various manifestations. It has expression, for example, in the ancient preoccupation with genealogy. In the United States of America, the concern has of late reached a peak in the widespread search for "roots" and the new ethnic awakening. The fact is a good reminder of the interplay of biology and culture. The people of the USA have probably become exhausted by two centuries of democracy, or more precisely by two centuries of an ideology that tended to undermine behavior issuing from predispositions of hierarchy. At the same time, the ideology, true to form, was highly selective of the areas in which it sought to impose its force. If, on the one hand, it tended to condemn aristocracy, especially of the hereditary variety, on the other hand, it allowed distinctions whereby it was better to be white than black; better North European than South European; better Protestant than Catholic; indeed better Presbyterian than Lutheran; and so forth in innumerable ways. In reaction, countless US Americans are avidly searching for their family trees. And it is little wonder that, instead of finding humble peasants and craftsmen as well as an occasional murderer or prostitute on them, they so often find African princes, European barons, and even such luminaries of art and science as Newton and Michelangelo.

When the need for recognition is not satisfied, it breeds discontent, often highly destructive, that can have societal as well as personal consequences. There is reason to believe, for example, that the current search for roots is a genteel tail end of the various protest movements of the 1950s and 1960s. Those were complex phenomena to be sure, but it is interesting to note that they were largely middle-class phenomena (e.g. Lipset, 1967; Skolnick, 1969). Indeed, they were disproportionately energized by the grandchildren and great-grandchildren of immigrants who entered this country toward the end of the last century and the beginning of the present under a social atmosphere that is best described as social rejection. To a considerable extent, the movements constituted a loud affirmation that "the new Americans" would suffer social humiliation no longer; that they had arrived; that they had earned their social credentials if they ever lacked them—and damned be those "WASPS" for ever doubting their worth.

In the absence of adequate vehicles for the redemption of self-esteem, such behavior is sometimes replaced by what Nietzsche, and later Max Scheler, termed *ressentiment*: an intense feeling of envy and hate associated with a recurrent fear to actively express the feeling against the person or group evoking it (see McGill, 1942). Often, however, it becomes extremely destructive. Turning again to literature, a brilliant case in point is provided by the immortal Dickens, *primus inter pares* in the art of constructing personality types. The character is Uriah Heep in *David Copperfield*, legal apprentice to Mr Wickfield. Uriah's family background has bred a terrible grudge against those to whom the "umble" feel constrained to bow, and he copes with his rancor with trickery and deception that bring ruin to his benefactor, and with romantic attention improbably directed toward the gentle Agnes, the victim's daughter. In the end, Uriah Heep swallows his own medicine, but in the process he has sown great misery and indignities.

Closer in time, I am reminded of one inside view of the Nixon administration. Charles Colson, Special Counsel to President Nixon, relates a life history of resentment and wavering pride for bearing an ambition that did not find ready recognition. A "Swamp Yankee," he could not tolerate "the condescension of aristocratic men to those who came out of less fortunate backgrounds" (Colson, 1976), and at one time cherished the chance to reject admission and full scholarship to Harvard in favor of admission to Brown University. According to Colson, Nixon's White House was populated by people essentially like him: proud of current accomplishments but unforgettingly resentful of slights, real or imaginary, and therefore past and ever-recurring. By 1970, "a siege mentality" was settling in the White House. "It was now 'us' against 'them.' Gradually as we drew the circle closer around us, the ranks of 'them' began to swell" (Colson, 1976: 38). "Them" were the intellectuals, the Press, the Democrats, those who opposed the war in Vietnam, indeed anyone who perceived the drawing of the circle itself. Recognition unfulfilled not infrequently sours into vengeance.

The affairs of state of a nation owe a lot more to the historical conditions of its becoming and to the challenges that are posed by its environment, including other nations, than they do to the personality structure of one segment or another of its ruling class. Still, if the reins of state fall in the hands of individuals in whom the predispositions of self-enhancement are not sufficiently fulfilled and tempered by forces of sociality, those individuals may play in political history the role that the trigger plays in a destructive weapon.

In conclusion, I wish to cautiously reinforce the hypothesis that the need for recognition may in principle become devoid of any adaptive value in the biological sense in which I have been using the term; it may become a source of self-expression that provides purely intrinsic satisfaction. The psychologist Abraham Maslow (1954) has offered a theory of motivation that has the advantage of seeing human needs organized in a hierarchy. Not all needs are equally basic and, therefore, simultaneously operative. Their relative expression depends on the prior satisfaction of more fundamental ones. Thus, the physiological needs are the foundations of all others. They are followed by the need for safety, the need to belong and to be loved, the need for esteem, the need for self-actualization, the desire to know and to understand, and the aesthetic needs, in this order. Each subsequent need is triggered off to an urgent degree only after prior needs have already been fulfilled. It follows that the more completely satisfied the more basic needs are, the more urgent the higher-order needs become. Hence, if the theory is valid, one would have to reason that with the advent of the horticultural revolution and the dawn of civilization, the satisfaction of the more basic needs became less problematic, and such other needs as self-actualization, knowledge, and aesthetics became permanent and possibly growing traits of human populations, though not of all individuals therein. From the perspective taken in this volume, it is a little difficult to conceptualize behavior that is not motivated by biologically adaptive predispositions. But it is not impossible, and indeed the effort may be the better part of a biocultural orientation. Chapter 6 will be specially devoted to such an endeavor.*

---

* For further consideration of the self-actualization concept, see Maslow (1954 and especially 1968: ch. 10) as well as the related work by Marx (1844), Fromm (1947, 1961), and Argyris (1973), among others.

5 Predispositions of Sociality, I

"There is always a rivalry between the spontaneous definitions of the situation made by the member of an organized society and the definitions which his society has provided for him" (Thomas, 1923: 43; also Freud, 1930). In our terms the individual tends toward self-enhancement; the society requires a degree of self-sacrifice, subjection to group pressure, or at least an inclination to pursue one's goals within a context that is animated by a give-and-take rule. Accordingly, in addition to predispositions of self-enhancement, we must be able to conceptualize predispositions that motivate behavior at least in part consonant with the demands of group life. Such demands are both a cultural reflection of an innate sociality and the tools that in varying degrees stimulate the predispositions of sociality. That is a statement of coevolution. We are also aiming at the statement that to an extent predispositions of sociality mitigate the operation of self-enhancement forces. But they do not negate it. Indeed, to be at all strict, we must view the sociality forces and their sociocultural manifestations as the evolutionary context within which the forces of self-enhancement work themselves out.

Sociality is a matter of degree. At one logical extreme, we may envision individuals behaving entirely for the benefit of others. At the other end, we may hypothesize behavior that self-servingly "calculates" the maximum benefits derivable from participation in a social contract. In the concrete, and in the typical case, individuals are subconsciously driven to pursue strategies that yield for them a net balance, large or small, of benefits, as compared to costs. In this sense, then, predispositions of sociality and their manifestations may be said to be adaptive, though in the relative sense underscored in this volume.

5.1 Reciprocation

Consider one of the most widespread phenomena of the social scene, both human and animal. The very name given to it, "reciprocity," conjures up relationships of symmetrical value. Thus, Marshall Sahlins (1976: 87— first emphasis mine), criticizes a very useful sociobiological theory, to be discussed presently here, as follows: "Trivers becomes so interested in the fact that in helping others one helps himself, he forgets that in so doing one also benefits genetic competitors *as much as oneself*, so that in all moves that generalize a reciprocal balance no *differential* (let alone optimal) advantage accrues to this so-called adaptive activity." That is, reciprocity interferes with the sorting process of natural selection. Well, not quite. Indeed not by a long stretch. But Sahlins's error flows from a degree of ambiguity associated with the reciprocity concept that is only partly of his own making.

A critique by Schulman (1978) effectively disabuses Sahlins of one misconception but leaves an even more fundamental problem untouched. "What Sahlins has overlooked," according to Schulman, "is the fact that although the benefits of reciprocity may be nullified among altruists, their fitness is greater than the mean fitness ($\bar{w}$) for the population, which presumably includes nonaltruists as well as altruists." In fact, however, the principle of reciprocity, a sociocultural universal, in no sense guarantees that those who apply it derive equal advantage from it. Some individuals are more conscientious, more diligent, more compelled—or whatever—than others in the application of the principle. Conversely, others are more adept at brazen cheating, or they are less required to reciprocate, and so forth. As a result, rules of reciprocity notwithstanding, some give much more than they receive, and vice versa. The fundamental reason lies in the probable fact that reciprocity is based on a behavioral predisposition that may be termed *Reciprocation*, and must be viewed as a product of natural selection. Consequently, we may expect to find that, with some exceptions, whenever we can get away with it, we are inclined to use reciprocal opportunities to our own advantage. This is a central theme of the discussion that follows.

Philosophers and social scientists have long treated reciprocity as a cornerstone of social life. For example, Hobbes's theory of the constituted public authority and Kant's theory of duty are both essentially theories of reciprocity. Within social science, it is worth noting that economics deals fundamentally with systems of reciprocal exchanges. Anthropologists and sociologists have likewise shown a lively interest in phenomena of reciprocity and, relatedly, of exchange. To-date, the most formal theory of the evolutionary selection and retention of reciprocity, and indirectly therefore of reciprocation as a biological predisposition, has been stated in an interspecific key by biologist Robert L. Trivers (1971), although, to be sure, this elegant theory does not contain the richness of data and considerations that may be found in the writings of many philosophers and social scientists.

Trivers's theory, known as a theory of "reciprocal altruism," is best understood against the background of a theory of kin selection by W. D. Hamilton (1964, 1972), for kinship relations and reciprocity are very closely related. A central proposition of Hamilton's theory states that organisms may be expected to engage in acts of beneficence toward others (to be "altruistic") in direct proportion to their degree of genetic kinship to them, and to the degree in which the cost incurred through altruism is lower than the benefit accruing to the recipient(s) of it. The theory, thus, underscores the maximization principle. For example, doing a life-saving turn to a close relative may serve to protect the share of genes that we hold in common.

Trivers's theory expands kin selection theory by encompassing reciprocally beneficent behaviors between unrelated individuals, indeed even between members of different species. A rather astonishing case in point is represented by the phenomenon of cleaning symbioses among fish (Trivers, 1971). At least forty-five species of fish, including six species of shrimp, perform a service for many other species of fish (the "hosts") that is somewhat reminiscent of medicine and dentistry in human society. They are the "cleaner" fish.

Briefly, the cleaners enter the mouth and gill chambers of the hosts, usually much larger individuals, and rid them of parasites (see Feder, 1966, for a review of the literature on cleaning symbioses). Cleaners have been observed to take up rather permanent stations, setting up thereby a sort of clinic, while the hosts have sometimes been observed waiting in a queue in order to obtain cleaning service. The hosts' behavior has apparently evolved in response to the problem of ectoparasites, for the more severe the need for a cleaning, as when a fish has a significant infection, the more frequent the trips to the cleaner. By the same token, selection has worked so as to preclude anything but a sporadic accident whereby the cleaner ends in the stomach of the host.

The cleaner's behavior seems to have considerable adaptive value, too. Some species of cleaners subsist nearly entirely on a diet of ectoparasites. Furthermore, their good service has at times been observed to be rewarded by their host fish with defense against predator fish. It is hard to imagine a more obvious case of mutually benefiting reciprocity.

The selection of reciprocal altruism in the evolution of the species, our own included, is allegedly a function of three major conditions: (1) a large number of situations in which given individuals can be of mutual help; (2) repeated interaction in a small group; and (3) fairly symmetrical exposure to situations inviting mutual aid. These broad conditions are elaborated by Trivers into six sociobiological parameters: (1) longevity; (2) low dispersal or migration rate; (3) mutual dependence (as for purposes of defense against predators); (4) prolonged parental care; (5) a weak dominance order; and (6) exposure to combat situations (which makes aid particularly valuable). These, presumably, were exactly the conditions that prevailed among the hominid species of the Pleistocene, and probably even before (Lee and DeVore, 1968).

Social science is rich in studies of reciprocity, social exchange, and related behaviors, so that here is an area of science where collaborative effort is both inevitable and very promising. One of the classics is Marcel Mauss's *The Gift* (1925). Mauss theorized that reciprocal gifts in primitive societies predominated over economic transactions as a form of exchange. Consequently, Mauss concluded, reciprocal gifts in such societies were more basic to the moral order than is the case in modern society. Indeed, the reciprocal gift allegedly transcended the economic dimension in the sense that it was "a total social fact"; that is, it was an event that permeated the entire institutional framework: religion, magic, economy, law, morality, marriage, and so forth. Reciprocity was a rule that pertained to all types of relations, even those formed in the family. This position led Mauss (1925: 71f.) to criticize an earlier theory by Malinowski (1923), according to which a man's gifts to his wife and children were "free" or "pure gifts" rather than reciprocal gifts—there was no obligation to reciprocate for such gifts.

The ubiquitous nature of reciprocity apparently was so compelling that Malinowski (1926b: 39–49) later accepted Mauss's criticism and indeed observed that he had arrived quite on his own at a correction of his previous error. The correct position to have taken, allegedly, is that in the ideas of the natives themselves the entire system of exchanges, including those between a man and his immediate family, "is based on a very complex give and take, and that in the long run mutual services balance." Whoever the recipient, the failure to reciprocate places a person in "an intolerable position, while slackness in fulfillment covers him with opprobrium." Test cases show that when, through laziness or eccentricity, persons fail to conform to the rules of reciprocity, the deviants become automatic "outcasts and hangers-on to some white man or other."

Malinowski's principle of "give-and-take," like Mauss's principle of reciprocal gift, was alleged to permeate the entire social order, involving ceremonies at marriage and at birth, rites of death and mourning, the worship of spirits, and so forth.

> When, for example, at the annual return of the departed ghosts to their village you give an offering to the spirit of a dead relative, you satisfy his feelings, and no doubt also his spiritual appetite, which feeds on the spiritual substance of the meal; you probably also express your own sentiment towards the beloved dead. But there is also a social obligation involved: after the dishes have been exposed for some time and the spirit has finished with his spiritual share, the rest, none the worse, it appears, for ordinary consumption after its spiritual abstraction, is given to a friend or relation-in-law still alive, who then returns a similar gift later on (Malinowski, 1926b: 44–5).

The custom is reminiscent of the widespread practice in modern society whereby at the death of an individual the culinary activities of the immediate family are temporarily taken over by friends and relatives who typically bring over a prepared meal. In some parts of southern Italy, for example, such a gift of food, typically much more lavish than an ordinary repast, is known as the "ricunsulu." It is a major element in a complex network of exchanges among friends, relatives, and godpersons encompassing ceremonies of birth, baptism, marriage, confirmation, illness, departures and arrivals, as well as death. The gift cements the group at a crucial juncture of the life cycle. Conversely, the failure to return a *ricunsulu* upon the proper occasion is almost invariably a sign that old ties will be broken, if they have not been already ruptured.

Some forms that reciprocity takes are especially good at stressing the biological substratum and the operation of natural selection. Lévi-Strauss (1949: ch. 5) proposes that the exchange of presents at Christmas among Euro-American peoples is "nothing else than a gigantic potlatch." It often entails rich decorations and attempts to impress with gifts that may have little or no use-value but may leave one economically debilitated for some time to come. That the Christmas exchange of gifts can be a matter of the utmost severity is probably dramatized by a report I seem to have read not long ago that around Christmas time the suicide rate in a country like the United States of America increases by a significant percentage.

The potlatch is most typical of the Indians of Vancouver and Alaska but is "a universal mode of culture, although not equally developed everywhere" (Lévi-Strauss, 1949). Mauss (1925), for example, has noted its practice in Melanesia and Polynesia. In its most characteristic form, it entails a feature that is reminiscent of one of our banking transactions. Gifts between the parties are either exchanged immediately at an equivalent value or received by the beneficiaries with the understanding that on a future occasion they will reciprocate with gifts of superior value.

The potlatch performs several functions. Three are most striking from the perspective of biocultural science, and clearly suggest its differential adaptiveness for different people. In Lévi-Strauss's (1949: ch. 5) words, they are "to give back with proper 'interest' gifts formerly received; to establish publicly the claim of a family or social group to a title or privilege, or to announce a change of status; finally, to surpass a rival in generosity, to crush him if possible under future obligations which it is hoped he cannot meet, thus taking from him privileges, titles, rank, authority and prestige" (see also Benedict, 1934). Again, as Deacon (1934: 637) notes, "one does not gather riches [or give them] except in order to rise in the social hierarchy."

The potlatch is especially pronounced at the tribal level as an interclan phenomenon. It is the center of an intense struggle between chiefs or noblemen in which sometimes gift-giving is replaced by conspicuous destruction of wealth. The practice, thus, can be dangerous in the extreme, for the destroyed wealth occasioned by the competition is never, unlike exchange, redeemable through restitution. But the competition for self-enhancement is unambiguous. As Mauss (1923: 4–5) put it for the Indians of Vancouver, the potlatch is "above all a struggle among nobles to determine their position in the hierarchy to the ultimate benefit, if they are successful, of their own clans," that is, of themselves and their closest relatives.

The behavioral predisposition of reciprocation works in complex ways, and its manifestations are intricately interwoven with myriad behaviors. Under the circumstances, the topic is naturally surrounded by a certain amount of controversy. Alvin Gouldner (1973: ch. 9), for example, criticized the heavy theoretical emphasis placed on the norm of reciprocity on the basis that it works poorly in certain social relations, for example, those involving children, old people, and the handicapped. Accordingly, he sought to demonstrate "the importance of something for nothing," or the existence of behavioral rules that go "beyond justice and reciprocity." The rule of reciprocity presumably implies justice or the right to receive what is one's due. Gouldner's central point was that justice may be a necessary but not sufficient condition for "social system stability." "For even when a man is given what is deemed rightfully his he may still have woefully less than he either needs or wants."

The reader may recognize in this statement a version of Karl Marx's (1875) intended correction, in the *Critique of the Gotha Program*, of a complex of moral rules that, as a subject of theorizing, goes back, through such scholars as Augustine, Cicero, and Aristotle, at least to Plato. The ancients argued that the survival of society required just relations, and justice existed when persons asked only, and received, what was due to them. Marx recognized that the communist society of his predilection would unavoidably be a society of great sufferings if the meter of equality were to be justice in this ancient sense. What of those whose contribution to the commonwealth, because of disease and other handicaps, would fall short of their needs? The answer, Marx thought, lay in organizing social relations according to the already encountered principle, "from each according to his abilities, to each according to his needs."

Without what Gouldner later called the norm of beneficence, whereby "people must sometimes be given *more* than [is] legitimately theirs," social system stability is allegedly impaired. Indeed, even this much is not enough to hold societies together. The principle of beneficence typically comes to be sanctioned by supernatural commandments, giving rise to the principle of moral absolutism. Whereas beneficence requires one to "do good," moral absolutism requires that "one do his 'duty,' *even if others do not do theirs and even if they have not helped or are not needy*" (Gouldner, 1973: 287). Gouldner cogently argued that a stable social system contains all three principles, and that they, furthermore, both relieve each other's failings and create strains for one another by virtue of their respective imperious inclinations. As a result, "a moral code is a tensionful system of precarious values and fragile adjustments, in which conflict is not merely residual but nuclear" (Gouldner, 1973: 292).

Gouldner's discussion sometimes inclines a bit toward a morality that verges on moralism. Still, the next chapter will provide a modicum of support for this hypothesis of transcendentally buttressed altruism. But we certainly cannot depart very far from the position that, of all the moral principles, reciprocity is by far the most widely practiced. And that entails *quid pro quo* exchanges wherein the desired *quid* is often deliberately and self-servingly more valuable than the *quo*.

Malinowski (1926b: 173) wrote of the Trobriand Islanders that even "within the nearest kinship group rivalries, dissensions, the keenest egotism flourish and dominate indeed the whole trend of kinship relations." A more recent observer of the Trobriand Islanders writes that "Trobriand exchange creates and reinforces oppositions of self-interest at the same time that those oppositions are momentarily overcome." Trobrianders engaged in reciprocal behavior use magic to enhance self-interest, to fulfil the expectations of others, to gain or exert power, and to "coerce others into believing in one's sincerity and good intentions" (Weiner, 1976: ch. 9).

A study of pre-revolutionary China expresses rather clearly the dialectical tension existing between the forces of self-enhancement and the undercurrent, however slight, of individual self-sacrifice that nurtures the principle of reciprocity. Lang (1946) reported that reciprocity was most marked in the kin group. Thus, relatives often helped each other with gifts, and the value of the donations varied directly with the degree of relationship. Moreover, compensation or reciprocity was not always expected, although it was normally given. The expectation of reciprocity was allegedly least likely when doing favors, such as lending tools or giving farming assistance, to old or disabled relatives. On the other hand, we could argue that, at least in the case of old relatives, the gift took place not in the middle of a chain of exchanges but toward the end; that is, in view of the old age of the receivers, giving by the young could properly be interpreted as an act of returning a previous favor rather than as one intended to elicit a future one.

Still, Lang (1946: 168–9) wrote: "Everyone, rich and poor, banker and peddler, high official and worker, has a special 'gift book' in which the exact value of the presents received from relatives and friends is estimated. These entries are made so as to calculate the amount to be spent for return gifts. Not to reciprocate means a serious loss of face." In general, the institution of gift exchanges in China was so stylized that it was regarded as a sort of "insurance payment against the [heavy] expenses of weddings and funerals," much as people in industrial societies pay periodical premiums on insurance policies against future uncertainties. There are, thus, many alternative avenues to fulfilling a given need.

Lang questioned the prevalent idea that the extended Chinese family was "one for all and all for one" (for another society with a similarly questionable reputation, see Banfield, 1958; Lopreato, 1967). Unless they were members of the family nucleus, relatives in need were likely to meet with stingy relatives. "Many rich relatives fulfill their obligations with such ill grace that the recipients of their aid are full of resentment. One of our informants, for example, hated his uncle and brother-in-law who maintained him but constantly reminded him that he ought to be thankful. There are many cases of poor people working as servants in their relatives' houses" (Lang, 1946: 169–70).

Reciprocation and reciprocal behavior, then, attenuate somewhat the raw struggle for existence, but rarely if ever do they cease entirely to be at the same time mechanisms of that same struggle. That would mean, *inter alia*, that a fundamental problem confronted by reciprocation is cheating. Where social relations are asymmetrical in power and prestige, cheating may be expected to be rampant. By and large, the greater the power possessed, the more the cheating perpetrated. The statement implies that the more complex the society, the more widespread the cheating, for complex societies feature complex arrangements of power and social status. Little wonder that, as some recent reports point out, American society is an exceedingly litigious society. In a sense, the fact is readily suggested by the hordes of students who in recent years have flocked to law and business schools. Such schools train the personnel needed to sharpen up, on the one hand, the tools of successful competition among various individuals and organizations and, on the other, the means with which to cope with the resulting legal morass.

Sometimes cheating takes ingenious forms and extreme degrees. When, for example, the Roman Empire fell prey to the various barbarian invasions, and feudal society began its millenary vicissitudes, a manorial system developed that included specific reciprocal obligations between the various classes constituting the society. The masses of people who "commended" themselves to the protection of manorial lords, initially did so under contractual conditions that spelled out a complex system of mutual obligations. But as the feudal system matured, the humble saw their rights evaporate one by one, while the powerful accumulated privileges that reduced the masses to what was slavery in all but the name. One such privilege dramatizes the pertinacity of the maximization thrust in *Homo sapiens*. In many parts of feudal Europe, the manorial lord enjoyed, among his many rights, the *jus primae noctis*, whereby he could lay claim among his subjects to the sexual services of the bride on her wedding night (see, e.g., Pirenne, 1933; Bloch, 1940).

Trivers's theory contains the proposition that natural selection discriminates against cheaters if cheating has negative effects on the cheaters that outweigh their benefits—if, for example, they cannot bear the costs of vengeance. But sometimes it does pay to cheat—as when, for example, swindling goes undetected and no sanctions are brought to bear against the perpetrators. Cheating takes place in various degrees. Trivers (1971) distinguishes conveniently between a "gross" variety, in which there is little or no reciprocity, and "subtle cheating," in which the recipient of a beneficent act tries to get away with returning as little as possible. In general, however, reciprocation has evolved *pari passu* with a psychological system whose function is to regulate cheating. The system includes such traits as moralistic aggression, guilt, sympathy, friendship, and gratitude.* On the whole, selection favors subtler forms of cheating because they are more efficient in escaping the predisposition of vengeance. But such a selection, in turn, favors a more acute ability to detect the offense. Still, selection also favors mimicking the traits of the psychological system "in order to influence the behavior of others to one's own advantage," so that "individuals will differ not in being altruists [reciprocators] or cheaters but in the degree of altruism they show and in the conditions under which they will cheat" (Trivers, 1971: 47f.).

We can fairly agree with this analysis. Some doubt arises about the point that subtle cheating is more easily selected than gross cheating. The example of serfs in feudal Europe, among others, would seem to be quite a gross case of cheating. The reservation, upon scrutiny, verges on interpretation of one of the conditions specified by Trivers for the selection of reciprocal beneficence, namely, a mild dominance order. It may very well be that a low dominance order in the early evolution of the human species did favor the selection of reciprocal altruism and subtle cheating. But as reciprocal behavior persisted even with the emergence of a more rigid dominance order, cheating may have become easier and perhaps grosser as well.

Insufficient attention to the dominance order is probably one of the chief deficiencies of Trivers's theory. More formal recognition of the interdependence of reciprocal behavior and structural social inequality is available in social science. George C. Homans (1974: 215), for instance, has shown that the process of exchange (roughly similar to reciprocity) *gives rise to status*, and, once established, status "reacts upon the exchange process itself, by altering the rewards and the costs of the participants." Indeed, social status has achieved such importance in human society that

---

* The question arises as to whether this psychological system and cheating itself might represent behavioral predispositions in their own right and should, therefore, be included in the biogram. Most likely, as the theory of human nature develops, such traits will find some formal place within it. For the time being, I am interested only in a few major predispositions and shall treat such other traits only as appendages of those.

reciprocity, or degrees of it, and the reactions to it can hardly be understood apart from their connection to relative social status.

The phenomena of reciprocation and reciprocity entail not only the problem of cheating but also the problem of relative deprivation, or what Homans, and Aristotle before him, prefer to term "distributive (in)justice." The latter concerns the question whether or not one receives according to one's merits or entitlements. But justice and injustice are subjective and relative phenomena. A garbage collector, for example, may provide to a physician a service of health and hygiene that in the final analysis is no less important to the physician's life than the medical service provided by the latter is to the former. But not for that are there likely to be many garbage collectors who feel that the relative compensations should be of equal value.

In short, there are various questions associated with the problem of reciprocity and distributive justice that may be roughly epitomized by the notion of *relative merit according to relative social status*. There is theoretical justification for Aristotle's old position, and Homans's (1974: 249) more recent one, that relative justice exists—or better, is normally perceived to exist—when two persons of equal status receive equal rewards (see also Durkheim, 1897). The higher the status, the higher the reward. More importantly, the higher the status, the lower the felt necessity to reciprocate or to reciprocate at an equivalent level of value in relation to one of lower status. Status, moreover, is power. Thus, reciprocity takes place within a context of power differentials and with few exceptions can only operate to benefit some more than it benefits others. Given the character of natural selection, the Primeval Discriminator, matters could hardly be otherwise.

Gouldner's partly moral approach to reciprocal behavior is therefore noble, but by the same token it tends to shoot off the mark. Witness, as an illustration, the current controversy over the viability and fairness of the Social Security system in the United States of America, one of the outstanding forms of what may be termed an indirect system of reciprocity. There is fairly general agreement that the system is moving rapidly toward bankruptcy, threatening thereby the livelihood of countless people now depending on it and robbing current contributors of their future and just entitlements. Still, those now depending on it appear to prefer the assumption that the system will somehow be rendered viable without absorbing changes that in any way would undermine their present entitlements. Others who can count the large number of votes held by the elders are quite willing to coast along on the same assumption.

But what about those other politicians who would transform the system to make it more viable at the partial expense of current beneficiaries? I suspect that they behave in this fashion not out of a greater morality or a greater wisdom but because they have "younger" interests and sense that coasting along is expensive for future retirees, who may enter old age faced by an empty Social Security kitty. When the young awaken to the implications of business-as-usual with the Social Security system, we may expect to observe an effervescence of intergenerational conflict. The political reformers probably sense the coming of such stress, and are buying votes in their own special way.

In conclusion of this section, reciprocation and its corresponding cultural universal, the reciprocity principle, are crucial mechanisms of social and moral order which to an extent mitigate the forces of self-enhancement. But to an extent they also work in the service of those forces. In particular, in view of the logic of natural selection, reciprocation and reciprocity benefit some more than they do others. As a general rule, they tend to benefit most those who are already winning the evolutionary race, namely, the powerful and the prestigious. The fact is more obvious in some places and times than in others. In a study of Thailand, for example, Ruth Benedict (1943: 5) noted that land ownership was distributed according to title and honor: the greater the title, the greater the ownership. "This system not only fixed the relative rank of every man in the kingdom, but it actually placed a value on him. He was literally 'worth' so and so much. If he had to be fined for an offence, the fine was graded according to his Sakdi Na [his place in the title-wealth-status system], and if compensation had to be paid for his death or for any injury, this was likewise computed on the same scale."

5.2 Predispositions of Domination and Deference

We have been moving toward consideration of predispositions that may underlie what is severally referred to as a stratification system, a dominance order, a social hierarchy, and so forth. My preferred terms are "dominance order" for the sociocultural universal, and *predispositions of Domination and Deference* for the biological forces more directly associated with it. I use these terms not to direct attention to what some social scientists in a narrow historical sense call class structures—and certainly not to refer to any particular form that a system of institutionalized inequalities may take in given times and places. I employ them to account for the fact that human communities and societies feature distributions of power and related phenomena, for example, authority and prestige, whereby some individuals and aggregates have higher or lower life chances than others, *regardless of their contribution to the economic vitality and historical viability of the society, and regardless of their loyalty to the rules of reciprocal behavior*.

Life chances are here understood not in the purely economic terms suggested by Max Weber (1922) to sociologists, but in the terms described in the discussion of self-enhancement forces. That entails grasping life chances from the perspective of inclusive fitness, so that predispositions of domination or deference along with dominance orders are mechanisms of natural selection and cultural evolution whereby some individuals and aggregates contribute more units of heredity to future generations than their random share. A basic feature of a dominance order is that some individuals can impose their will over others and interfere with their reproductive success. This power, as I have called it, expresses itself in the form of imperious, often monopolistic, rights over material resources, sexual activities, group-relevant decisions, and related privileges. It follows that the dominance order manifests itself in networks of aggressive–submissive relations, more widely known in social science as relations of superordination and subordination. That is the point behind singling out the predispositions of dominance and deference. But in keeping with our systemic view of the human nature, such predispositions are typically accompanied by other behavioral predispositions. One faithful companion especially worthy of mention is territoriality. As already noted, hierarchy and territoriality work closely together (van den Berghe, 1974). It is fairly easy to see why: it is in the "territory" that the fitness-enhancing but scarce resources are to be found, and hence it is there, too, that the struggle for supremacy, and thus the predispositions of domination and submission, are most clearly observable through a dominance order.

Another, even more closely associated, predisposition is the climbing maneuver. Indeed, inasmuch as an evolutionary approach by definition dictates a dynamic conception of phenomena, the climbing maneuver is never absent where predispositions of domination and deference are at work. That entails understanding dominance and deference as being in a largely precarious state of mutual tension, with the result that a dominance order is ever in a state of flux. Still—and this is a central point of this section—human beings also seem predisposed to perpetuate systems of dominance and submission even as they are constantly attacking them for self-serving purposes.

Let us first dwell a moment on the consequences of the dominance order for differential inclusive fitness. Among other animals, the fitness payoff of dominance is unmistakable. Wilson (1975b: 287; see also Wynne-Edwards, 1962) bluntly writes that "to dominate is to possess priority of access to the necessities of life and reproduction. . . . With rare exceptions, the aggressively superior animal displaces the subordinate from food, from mates, and from nest sites. It only remains to be established that this power actually raises the genetic fitness of the animals possessing it. On this point the evidence is completely clear." At least for the males. Thus, in many species, nearly all females mate, whereas only a minority of males do most of the breeding (e.g. DeFries and McClearn, 1970;

DeVore, 1971; Geist, 1971; Schaller, 1972; Wiley, 1973; LeBoeuf, 1974).

Among humans, the story becomes much more complicated, although the ultimate consequence is essentially the same. Dominance-order behavior in human society reflects an extremely keen imagination and an enormously complex sociocultural system giving vent to it. The custom of the potlatch encountered earlier is indication enough of that. Indeed, human beings can wreck each other's families and even drive each other to suicide with mere invidious comparisons of honor or with competition for the possession and consumption of means that are largely superfluous to their basic needs. There is rarely a greater despair than that suffered by a filthy-rich man who, in the vain search for acceptance into the mysteries of family vintage, "drowns" his haughty guests in champagne and constantly advertises his wealth in one vulgar fashion or another. Humans specialize in conspicuous consumption (Veblen, 1899).

This fact may indicate that the functions of our dominance orders have to an extent evolved beyond those that serve the process of natural selection. But the fundamental function remains. We may recall the reproductive privileges of the European feudal nobility. More broadly, as we shall see in a later chapter, hypergamy is rather the rule in human society, with the necessary consequence that a certain number of low-status males simply remain nonreproductive. They are often forced to languish in jail into the bargain. The same stratum of the population receives first call to the front line in times of war. Likewise, polygyny is widespread in most preagrarian societies. The harem is still a fact in some parts of the world. The European aristocracy was until recent times notorious for the number of bastards it produced among the commoners. Indeed, powerful men have everywhere tended to have an exorbitant number of illegitimate children (e.g. Darlington, 1969). Official data on birth rates by social class are quite unreliable in this respect. Among the Yanomamö of Brazil, for example, dominant males father a disproportionate number of the children in their villages (Neel, 1970).

Human dominance orders are in part contrived results of coercive institutions like the army and the police. Furthermore, inherited wealth and social as well as family connections often confer a position of dominance even where the inborn capacity to achieve and retain it is entirely absent or greatly limited. The case of the child emperor or the idiot king is but an extreme example of a rather widespread phenomenon. More subtly, studies of the United States Senate show a direct relationship between social status and the probability of being elected to the Senate (Matthews, 1954, 1960). Again studies of occupational mobility in national populations show consistently that there is a large, nonrandom, degree of occupational, and thus status, inheritance from one generation to another (e.g. Lipset and Bendix, 1959; Miller, 1960; Jackson and Crockett, 1964; Blau and Duncan, 1967; Lopreato and Hazelrigg, 1972). The latter study, for example, estimated for Italy an expected total mobility rate (in either direction) of nearly 80 percent at the intergenerational level.* But the observed rate proper was only a little more than one-half as much, and of this only about two out of three cases could be attributed to a free movement mechanism, the rest reflecting a "forced" movement due to structural expansion in the higher reaches of the occupational hierarchy (Lopreato and Hazelrigg, 1972: ch. XIII).

Status inheritance is also observable among animals, especially primates. Biologists distinguish between basic rank and dependent rank (Kawai, 1958). The latter, unlike the former, is affected by kinship ties. Several studies of primates show, for example, that the young begin play-fighting early in life, and the outcome is determined by mother's status. Donald S. Sade (1967: 113) puts it as follows about rhesus monkeys: "Offspring begin to fight as old infants or young yearlings. They defeat their age peers whose mothers rank below their own and are defeated by their age peers whose mothers rank above their own." The hierarchy thus established persists for several years and in many cases for the long haul, although physiological differences do seem to come into play at puberty, and at this time there are losses and gains in rank (see also Marsden, 1968; van Lawick–Goodall, 1971; Missakian, 1972). Speaking more generally, E. W. Count (1973: 55) has stated that "an old or even ill individual may continue to dominate vigorous associates long after it has lost the power to defeat them in actual combat."

It would thus seem that the workings of dominance orders are to an extent the result of environmental, including, for us, sociocultural factors. But the wide diffusion of the institution in human societies as well as in the animal kingdom (see, e.g. Wynne-Edwards, 1962; Wilson, 1975b: 282f.) leaves little doubt about its biological, evolutionary origin. There is no doubt about this proposition with respect to complex societies, where we find elaborate class structures interlaced with systems of stratification along ethnic, religious, educational, and other such dimensions. Some doubt may arise concerning the hunting and gathering society that predominated until a few thousand years ago. Archaeological evidence, to my knowledge, is of little help in this respect. Ethnographic evidence, on the other hand, reveals virtually without exception (e.g. Turnbull, 1963) a dominance order that is associated with the presence of leadership functions in society (see, e.g., Schapera, 1956; Lenski, 1966; Murdock, 1967; Lee and DeVore, 1968). In his survey of hunter-gatherers, G. Lenski (1966: ch. 5) notes that in such societies an appreciable economic surplus is typically absent, but the unequal distribution of prestige is the rule, and "prestige usually goes hand in hand with political influence."

---

* The estimate, however, is based on the questionable assumption that ability was randomly distributed among the occupational classes.

As we might expect, however, in view of the interplay between genetic and environmental factors, hunting and gathering societies vary considerably in the intensity with which they manifest a dominance order. At one extreme, a hierarchy is observable only by virtue of the fact that certain individuals perform such functions as the termination of quarrels and fights or the adjudication of serious disputes. At the other extreme, some 12 percent, according to one sample, feature slavery and/or a class structure (reported in Lenski and Lenski, 1978: 100–1).

Rather compelling evidence of the universality and evolutionary origin of dominance orders derives from the readiness with which they arise and, conversely, from the seeming impossibility of doing away with them for any sustained period of time. Sociobiological scientists have observed that initial encounters between animals are often characterized by repeated threats and fighting. One result of this behavior is the social hierarchy. If crickets, for example, are brought together for a time in a closed group, a series of fights is likely to break out which subsides with the emergence of a dominance order (Alexander, 1961).

Something similar happens in human societies. Streetcorner boys, for example, are quick to establish a dominance hierarchy, and the gangs they form are recurringly engaged in achieving predominance over a given territory, or "turf," and the resources valued in it (e.g. Whyte, 1943). At a macro level, the tendency is abundantly illustrated by the history of immigration in the United States of America. Wave after wave of immigrants, and cultural group after cultural group, landed in a context of a tremendous, often violent, competition for space and housing, for jobs, marriage mates, and a great variety of other scarce and desired resources. The result was a sort of pecking order, definable in terms of relative socioeconomic and political success, colorful but damning epithets, and discrimination of various sorts. Prestige, influence, and general success in the order were a function of relative ethno-religious similarity to the early English settlers combined with the length of experience in the country. By and large, the more phenotypically similar a group was to the English, and the longer its history in the country, the higher its position in the various economic, political, religious, and social organizations (see, e.g., Wittke, 1939). The pecking order was so refined that a large portion of American literature and film activity was devoted to describing its composition. The practice persists to-date but little abated (see, e.g., Glazer and Moynihan, 1963; Gordon, 1964; Lopreato, 1970; Greeley, 1971, 1974; Novak, 1971; McLemore, 1980).

The phenomenon is only superficially different from what may be observed when a strange individual is introduced into an established group of animals. Aggression immediately rises as subordinate individuals seek to establish themselves above the newcomer, and the dominants are concerned to maintain their superior status. The scarcity of territory and resources occasioned by the new arrival exacerbates competition and aggression, and the preexisting dominance hierarchy usually accommodates the newcomer at the bottom of the ladder (Barash, 1977: 219).

As we noted in Chapter 4, the most famous theory of social stratification in human society predicted the end of the dominance order, and partly on its basis sociopolitical experiments have been made on so large a scale throughout the world, and for nearly three-quarters of a century, that the accretions of the experiments and the reactions to them may be considered the major political international drama of the period (Marx, 1867, 1875). The classless society, to the extent that Marx was tempted to describe it—reluctantly, to be sure, as Karl Mannheim (1936: 126) justly noted—was to offer a variety of unique features. Among these were: (1) the equality of all labor, that is, evaluation of labor only in terms of hours worked; (2) free circulation of talent and skill, that is, equality of opportunity plus periodic and publicly funded retraining and reemployment of the work force; (3) public leadership chosen on the basis of the highest qualifications, and subject to ready recall by referendum; and (4) abolition of private property, that is, the establishment of "collective ownership" combined with the principle of "fair distribution" according to that famous rule: "from each according to his ability, to each according to his needs!" (e.g. Marx, 1867, 1875; Marx and Engels, 1845–6; also Lopreato and Hazelrigg, 1972: ch. 1).

If the validity of theories depended only on logical possibilities and on heartfuls of love and compassion, today the social sciences would be much richer, and the human race far less afflicted by the baneful features of dominance orders. Those societies whose governing classes claim to have applied Marxist theory, to varying degrees, have made a shambles of it; and, to varying degrees, they have atrociously distorted the psychological-political principle that was dear above all to Karl Marx himself. It goes by the word of freedom—"freedom *to* as well as freedom *from*"—and is in part incorporated in his very motto: *de omnibus dubitandum* (one must doubt everything).

The classical statements on what has happened in societies that claim to have experienced Marxist revolutions belong to such scholars as Milovan Djilas (1957) and Solzhenitsyn (1973), respectively, political scholar as well as former high functionary of the Yugoslav Communist Party, and Russian litterateur and experienced inmate of the Gulag (see also Fischer-Galati, 1979). There are ups and downs, of course, in all things. But the rhythm that in both reports leaps quickly to the eye bears the name of repression—repression of the individual capacity to compete, to move, and to think freely; repression of ethnic minorities, repression of whole nations that have fallen under the millenarian heel. Hungary, Czechoslovakia, Afghanistan, and Poland are reminders enough.

As we have already seen, according to Djilas, everything has happened differently in the USSR and other communist countries from what theory predicted. State, classes, and private property were to wither away; but they have grown gradually stronger. A new class of owners and despots has evolved with special privileges and economic preference based on the administrative monopoly they hold. They need make no concessions, for workers' strikes are forbidden and parliamentary action is entirely under the control of the supreme communist party.

The irony is that nearly absolute domination by the political bureaucracy comes precisely from "collective ownership." The new class administers the public affairs in the name of "the society," but in name only. A chasm has developed between the humble masses and the governing class not only in power and freedom but also in style of life. The latter have country homes, special quarters, access to special and inexpensive stores, freedom of movement within society and between societies, chauffered automobiles, and countless other privileges, including superior medical attention and children who attend the elite schools and are relatively immune to the hard-boiled rule of "proletarian" law. The masses spend much of their time worrying about the endless queues that lead to the scarce staples and other basics, about buying off one petty bureaucrat after another in order to solve the simplest problems, and of course, about the watchful eye of the secret police. If they chance to be scientists, artists, or litterateurs who question absolute domination or produce works of art that do not obey party standards, they have a good chance of landing in jail, the Gulag, or the psychiatric ward.

The literature on destratification or attempts to prevent the emergence of dominance orders is quite substantial (see especially Lenski, 1978). Consider the experiment initiated around 1910 by certain Jewish intellectuals and socialists in what is now the state of Israel. They organized a series of collective agricultural settlements (kibbutzim) with the express purpose of realizing something similar to the Marxist vision of the classless society. At least as recently as the late 1940s, they were relatively small communes, numbering an average of 500 people. Eva Rosenfeld (1951), among others, has traced the history of the kibbutz. Among the social principles originally instituted were: (1) collective ownership of all property; (2) central distribution of all goods; (3) communal eating; (4) a form of participation essentially inspired by Marx's "from each according to ability, to each according to need;" and (5) a keen awareness of the insidiousness of power, which provided for frequent turnover in managerial positions and for the rotation of the less pleasant tasks.

In principle, then, according to Rosenfeld, the social structure of the kibbutz prevented the rise of social classes. But principle did not become fact. First there emerged a distinct system of differential prestige whereby some individuals received deference while others humbly gave it. By the time of Rosenfeld's field study in the late 1940s, a class structure was clearly in the making. Managerial work had been originally invested with a negative premium. Now it received the highest honor, and the reelection of certain individuals to managerial positions became a matter of course. Two distinct social strata developed. One, the managers, received special favors from the distribution officers, experienced intrinsic satisfaction from work, and fathered the children who grew to receive special educational opportunities under the pretext that they showed great promise and special talent. The other stratum, the rank and file, was frequently frustrated and humiliated by the distribution officers who, for one thing, had no special interest in matching a body with a suit of clothes; it also rankled under the suspicion that hard work did not benefit adequately or fairly; and it stressed a higher level of living and greater freedom from institutional encroachment (see also Cherns, 1980).

Wondrous indeed are the ways in which the forces of dominance and submission somehow manage to mushroom through the thin veil of forbidding ideologies. In a study of a calendrical ritual honoring an important saint in a highland Maya community, Renato Rosaldo finds that the ritual is a complex mechanism for expressing differences in social position in a community that would forbid such expression in secular activity. "A metaphor for ascribed status (age) is translated into one for acquired status (wealth). Not only does [the ritual] reflect and confer prestige within the community, but also . . . it gives men a culturally appropriate way to act out a social order based on acquired status, an order forbidden in hamlet life" (Rosaldo, 1972: 368–9).

Some twenty years ago, V. Wynne-Edwards (1962) published a book that tied the dominance order among animals to the Malthusian flavor of natural selection within a context of group selection. The volume stirred such a lively debate in biology that some scholars (e.g. Sade, 1975) have equated it to E. O. Wilson's (1975b) famous volume on sociobiology for its ambition and attention received. Wynne-Edwards marshalled a wealth of evidence to support the thesis that species of animals, especially among the higher phyla, have evolved various mechanisms whose function is to adjust population density to fluctuating levels of resources. Where, for example, the resource in question is food, as it usually is, mechanisms operate to prevent overcrowding of habitat and, therefore, depletion of future yields of food supply. In this sense, they have an equilibrating, or homeostatic, character. The homeostatic mechanisms may be conceptualized as a system of "conventional prizes," won through contests, whose material value is measurable in terms of priority of access to the scarce resources. Contests for dominance and for appropriation of territory, for example, are really devices whereby only the winners gain "the right" to reproduce. According to one of the most daring features of Wynne-Edwards's thesis, all social behavior has its origin in the competition for scarce resources, and much such behavior constitutes precisely a set of mechanisms whose function is to regulate population and keep under control the Malthusian problem. Thus, hierarchies "commonly play a leading part in regulating animal populations; not only can they be made to cut off any required proportion of the population from breeding, but also they have exactly the same effect in respect of food when it is in short supply" (Wynne-Edwards, 1962).

Hierarchies have provided the impetus for an entire field of study, which, under the name of "hierarchy theory" (e.g. Pattee, 1973), brings in cooperative effort physicists and biologists, among other specialists of system analysis. The area provides a complementary explanation of the emergence of dominance orders and a rationale for the "emergent properties" that are so fundamental to sociological theorizing. Central to hierarchy theory is an idea based on the physical concept of "pion," which is roughly conceptualizable as a binding force. The crucial idea, specifically, is that the multiplication of units constituting a given set gives rise to levels of order that arrange themselves hierarchically. Thus, in its simplest sense, a hierarchical order refers to a complex of successively more encompassing sets, wherein to "encompass" is understood to mean that the properties of a set at a lower level are derived at least in part from the properties of a set at a higher level, just as the information that generates the higher level of order is translated and transformed from information existing at lower levels of order. In sociology, Talcott Parsons's (1951) conception of the hierarchy of control obtaining among the systems of the general theory of action may be viewed as recognition of, and at least partially successful effort to apply, hierarchical theory at a macro level.

The emergence of new sets or systems has been referred to as "neogenesis" (Grobstein, 1973). "Enzyme activity of proteins," for example, "is an emergent property dependent upon amino acid sequence but manifested only after establishment of a specific configuration and in the presence of a suitable substrate. Hierarchical organization in biological systems thus is characterized by an exquisite array of delicately and intricately interlocked orders, steadily increasing in level and complexity and thereby giving rise neogenetically to emergent properties" (Grobstein, 1973: 46–7).

In the next chapter I shall argue analogously that one type of true altruism in human society is an emergent property dependent on the psychological emergence of self-deception but manifested only after the establishment of a specific symbolic configuration, the soul. Here, I shall conclude this brief discussion of hierarchy theory by noting that hierarchies or dominance orders are steadily generated and enhanced by growing living systems. In the process, new properties are continually emerging. Emergence is a term that may be attributed to an association of old and new components of systems such that the association assumes the function of a set or component in its own right; that is, it contains information that is not found in the associated components when viewed individually. The dominance order that has been the topic of this section would, thus, seem to be, at least in part, an emergent property of social relations whose historical viability is dependent on a higher order of information and regulation of activity.

This statement may be explicated further by going into what may be termed the solidarity function of dominance orders. In so putting the matter, I in no way lose awareness of the former argument that, from another perspective, we may view dominance orders as arrangements whereby some individuals rob others of the wherewithal for their livelihood and genetic fitness. Indeed, as noted, that is the ultimate viewpoint enjoined by natural selection. My goal here is to focus on what may be a secondary adaptation of social hierarchies. Biologists and social scientists have observed that under normal circumstances dominance orders provide for a regularity of social relations that often redounds to the benefit of more than just the dominant individuals (e.g. Kummer, 1971; Brown, 1975; Wilson, 1975b; Eaton, 1976). In some primate societies, for example, dominant individuals use their influence to prevent or terminate fighting among subordinates that could otherwise cause harm, sometimes fatal. As a related phenomenon, it has been observed that the removal of dominant individuals sharply increases aggression in the lower ranks (e.g. DeVore and Washburn, 1963; Bernstein and Sharpe, 1966; Eisenberg and Kuehn, 1966; Tokuda and Jensen, 1968).

The observation seems to apply to a variety of species (e.g. Schaller, 1965, 1972; Guhl and Fisher, 1969). Allee (1938: 135–6) compared stable flocks of hens with flocks in which a hierarchy was not allowed to stabilize. Individuals in the former "pecked each other less, ate more, maintained weight better, and laid more eggs than did their fellows in the flocks steadily undergoing reorganization." In a cross-species study, social scientist Mazur (1973) has noted that "high ranked members—particularly the leader—perform service and control functions for other members and for the group as a whole." Indeed, there is convincing evidence that aggression is elevated to pathological levels by a breakdown of hierarchically organized systems (Scott, 1958).

A lesson that would seem to strain through these facts and considerations is that to some extent dominance orders are group-wide adaptations, and are thus held together by both consensus and coercion. "In fact," to quote the sociologist Pareto (1916: 1154), "only when it is crumbling and is about to give way to another does a hierarchy cease to be spontaneous through becoming preponderantly or exclusively a matter of force." Or, again, in Georg Simmel's (1950: 183) words, even where authority seems to "crush" us "it is not based only on coercion or compulsion to yield to it."

On the surface of it, we are faced with a paradox. How can a dominance order both foster and reduce conflict? The answer is surprisingly easy. Groups are historical entities; hence, the quality of their interpersonal relations varies in the course of time. Power is more afflictive some times than others, more justifiable this year than next, better defended today than tomorrow. Without relations of dominance, available evidence suggests that the Hobbesian problem of order would go unresolved; internal warfare would be continuous. A dominance order regulates relations if for no other reason because it reveals whom we are likely to win against and who instead is likely to beat us. In practice then, it tends to establish a convenient order of access to resources (van den Berghe, 1974) by economizing energy expenditure. Moreover, as we have seen, dominant individuals (and groups) regulate conflict by refereeing and/or terminating fighting. In recent human society, the law, for example, has often been invoked by governments to settle protracted strife between labor and management, between one religious or racial group and another, and so forth.

In short, with the establishment of the dominance order, conflict is *in principle* eliminated because everyone knows one's own place in the group and to a degree controls the climbing maneuver, the drive to rise above one's own station. In this sense, hierarchies tend to achieve internal consensus. But this tendency has other tendencies to contend with. The climbing maneuver, for example, should not be expected to remain permanently quiescent. Individuals remain decidedly interested in being on top rather than at the bottom. That means that they are constantly on the alert for signs that the dominance structure can be transformed to their advantage. Hence, the hierarchy is constantly being challenged and defended at the same time. In this other sense, dominance relations are conflictual relations. It follows that when the dominance order is the least stable, the group also features the most conflictual relations, and vice versa. The order is crumbling, and the death rattle is noisy.

In human societies we often observe the most proper behavior in the relations existing among partisan politicians. But come the time to elect a new government, which is in principle an act of destruction and reconstruction, and the candidates go at each other's throats with a vengeance. They become mudslingers, and often ruin permanently each other's careers. What is even more revealing is that this same sort of behavior is observable also among the members of one same party. The honorable MP or Senator X retires, or piously takes leave to join more heavenly hierarchies. Bedlam breaks loose. Dozens of pretenders, who only yesterday had massaged each other's egos at a fund-raising gathering, now start noticing, vociferously, the most awful deficiencies in each other's intelligence and competence, to say nothing of their character. On the international scene, it happens that, in appearance at least, former colonists have some basis for saying "I-told-you-so" about one or another new nation that cannot "decide" whom to trust with the affairs of government. Africa today features a tremendous degree of elite strife. But the fact does not reflect a problem of political competence. It is rather what happens when one master is removed and another must be found. Consensus is typically lowest at such junctures.

The dominance order may have a variety of causes, but since it is expressed through acts of dominance and submission, ultimately it necessarily rests on the behavior of individuals who have evolved to act out relations of dominance and subordination. Accordingly, it is reasonable to postulate a pair of correlative predispositions that, interestingly enough, have had wide diffusion in social science. Thus, for example, Spencer and Pareto wrote, respectively, of "fitness for ruling" and "sentiments of superiors." On the other side, they wrote of "reverence for authority" and "sentiments of inferiors," respectively; others, for example, Michels (1914), wrote of "the cult of veneration."

These behavioral forces—*Domination* and *Deference*—may frutifully be thought of as adaptations that ritualize behavior through social status. That is another way of putting what I have already stated: that such predispositions help to order activity and economize it within a competitive context. It is dangerous, for example, to challenge at random when competing for sexual services. One may end by never having another chance at successful competition. Putting the matter this way, note, suggests the hypothesis that the two predispositions may be found side by side in the same individual, and that the manifestation of one or the other is a function of varying intensities associated with varying life cycle and environmental factors.

The propensities to dominate and defer are not constant in their intensity. They vary as the social context varies. It often happens, for example, that a football team that wins regularly against a second, also loses against a third team that is ordinarily beaten by the second. Likewise, dominance hierarchies feature dominance triangles (e.g. Count, 1973) whereby individuals dominate some and defer to others, without necessarily suggesting any syllogistic logic. Again, subordinate individuals may ultimately have a bright evolutionary future (Barash, 1977). In human society, the rise and fall of the elite—for example, through revolution—illustrates clearly the point. Today's underdog becomes tomorrow's top dog.

There are striking differences between the behavioral profiles of dominant and submissive individuals. The dominants are characterized by expensive tastes, pride, and arrogance. To balance things off, they also practice patronage and paternalism. Upper-class people are found in the most exclusive social clubs, in positions of political power, in the higher rungs of the military, religious, and business worlds. Their life is ruled by a highly formalized code of etiquette. The code is the external symbol of a synthesis of attitudes that presents their "superiority" as legitimate. The more stable the dominance order, and thus the more secure the positions of dominance, the more rigid the code (Weber, 1922; Amory, 1947; Mills, 1956; Kahl, 1957; Baltzell, 1958).

In a useful, partial elaboration of the great Marxian synthesis, Max Weber (1922) argued that periods of rapid economic and technological transformations tend to accentuate the economic phenomena of class; thus, economic acquisition and consumption, competition, and raw exploitation come to the fore. Conversely, periods of relative stability tend to give vent to the ritualistic aspects of social status. Style, noblesse oblige, and aristocratic demeanor are the tokens of high position. The position of privilege is precarious in the former case because the struggle for the top is furious. It is relatively secure in the latter because the struggle has abated.

The superior, arrogant, and self-assured bearing of the dominant human being has its counterpart among animals. The dominant wolf, for example, holds his head, ears, and tail high. His face is thrust forward in a sort of challenge when he meets other members of the group (Wilson, 1975b: 280). An analogous syndrome is featured among leading rhesus monkeys (Altmann, 1962). The extreme expression of dominance is found in some species, for example, baboons and macaques, where dominant males assert their power by a ritualized form of homosexual mounting, while subordinates present their rumps in the female posture as a sign of submission and conciliation (e.g. Maslow, 1940; see also van Lawick-Goodall, 1971, for chimpanzees).

One reason status is relatively secure during periods of relative economic stability is that at such times the environmental opportunities to successfully challenge the existing dominance order are relatively few. That implies, in turn, that the feeling of deference is relatively intense. Manifestations of this predisposition, which we may otherwise call *the sense for authority*, are endless. The image of the poor peasant, hat in hand, begging a favor of the landlord, or merely bowing to the local gentry, comes immediately to mind. In traditional India, the lower castes granted the Brahman, standing at the top of the social hierarchy, a deference worthy of a God. Like God and the cow, the Brahman was inviolable. His murder was a cardinal sin; and he could not be beaten, put in irons, fined, or expelled from the group (Dumont, 1966). In a day gone by, the illiterate revered the person of knowledge. One who spoke only a modern language felt inferior to one who spoke Latin or Greek.

Today feelings of deference may be weaker, but they are still here; thus we pay less deference to educated people, but we have "stars" who compete successfully with Brahmans for the amount of respect and veneration received. Without the predisposition of deference, politicians would not land in airports where masses of humanity push and shove to catch a glimpse of them. Upon running for office, they would not have any number of helpers sometimes willing to give up job, money, and even family for the honor of serving in the campaign. Singers, athletes, actresses, and other stars would be without fan clubs, and few if any people would care to have their autographs. Sometimes one is surprised to observe normally skeptical, perfectly reasonable, even congenitally "republican" individuals who go to any length to experience firsthand the arrival of a king, a queen, a princess, or merely a 14-year-old "messiah." In modern society, kings and queens and princesses and princes are no longer à la mode, though their marriages still cause orgies of obsequious behavior on the international scene. But have you noticed how much nostalgia we suffer? Now we crown Miss Universe, Miss America, Miss Texas, Miss Wool. . . . *Vin ordinaire* sometimes inebriates more easily than vintage wine. Catholics would be at a loss to obey the dictates of an hierarchy that regularly changes its mind on matters of eternal salvation and bears no resemblance, in its rich and bejeweled attire, or in its periodically pompous existence, to the divine being whom it presumes to represent and who allegedly subsisted on little more than bread and water. These are not purely cultural phenomena. They originate in instinctive behavior of humility that facilitates the privileges and the historical role of dominant individuals.

In the animal realm, too, the demeanor of subordinates correlates pathetically with the superior behavior of dominants. Frequently, especially after a nasty encounter, the former slink about and try to appear as inconspicuous as possible. The head and tail are held low. Feathers or furs blanch in contrast to the flushing of the dominants. After a defeat, a fish may hang limply in the water (e.g. Count, 1973). Indeed, the behavioral repertory of the humble animal is rich. There are hormonal effects of subordination whereby some scholars speak of "psychological castration," at least among males (e.g. Barash, 1977). Nonhuman primates, for example, may experience a large reduction in size of the testicles and in hormonal output, which is apparently reversible when subordination turns into dominance. Subordinates may also experience enlarged adrenal glands, which are, in turn, associated with a variety of other dysfunctions, for example, renal failure, impaired antibody formation, greater susceptibility to disease, and even death (Barash, 1977).

Erving Goffman (1959) has argued that modern human beings are surrounded by an aura of respectability reminiscent of tribal sanctity. Indeed, it appears that "the sanctity of the individual" is one of the fundamental principles established by the secular rise of the bourgeoisie in its struggle against the *ancien régime*, and by the ideology of democracy that accompanied it. The relative respectability and esteem of the common people in modern society have been achieved, to an extent, by the organizational and economic prowess conferred especially by political parties and trade unions. But it is probably premature to envision, as some seem to do (e.g. Collins, 1975: 210f.), "the evaporation of deference cultures." Modern society has multiplied the bases, or organizational contexts, of deference, and provided endless substitutes for the monarch, the warlord, the chieftain, and the "superior" employer. Some scholars argue, moreover, that the power of the dominant sectors, or elites, over the masses of a society like the United States of America has grown greater rather than weaker (Mills, 1956). Beyond that, princes, warlords, and the like have been replaced by the likes of Elvis Presley, John Lennon, Roger Staubach, George Best, Robert Redford, Alan Alda, and Brooke Shields, to mention but two classes of "stars" to whom countless millions would humble themselves in multiple and sundry ways.

Moreover, aside from the fact that most of the world is still far from being modern (if by that is meant, roughly, industrial), modern society has managed to spawn a class of people, representing as much as 10–15 percent of the total population, who live largely at the margin of society and must do a lot of bowing for their daily bread. Their dependence on the state and its petty bureaucrats is so profound that their subordination and humiliation may be said to be total. The anthropologist Oscar Lewis (1966) saw them as suffering from "the culture of poverty": from a precarious style of life that is perpetuated from generation to generation and at the heart of which are apathy, a sense of hopelessness, a humble awareness of dependence on the "high brows" for their livelihood, and constant fear of the morrow and the rainy day. Psychological castration on a large scale! Dominance and submission are still very much with us (on the culture of poverty, see also Miller, 1968; Miller and Reissman, 1968; Valentine, 1968; Rainwater, 1970; Gans, 1973; Feagin, 1975).

In conclusion of this section, a comment is in order about the dominant current conception of social stratification. The topic would deserve much more detailed attention from an evolutionary perspective, but this is not the place to confront the task in full. In a nutshell, it may reasonably be said that most of the prevalent theory of social stratification fails to rise above the uniqueness and peculiarities of time and place (but see, e.g., Pareto, 1906, 1916; Weber, 1922; Davis and Moore, 1945; Davis, 1948; Lenski, 1966; Collins, 1975). Indeed, in some quarters, especially in Western Europe, where Marx's anti-capitalist theory still holds sway, the very term "social stratification" is abhorrent, inasmuch as it presumably conveys an image of gradation of inequalities rather than the preferred conception of interest groups in conflict, *qua* organized classes, against one another.

It is as if either social stratification were less than 300 years old or the task of social science were to explain it only as a particular feature of 200–300 years of industrial society. The irony is that new developments in stratification theory continue to be variations on a Marxist-industrial theme (e.g. Marx and Engels, 1848; Marx, 1850, 1867) even as many scholars are busy showing that the social, economic, and political developments of industrial societies since Marx's time have outpaced the theoretical reach of

Marx's work (e.g. Geiger, 1949; Dahrendorf, 1959; Bell, 1973; Giddens, 1973).

Few things in social science are intrinsically as admirable and informative as Marx's great work and the later developments that are anchored therein, whatever their limitations. So, it is not at all my intention here to disparage their value. Nor has it ever been (e.g. Lopreato and Hazelrigg, 1972). Rather, my aim is to note that if the focus of social science is to fall on the *science* part of its name, it would be more appropriate and more fruitful to seek to explain forms of social stratification in time and place (1) by reference to fundamental forces that are universal in time and place *as well as* with the aid of historically and spatially particular circumstances, (2) with keen awareness of the wide diffusion of systems of social inequality in the animal kingdom. In sum, I think we would stand to gain from a more profoundly historical and comparative approach to the problem at hand.

It follows that post-Marxian theories of stratification may be said to have had varying degrees of formal success, depending in part on whether their focus inclines toward the historically specific or the historically general. Thus Dahrendorf's (1959) theory of class conflict in dichotomous, imperatively coordinated associations encounters many of the same problems faced by Marx's theory (Lopreato and Hazelrigg, 1972: ch. XVII) and is called into question by the developing features of "the post-industrial society" (e.g. Bell, 1973; Giddens, 1973). On the other hand, theories that have to any degree logically crashed time barriers show greater promise of scholarly durability. Max Weber's (1946: 180–95) famous essay is limited in scope but nevertheless very valuable. It reckons, more explicitly than Marx's work, with the multidimensionality of structured inequality; and above all, it emphasizes invidious considerations of status as human uniformities, and the distribution of power as ineluctably asymmetrical and both derivative and promotive of selfish interest. Weber wisely notes also that the quest for economic success, social honor, and power have varying expressions depending on varying historical circumstances and varying degrees of technological transformation.

Likewise, the theory by Kingsley Davis and Wilbert E. Moore (1945; see also Davis, 1948), while still problematic and controversial, does have the virtue of enjoining consideration of the basic causes of social stratification, whatever its particular form in time and place. In the process, it suggests assumptions about human nature and the evolutionary character of society that are promising for a general theory of social stratification. I have no doubt that as the behavioral sciences effect their evolutionary rapprochement, there are greater promises still for their cooperative pursuit.

5.3 The Need for Conformity

With considerable but convenient oversimplification, it may be said that a dominance order is a social arrangement whereby individuals pay obligations of reciprocity more or less asymmetrically. The asymmetry is associated with differential opportunities to fulfil the drive toward genetic fitness as well as to enjoy a comfortable and honored existence. Thus, at various proximate levels, it is power that overlays a reciprocity system with a dominance order (see Adams, 1975: 27–8, for a related idea). At an ultimate level, reciprocity and dominance are joint results of the struggle for existence inherent in natural selection.

From either perspective we can glimpse the hypothesis that, to a degree, the viability of reciprocity and dominance orders depends necessarily on some tendency to conform to certain rules and to demand more or less similar behavior from others. Such rules cannot be founded merely on reason, free will, and good intentions for their perpetuation. These are chancy factors, and the universality of reciprocity and dominance orders counsels rejection of chancy factors as basic explanatory devices. The hypothesis, therefore, may be extended to conceptualize a behavioral predisposition that I shall term the *Need for Conformity*. The predisposition is basically a behavioral force that sustains social orders by supporting reciprocal behavior as it is filtered through the workings of dominance orders. To a large extent, therefore, the degree of conformity existing in a social aggregate is a measure of the viability of the social contract and the distribution of resources that it tends to specify. The question of conformity, therefore, concerns, *inter alia*, the previously encountered problem of cheating. The predisposition in question minimizes deliberate failures to reciprocate among peers, and to an extent attenuates the problem inherent in the asymmetrical obligations associated with dominance orders.

The evolution of the need for conformity is, thus, most probably linked systemically to the evolution of reciprocation and of the predispositions of domination and deference, which are fundamental features of group life. This hypothesis differs somewhat from one advanced by Wilson (1978: 184–7) according to whom voluntary conformity is "a neurologically based learning rule that evolved through the selection of clans competing one against the other." Such an hypothesis does not appear to be sufficiently historical. Ultimately, the tendency to conform is a mechanism associated with the utility of group life per se, which is much older than clan organization. There is truth in the old adage that two pairs of eyes see better than one. Where food is scarce, pooled knowledge bestows upon the conformist an advantage over the nonconformist. The same principle applies when the environment is fraught with dangers. A large literature in biology shows that predators have a strong preference for deviant individuals, whether deviance be expressed in terms of movement, spatial positioning, color, or a variety of other factors. In his work on sparrow hawks, Mueller (1971), for example, demonstrated the working of the "oddity factor" whereby oddly colored mice were preferred as prey by the hawks. In humans, the phenomenon of the scapegoat, or more simply of the "goat" in small group research, seems to be just of that nature. The "goat" is constantly ridiculed and picked upon.

Little wonder that, as Wilson (1975b: 562) has remarked, human beings "are absurdly easy to indoctrinate." The tendency performs many crucial functions. It helps us to participate in systems of reciprocal behavior and to avoid the punishment of others who object to our transgressions and have what it takes to enforce given types of behavior. In this sense, willing conformity is an adaptation away from the costs entailed by forced conformity.

The predisposition to conform is most directly manifested by imitation, or observational learning. Thus, it is a sort of teacher built into each of us, and facilitates the process of socialization (Piaget, 1951; Bandura, 1977). Imitation is "the most elementary way in which tradition is preserved" (Lorenz, 1977: 200). Voluntary conformity grants us a learning shortcut, quicker and more efficient than individual trial-and-error behavior. Biologically, it appears to be a complex phenomenon, for successful imitation would seem to require the anterior incorporation of genetic information that entails the ability to recognize what, when, where, and whom to imitate (Campbell, 1966; also Bandura, 1977; Bandura and Barab, 1971).

None of this is to deny the importance of group competition and defense for the retention, if not the evolution, of the need for conformity. The aforementioned preference by predators for odd individuals may also be placed within the context of group defense. When two opposing armies wage battle, the most vulnerable individuals are also those who are the most exposed and the farthest removed from the core. The first line of attack, the scout, the sentry, and the like fall under this category. Odd individuals in the sense encountered above may very well be the behavioral analogs of exposed warriors.

Observational learning, the immediate cause of voluntary conformity, is widespread in animal species. For example, Pfeiffer (1969) reports that many practices are spread and maintained among chimpanzees through imitation, notably the use of crumpled leaves to sponge inaccessible water and the use of probes to catch termites (see also Kawai, 1965; Hediger, 1970; van Lawick-Goodall, 1971). Hinde (1970) refers to such practices as "true imitation"—copying an otherwise improbable act or utterance. Other forms of imitation are "social facilitation," which stimulates the tendency to perform an action already in one's repertoire; and "local enhancement," in which an individual becomes sensitized to the value of a particular site on the basis of observing another's experience there. All three forms have obvious adaptive value—the first, in acquiring new and successful behaviors, the second in triggering individual abilities and coordinating group activities, and the third in receiving information economically and efficiently. In various ways, then, the conformist takes advantage of the pooled knowledge of the whole group.

If the hypothesis that imitation is adaptive is valid, we should also expect individuals to be inclined toward preferring certain "models" over others. Under normal circumstances, dominant models, for example, should be preferred over subservient ones because the former are more successful actors. Likewise, relatives should receive preference over strangers, for the former have an investment in the observers and are less likely to lead them astray. The evidence is overwhelming from humans, where children have a strong preference for their mother. But kin selection clearly suggests that such behavior is widely diffused. Cats, for example, not only learn by observing; they also learn best by watching their mothers (Pulliam and Dunford, 1980).

Among social scientists, few have been as keen to the role of conformity in group life as Durkheim (1893, 1897, 1912), Sumner (1906), and Pareto (1916); and only the latter scholar has specifically identified the trait as an item in a theory of sociocultural universals. According to Pareto, "in human societies the uniformity desired may be general throughout a people, but it may also differ according to the various groupings of individuals within the people." Thus, there are many centers of similarity in a given society—as, for example, when the society contains diverse religious or ethnic groups. This circumstance is again related to the previous problem of the inseparability of consensus and conflict in society, for if the need for conformity leads to consensus within each center, and sometimes in the society as a whole, it also leads to conflict when one group finds resistance in seeking to impose its own particular rules upon others.

Sociality forces are basically disciplinary in nature. Discipline or social control can be internalized, or it can be imposed from the outside. Accordingly, we may explicitly distinguish between two phases of the need for conformity: one works on the self; the other demands acquiescence from others. The two are reminiscent of Durkheim's (1953) two dimensions of morality: desirability and obligation. The former relies on techniques of learning, such as imitation, and finds expression in virtually pure form in such phenomena as fashion and the shared liturgy of a congregation. The latter is most evident in hostile reactions to violations of social rules.

An illustration of the need for conformity may be useful at this point. In recent years, as it has often happened in many societies in times following periods of sociocultural upheaval, some Western societies have experienced an intensification of the religious spirit and an accompanying proliferation of religious cults and doctrines, some of which represent moral invasions properly speaking (see, e.g., Wuthnow, 1978). Examples of these are the Hare Krishna sect, founded in the United States of America in 1966 by Indian guru A. C. Bhaktivedanta Swami Prabhupada, and the movement founded by the Korean "messiah," Sun Myung Moon. The conversion of thousands of US Americans to these religions has been met with great apprehension in the Judeo-Christian tradition, and several years ago some parents of the converts resorted to both legal and strong-arm methods to win back their wayward children, some of whom were quite past majority age. Ironically, many parents were at best nominally religious, so that it would appear that they would rather have areligious children than communicants of a religion different from the one to which they nominally belonged.

The terrific revulsion to the foreign religions is illustrated by "deprogramming," a technique in which the wayward converts were subjected to threats and arguments until they gave up their new faith. Before that took place, however, deprogrammers often had to abduct the recalcitrant converts and place them in special deprogramming centers. One such center was titled, with more than a touch of unintended irony, the "Freedom of Thought Foundation." The Foundation led the parents of converts to sue for a thirty-day "conservatorship" over their offspring. This stratagem was made possible by dubious appeal to laws designed mainly to protect the senile from fraud. Legal hearings were held, and parents were usually granted custody of their convert children without knowledge of the latter, who, however, could then be legally seized and subjected to deprogramming in given centers. All this was supported by many categories of people, including social scientists testifying that conversion entailed brainwashing and "mind control." In March 1977 finally, the New York State Supreme Court ruled that abduction and deprogramming constituted a direct and blatant violation of constitutionally guaranteed freedom of religion.

Many theories have been advanced to explain the resistance that these new religions have met in the United States of America. A recurrent theme is fairly summarized by the following statement offered by Harvard theologian Harvey Cox (1977—emphasis provided) in a key remarkably congenial with mainstream social science: "Some oriental religious movements bother us because they pose a threat to the values of career success, individual competition, personal ambition and consumption, on which our economic system depends. *We forget that Christianity, taken literally, could cause similar disquietude.*" Remarkable sophistry. On the one hand, it is a splendid example of the human tendency to find logic where there is none. On the other, it reflects the tendency among intellectuals to deny rationality to the populace when it is convenient for their theories. Christianity was never taken literally, if for no other reason because it is too complex and heterogeneous a phenomenon to have a literal dimension.

Why, then, should Americans fear other religions for their literal economic implications? The new religions are offending the need for conformity, and the religious offense to the need is always one of the most abominable. What is somewhat surprising is that the reaction against the religious intrusions has not been more violent. On the other hand, it is also true that the challenge so far has not been widespread. Thus, it can be rather easily countered with a few legal and paralegal skirmishes, like the taxes accusations against Mr Moon, and the reactive strength of traditional religion itself. Should there be more than a few thousand converts to the movements of Hare Krishna, Sun Myung Moon, and the like, the reaction would be greatly more violent, whatever the economic system.

In his attempt to show how learning is often overridden by natural selection, Barash (1977: 35) appeals to an interesting case of animal behavior. When a cow is bitten by a stable fly or hears its buzz, it almost nonchalantly ripples its skin or flicks its tail, despite the fact that the fly can inflict immediate and sharp pain. By contrast, the cow reacts furiously to the sound of a warble fly, which inflicts no pain but deposits eggs that will later develop into larvae and burrow into the cow's skin, causing serious harm and even death. The cow, in short, has been selected to tolerate an innocuous bite, even if painful, but to do all it can to avoid a parasite whose injury is nil in the short run but enormous in the long run.

There is an analogy to this behavior in the Americans' strong reaction to the Eastern cults when this behavior is compared to the same people's reaction, say, against the demonstrations of Iranian student guests in favor of the Khomeini regime even as Khomeini was humiliating "the great Satan" by holding hostage fifty Americans in Teheran. Opinion polls at the time reported that Americans were annoyed at the taunting behavior of the Iranian students but, with few exceptions, believed at the same time that it was a nuisance that they should properly cope with. The contrast to the abduction and deprogramming of American youngsters who, after all, showed every indication of being lawful, humble, and upright is thus dramatic. Why? The conduct of the Iranian students was painful, but it was seen as coming from a small minority of transients, and was bound to be ephemeral as a consequence. By contrast, the converts to the foreign religions were experienced as being borers from within. They threatened the sacred principles on which the American social order was largely based and, if unchecked, might have succeeded in destroying sacred principles and social order alike.

Generally, the more strongly held is a rule, the greater is the tendency to defend it by resort to force. From a biocultural perspective, there is almost a tautology in that statement, for the rule occasions vehement adherence precisely because it is tied to a need for conformity so intense that it demands acquiescence at all costs. Herein lies the danger of the true believer, the Democrat of democrats, the True Communist, the ethically pure. The true believer stops at nothing in order to attain general purity, and when the power of organization is on his side he likes nothing better than to bathe in blood his offended morality and his instinct of vengeance.

The association between the use of force and the intensity of predisposition and belief helps to explain the frequent association found in history between religion and the use of force. Many a Christian has been at a loss to explain how the church of the humble Christ, who preached to his followers that they turn the other cheek, could have produced the Crusades, the Inquisition, and in general that violent spirit that has meted out most foul and cruel punishment to people in chains, on the burning pyre, on the torture rack. More astonishing still is the fact that the most cruel penalty has often been reserved for the most pious: the Savonarolas and the Companellas. From an evolutionary perspective, there is no mystery to such apparent inconsistency. Religion is a fundamental human institution. Hence, except when it is undergoing radical and immanent changes, it exercises the belligerent predispositions that preserve the existing social order (Durkheim, 1912; Pareto, 1916; Weber, 1922, especially II: ch. VI). Any serious challenge to the religious uniformities of the times is bound to provoke a strong reaction—however peaceful the religious philosophy of the founders. Humility generally goes only with a minority status. That explains also the piety of the challengers: the likes of the Companellas are harbingers of things to come and, therefore, by definition, members of a minority.

An exacerbated need for conformity creates fundamental problems of order for the mammoth nation of our times. The human capacity to identify at a gut level with a large group is still in its infancy. Intra-national heterogeneity along racial, ethnic, linguistic lines is but the surface appearance of deeper, basically religious, differences reflecting diverse sociocultural directions that the need for conformity can take. The leviathan nation is mostly a result of the great acceleration in recent millennia and centuries of sociocultural evolution, especially in its dimensions of technology and communication. But biological evolution is much slower, and that creates the absurdity of the "cave man" strolling on Broadway or Piccadilly Square and sniffing the global conceptualizations emanating from their processing nuclei of civilization.

That does not mean, of course, that sociocultural evolution necessarily inclines toward a final divorce from biological evolution. It is a remarkable fact that if the agrarian revolution gave us the empire and the beginnings of the global society, it also endowed us with the universal, proselytizing, catholic religion. Buddhism, Christianity, and Islam mark the end of the jealously guarded and exclusively revered divinities and sacred principles of countless peoples at least since Neandertal man. One may reasonably hypothesize, therefore, that the rise of the global nation has a global religion puffing back there in its wake. Global may here be understood in the literal meaning of the term. It may very well be, for example, that the current struggle between "liberal democracy" and "socialist democracy," a feature truly worldwide in diffusion, is but a good skirmish of the battle that lies ahead and is destined to replace universal religions with *the* Universal Religion.

Space science and technology will most likely play a major role in this epoch-making evolution, for, as some science fiction suggests, "the war of the worlds" may be an inevitable by-product of space science and technology. And nothing brings a nation, or a world, together more effectively than wars between nations, or between worlds.

For the time being, however, the nations will continue to pay for their territorial hunger or privation with group-conflict indigestion. Witness the internal conflicts of Canada, Great Britain, India, Iran, the Soviet Union, the United States of America—of almost any existing nation, really. Some complex nations (I may call them "societies" for convenience sake only) achieve a quasi-societal unity through the political-ideological structuring of "pluralism," as is the case in the United States of America, for example. But, as we shall see more fully in Chapter 9, "ethnic affirmation" and attacks against "the melting pot" achieve recurrent salience in that society. Moreover, it was only yesterday that religion raised its authoritarian head and a Catholic president showed up clearly as an historical anomaly. Today one may even hear from the guru of a large Christian body that Jesus does not hear the prayers of the Jews.

All of this is to say that parochialism or tribalism is inherent in the need for conformity and is still very much with us, although complex modern organization makes adjustments that sometimes hide the primeval tendency. As a corollary, a nation may be seen as being constantly in a state of siege that is laid from within, the various social nuclei and interest groups tending to impose their several conceptions of right and wrong upon one another. Nations are, thus, to a large extent aggregates of "odd individuals," or more precisely compounds of "odd groups." The "enemy" is within.

If it happens that he is also without—as when an internally riven nation is at odds with another—we may expect to observe an increase in enforced conformity, *even when the action is objectively superfluous*. Classical illustrations are provided by the "Jewish question" in World War II Germany and, at a much less destructive but not necessarily less ignominious level, by the concentration camp for American *citizens* of Japanese descent in the United States of America during the same war.

It also follows, conversely, that the more solidary the nation—the more societal the mass of people—the more voluntary the conformity, at both group and individual levels, precisely during periods of national crisis. Thus, many scholars have noted that voluntary conformity is greater in periods of intergroup conflict: for instance, in times of war and religious strife (e.g. Sumner, 1906; Park, 1941; Stouffer *et al.*, 1949). These are periods in which, depending on their mutual solidarity, the nation-groups are pooled together by a set of "core values," and experience what Sumner (1906) termed "ethnocentrism," namely, a feeling that their association and their cultural complex (the "in-group") as a whole are so far above those of the enemy that, whatever their mutual differences, they are still the proper yardstick for assessing the contemptibility of the "out-group." The phenomenon has adaptive value. *Ceteris paribus*, the greater the ethnocentrism in times of international strife, the greater the probability of success. Indeed, it is more than a matter of light interest to remember that cohesive North Vietnam prevailed over riven South Vietnam, despite the much greater war arsenal of the latter.

The importance of conformity for group life may be grasped from the viewpoint of how group members relate to each other when conformity is problematic. Festinger and his associates have shown, for example, that the more cohesive a group is (the greater the proportion of voluntary conformity, according to our terminology), the greater the frequency of interaction among its members. But frequency of interaction varies in an interesting way in relation to conformity and nonconformity. The more cohesive a group is, the greater the attention paid to a deviate in the form of bringing him into line—up to a point. For, when such efforts to enforce conformity are perceived as failures, the deviate becomes an isolate, a rejected bad egg. By the same token, former members who defect are treated with greater hostility than are those who were never group members (e.g. Back, 1950; Festinger, 1950, 1957; Schacter, 1951; Homans, 1974). Thus, it would seem that, at least in an experimental small group context, conformists tend to be taken for granted, while deviates are given the benefit of the doubt as long as there is a chance that they can be brought back into the cohesiveness of the group. When, however, they become lost causes, they are given up and ostracized. Likewise, the conformity of the group is experienced as so valuable a dimension of social life that the traitor receives greater abomination than the mere stranger. That helps to explain the fact that heretics are meted out more severe punishments than mere nonbelievers.

Returning to a developmental key, C. H. Waddington (1960) has cogently argued that evolution has predisposed children of the preadolescent period to be intensely gripped by orthodox imitation. These are the formative years. Later, those same children may be expected to use their basic information in a more flexible and even creative fashion. The developmental problem, however, concerns aspects of child psychology, and frequently these are arrived at through verbal tests that introduce the problem of sophistry, what may be termed the problem of confusing the sign of a thing for the thing itself. Nevertheless, such data are certainly not the worst at our disposal. An experiment by educational psychologist

Lawrence Kohlberg (1969) based on the interpretation of children's verbal responses to questions about moral issues is particularly worthy of note. Kohlberg appears to have found that the child moves in time through a series of stages as follows: (1) conformity to authority to avoid punishment; (2) conformity in order to receive rewards and participate in the exchange of favors; (3) conformity to avoid rejection; (4) conformity, through a sense of duty, to avoid disruption of rules and avoid guilt feelings; (5) legalistic orientation and recognition of the value of contracts in which a degree of flexibility in rule formation is accepted as positive; (6) principle orientation, in which one's conscience is in principle given precedence over formalistic rules. Most individuals seem to attain only stage four or five, suggesting that conformity is to a large extent a function of the emotional controls of the brain, and that reciprocal behavior may be its fundamental "aim."

E. O. Wilson (1975b: 563) has presented an argument that, to put it mildly, has amused some philosophers and cultural scientists. Ethical philosophers allegedly "intuit the deontological canons of morality by consulting the emotive centers of their own hypothalamic-limbic system." This system incorporates and expresses the individual's particular allocation and combination of species traits and, thus, is "programmed to perform as if it knows that its underlying genes will be proliferated maximally only if it orchestrates behavioral responses that bring into play an efficient mixture of personal survival, reproduction, and altruism" (Wilson, 1975b: 4). Wilson goes on to suggest that, as a result, the mind of the moral philosopher is taxed with ambivalence as it encounters stressful situations—situations which call for the emission of altruism at the expense of egoism, or vice versa.

On the assumption that ethical philosophers are not far removed in their biocultural constitution from the generality of human beings, Wilson's statement makes very good sense. The study of ethics is probably the most complex, and the least developed, of the behavioral disciplines. The fundamental reason is that its practitioners must place themselves in the unenviable if not impossible position of looking at moral behavior from the outside—apart from their own ancient emotional controls—at the same time that it is those very controls that lend them at least some of the tools with which to grasp the nature of morality. What we have seen indicates that ethics is a complex bundle of forces composed by phylogenetic, ontogenetic, and environmental contributions. Ultimately, inasmuch as the individual lives within the opportunities and constraints of group life, morality is necessarily confronted by a variety of conflicting requirements, chief among which is the necessity of balancing the good of the individual with that of the group. Moreover, the more narrowly focused the need for conformity, in the sense that it is tied to a small group type of sociality, the more problematic is morality in a complex-society setting, and the greater both the ambivalence and the formal problem of the ethical philosopher. Human nature, considered either as a whole or in its individual distributions, contains inherent contradictions. Few behavioral predispositions attest to the fact as clearly as does the need for conformity.

5.4 The Need for Social Approval

Three other predispositions of sociality will be dealt with. One, *Asceticism*, will form the basis on which certain biological and cultural selections will operate to yield genuine, non-self-serving, altruism, and will be treated in the next chapter. Two others will be dealt with in the remainder of this chapter. They are the *Need for Self-Purification*, and the *Need for Social Approval*. I turn to the latter first.

The need for social approval is widely recognized in social science, and has played a very considerable role in theorizing. It is a crucial, even if implicit, element in Durkheim's theory of "altruistic suicide" (1912), an explicit entry in Pareto's (1916) theory of the sentiments, and one of the four fundamental needs in W. I. Thomas's theory of the individual and society, which, under the label of the "wish for response," is considered the most social of the needs (Thomas, 1923). The fact of social approval is, in turn, a crucial variable of both exchange theory (e.g. Thibaut and Kelley, 1959; Homans, 1961, 1974; Blau, 1964) and cognitive theory (e.g. Festinger, 1957). My main interest in the present behavioral predisposition relates to its importance for the theory of genuine altruism to be presented in the next chapter; hence, we shall best appreciate the evolutionary significance of the predisposition in that connection. Here a few brief comments will suffice.

The need for social approval is distinct from the need for recognition, encountered in Chapter 4, although the two are sometimes confused in the literature (e.g. Pareto, 1916: 1160). The latter force is manifested primarily by the emotion of pride for given qualities, real or imaginary. It is thus seen at work in highly competitive situations, where self-enhancement is typically at stake, although, as we noted, it sometimes expresses itself in sublimated form and with nonadaptive consequences. By contrast, the need for approval is most directly expressed by our concerns for what others think of us as members of social groups and participants in their moral orders. In the terminology of George H. Mead's philosophy of sociality, the more fully developed the social self, the intenser the need for social approval. That entails the proposition that the more capable we are of taking the role of the generalized other—taking into account, simultaneously, the expectations of a whole group for directing one's own behavior—the greater is our need for social approval, and vice versa. For, under those conditions, our individuality and self-centeredness have been greatly attenuated and, conversely, our group orientation correspondingly enhanced. Festinger and his associates have found that social approval, together with communication, is a fundamental reinforcer of group cohesiveness (Festinger *et al.*, 1950; also Homans, 1974). Conformity to group standards, appropriate response to associates, reciprocity, and comportment according to one's own role in the group, these and other such forms of behavior are rewarded by social approval.

In short, the need for approval is the most social of the wishes, as W. I. Thomas put it, because it accompanies, if it is not altogether a basis of, all other predispositions of sociality. Most importantly, it is by social approval that reciprocity, conformity, and hierarchical orders are to a large extent sustained. Homans (1974) treats social approval as a "generalized value" because it can serve as reward for a wide variety of actions. Ethnographic reports show fairly clearly that, as F. E. Williams (1930) put it for the Orokaiva, "public reprobation" is the most important sanction of morality; and public approbation, conversely, is one of the most coveted rewards.

It follows that the need for approval is not devoid of adaptive value. There are times when group approbation is accompanied by rewards, such as election to a leadership status, that may have fitness-enhancing value. However, as Chapter 6 will argue, often social approval substitutes for material rewards and offers merely symbolic dividends. It states in effect that the recipients are respected members of the group, and as such they should feel that they are receiving their proper deserts. Some scholars have noted that the search for "respectability" is the typical social goal of the lower middle class (e.g. Kahl, 1957), where the difficulty of making headway in the division of labor invites behavior that stresses moral propriety as a substitute. This idea hooks up, we may recall, with Merton's (1968) theory of social structure and anomie, according to which it is precisely the lower middle strata that tend to resolve the goal–means disjunction with ritualistic behavior.

When social approval achieves the status of a purely symbolic reward, it ceases to have adaptive value. Hence, it becomes one of the peculiarities through which natural selection condemns some individuals to a genetic dead end. I would wager that the need for social approval reaches the highest levels of intensity among the least fit: those who reciprocate most, who are submissive rather than dominant, and who are most likely to conform voluntarily to group rules—in short, those whose lowly position subjects them most to the censure of their group. The behavioral predisposition in question is, thus, the coin that commonly pays the weak and the downtrodden—in brief, those who, in losing the Darwinian struggle for existence, tend to inherit the kingdom of eternal virtue laden with pats on the shoulders and humility.

There is, however, no relationship of necessity between low social status and an especially vigorous need for approval. It might be more correct to state that the intense need is a property of subordination, for that would leave the above point essentially intact, while at the same time allowing for an intense expression of the need in the higher reaches of dominance orders, where relations of superordination are also present. A useful illustration is provided by the suicide of the Japanese General Nogi, who killed himself with his wife on the day of the Mikado's funeral. In that case, as Pareto (1916) noted, the sacrifice of life had no adaptive value. It was a pure manifestation of predispositions of sociality, especially deference, combined with certain old-fashioned Samurai traditions and buttressed by an intense desire for the approbation of people sharing those sentiments. The next chapter will argue that the need for social approval within an hierarchical context can reach such an hypertrophic peak that its attainment becomes an end in itself and, thus, introduces self-denial in favor of others, namely, genuine or ascetic altruism.

5.5 The Need for Self-Purification

Related to the need for social approval is a feeling that one has done wrong conjoined with the need to undo the harm and restore one's sense of social and moral integrity. I call this the *Need for Self-Purification*. The predisposition is intricately associated with such emotions as guilt, remorse, fear of unworthiness, and, in the extreme case, even fear of ejection from the group contest.

Anthony Wallace (1966: 156–7) argues in a discussion of rituals that expiation, or penance and good works, are typical of individuals who experience severe shame or guilt for a variety of socially relevant reasons: for example, the violation of a taboo, the commission of a sin, the persistent failure to discharge a social obligation. The predisposition in question is, therefore, closely associated also with reciprocation. Expiation can remove the guilt, restore the sense of personal integrity and social belonging, and reintegrate the individual in the social order. As a consequence, notions of pollution and taboo, along with the purificatory rituals associated therewith, are widespread in human society. Inasmuch as the boundaries of conformity are problematic, as we have seen, taboos also represent rules that protect individuals and groups from ambiguity and dissonance with respect to the boundaries which constitute social and moral orders (Douglas, 1968). They thus aid group adaptation and, frequently through it, also individual fitness by reinforcing socially appropriate and remunerable behavior.

The subject is extremely complex, and my basic aim is to introduce it into a biogrammatic context, not to treat it with all the attention that it deserves. From this limited perspective, it seems that we may be dealing with a two-pronged behavioral predisposition that is probably rarely if ever observed in singular form. Briefly, the significance of the predisposition concerns the feeling whereby violators of a rule, a taboo, experience an alteration of their personal integrity, of their sense of inner equilibrium and worth, of their self as a relational, social entity. Associated with that feeling is another feeling: that the altered integrity can be restored by performing certain acts. As noted, we commonly understand the sense of altered integrity in terms of such emotive states as remorse, guilt, social and moral disorientation. Pareto (1916: 1241) puts it in a nutshell as follows:

> The person who violates a certain norm that it has been his habit to observe feels ill at ease from that very fact. He is conscious of being somehow less than he was before. To escape from that painful state of mind, he looks about for some means of removing the stain, of restoring his former integrity; he finds it and he uses it. The rites that are used to escape the consequences of violating a taboo illustrate the situation in fairly simple form.

At a basic level, the expression of such feelings directs attention to the formidable authority of customs, folkways, mores (Sumner, 1906). That, in turn, sensitizes us to the phenomenon of habituation, or the tenacity with which living things adhere to behavioral syndromes associated with adaptations. Hence, while we should consider that we are probably dealing with a uniquely human trait, it may not be amiss to consider suggestive phenomena among other animals. Konrad Lorenz (1966), for example, relates an engaging story about Martina, a young greylag goose pet who had become accustomed to mounting a staircase in the Lorenz house in a highly ritualized way. One evening Martina was let into the house a bit later than usual. She had become anxious, and upon entering the abode and mounting the staircase she deviated from her habitual path. After a few steps, however, she suddenly stopped, made a long neck, the sign of fear, and spread her wings as for flight. Her fear disappeared after she ran down the staircase and up again in her accustomed manner. Having done it "right," the greylag stopped, "looked around, shook herself and greeted, behavior regularly seen in greylags when anxious tension has given place to relief."

The Christian baptism has its counterpart in many other religions. Among Christians, and their social scientists, baptism is explained as a ritual associated with the concept of original sin, and specifically intended to remove that sin. The explanation is highly problematic if for no other reason precisely because a form of baptism, or ritual self-purification, is probably universal. Alternatively, we may entertain the hypothesis, rooted in Durkheim's (1912) work, that the evolution of human society has moved *pari passu* with the evolution of individual bearers of sacred principles, constituting in their ensemble a religion, that are at bottom mechanisms asserting the supremacy and sacredness of the society over the individual. Such principles are adaptive in the sense that they reinforce group defense and thus reciprocal aid. Self-purification would thus be a means of group selection through the individual's varying subjection to, and sanctification of, the group.

If we add the concept that altered integrity is extensible in time, we reach the notion that posterity is responsible for the transgressions of the forefathers (Pareto, 1916). The Christian doctrine of original sin and kindred notions—such as the Orphics' belief that the moral integrity of humankind had been primevally altered—would seem to flow logically from these considerations. Our ancestors, representing the society whose traditions we bear, committed an offense against the Sacred Principle, God, or the Sanctity of Society; more concretely they left a legacy of bloodshed and broken contracts *within* the society. Through abstraction of the need for vengeance, the offense required reparation. The story of Adam and Eve, therefore, may be a brilliant sophistry and fabrication, but it may also be a clue to the evolutionary forces that gave rise to the need for self-purification and to the cultural universal that Christians term baptism.

Consider, further, that through the centuries, the practice of baptism became excessively ritualized in the sense that behavior and predisposition became largely separated. That seems to be the case, for instance, when baptism is performed early in life, before the full maturation of the social self and without, therefore, the expiatory dimension that baptism entails. We should not be surprised, under the circumstances, by the rise of the idea of rebirth among adults. It is interesting to note that religious revivals typically follow sociopolitical upheavals (Sumner, 1906; Durkheim, 1912; Pareto, 1916; Weber, 1922). The recent increase in US born-again religion, for example, has taken place in the wake of the social turmoil that shook the very foundations of 1960s society (for a recent discussion of this topic, see Crippen, 1982). If societies have built-in mechanisms to control their population in relation to the resources available, it is hardly possible not to hypothesize that moral orders have also built-in mechanisms to control the forces that would break them up. I am suggesting that the recent rise in born-again religion is the manifestation of one such mechanism, whereby a large section of the population in effect proclaims *mea culpa* and surrenders its excessive individualism to the sacred principles of the constituted society.

In listening to a "testimony" in church or on PTL television, a Christian network that is international in its reach, one is typically treated to a public confession of "sins" that is extremely fervid and detailed. The themes recur with great regularity. At the core is the vision of a person who had navigated the ship of life in stormy waters with presumption but without reliable instruments. These are in the hands of Jesus, the Sacred Principle. Like a man drowning and gasping his last breath, he surrenders. It is the primeval call of the moral order. And so in the testimony and in the songs that accompany it, one hears statements of clearly theoretical value: "It is so much better to have Someone rule your life and control it." "What a mess I made trying to run my life!"—till Jesus took over. "We surrender ourselves, we surrender our hearts, our minds, our bodies to Jesus Christ."

It bears reiterating that baptism is not unique to Christian religion. Countless Hindus, for example, bathe every year in the Ganges River to wash away their sins. The act of self-purification is deemed to be particularly efficacious when it is committed during the Kumbh Mela, the pilgrimage festival that is held at twelve-year intervals, and attracts many millions of devotees.

Note that if the theoretical rationale given above for baptism has any validity, it is not necessary that those who turn to born-again religion (or other purifying cults) are also those who perpetrated the moral offense—produced the social turmoil. In one sense, we are truly all children of Adam and Eve. The offense of one is the offense of all. But neither should we be surprised to find a considerable number of actual, even famous, offenders who have "surrendered to Jesus." The cases of people like Eldridge Cleaver and Charles Colson are cases in point.

The latter person recalls the moment of the personal drama that embraces baptism and rebirth. After intense suffering for "the evil of pride" committed in the name of a true belief in the "sins" of President Nixon's administration, wherein he was special counsel; and after excruciating fear for the punishment that legal society was sure to mete out; one evening Charles Colson broke down and cried. "Something inside" him "was urging" him "to surrender." "Take me," he repeated over and over again. "And as I did, I began to experience a wonderful feeling of being released. There came the strange sensation that water was not only running down my cheeks, but surging through my whole body as well, cleansing and cooling as it went. They weren't tears of sadness and remorse, nor of joy—but somehow tears of relief" (Colson, 1976: 128–9). We need not worry about Colson's denial of remorse. The tears of relief are good enough for our purposes: relief can only be grasped in conjunction with some such property as guilt or remorse.

Another interesting application of the thesis I have embraced concerns the soul, a concept to be dealt with in greater detail in the next chapter. Here we may treat it as an abstract or imaginary offender. The idea underscores the moral character of the individual and, as we shall see in that chapter, the concept of immortality as well. The two, in turn, give rise to concepts of transcendental expiation. Among Catholics the type-case of self-purification at a transcendental level is the Purgatory. The Purgatory is allegedly a place where souls who sinned in their earthly lives atone those sins. Among the various redeeming penances devised by the church and inflicted upon the faithful, Purgatory is one of the most imaginative, for it immortalizes our earthly transgressions at the same time that it offers redemption through expiation. The need for self-purification also underlies, of course, the variety of liturgical devices implemented for the redemption of the souls in Purgatory: prayers, masses, acts of charity, and the like.

My aim has been to explain the practice of self-purification at an ultimate level. The label for this has usually been natural selection. But we are approaching an argument, concerning altruism, that to some extent permits us to transcend natural selection. So, we may prepare ourselves for the partial theoretical shift by saying here that the ultimate cause of the need for self-purification is that tendency, however partial and intermittent, whereby individuals subordinate their good to that of others. That was made possible, as again Chapter 6 will try to show, by the cultural selection of the concept of the soul, *inter alia*. The soul, as will be seen, is a property of the human concept of existence that mimics the biochemical property of DNA molecules through the idea of eternal salvation. Salvation, moreover, is usually attained through the good works, toward both deities and human beings. Accordingly, one major proximate cause of acts of self-purification resides in the individual's own capacity to find evil within himself (Bandura, 1977: 151–3)—evil that we conceptualize as an offense to the moral rules, often sanctified through transcendent principles, of the community.

The proximate mechanisms of transgression that call forth the need for self-purification are legion. Catholics, for instance, long believed that meat defiled when eaten on Friday and on certain other days, and the offense required certain rites, for example, confession, to cleanse them of the taint.

Among many peoples there have been purification customs in connection with homicide, voluntary or involuntary. Apollodorus seems to have thought that purifications were normal procedure after homicide. Thus the daughters of Danaus were purified by Athena and Hermes after slaying the sons of Egypt. The Argonauts were persecuted by the wrath of Zeus for slaying Absyrtus, until they were purified by Circe.

Sexual intercourse between a man and a woman has widely been considered a cause of impurity under some circumstances and at given periods of the year (e.g. M. H. Wilson, 1951). Menstruation, too, has often been considered defiling. The Greeks and the Romans, among other peoples, considered parturition a cause of taint. Unusual occurrences have been particularly frequent causes of a sense of transgression. In Gogo society the birth of twins is a sign that society in general is at sea and that the male ritual leaders in particular, namely, the keepers of the sacred lore, have failed in their duties. Rituals of purification ensue in which women practice violent, transvestite dances which, in a parody of the male's violent role, seek to effect a cure (Rigby, 1968; see also Wilson, 1951). But perhaps the largest categories of offenses requiring ritual purification concern, predictably, relations between persons and deities. The latter may be viewed as the symbolic referees of human relations. Hence, self-purification directed at divinities often calls forth action that provides formidable and just punishment.

Another category of impurities and taboos is extremely widespread, but it also seems to have implications for hygiene and, therefore, will be excluded from detailed consideration. I have in mind the common belief in funerary contagion whereby the relatives of the deceased are considered impure, and are often kept in isolation, for varying lengths of the mourning period. E. E. Evans-Pritchard (1956) writes of the Nuer that at "a mortuary ceremony there are a feeling of pollution and the need to cleanse men and things" (see also Hubert and Mauss, 1898; Hertz, 1907–9; Wilson, 1951; Wallace, 1966).

The multiplicity of proximate causes of ritual purification concerns, in our terminology, the cultural variants, just as the fact of purification itself is a cultural universal, and the feeling of transgression along with the need for purification represents the behavioral predisposition. Other variants that may be briefly touched upon refer to the means whereby the purification is attained. These, too, are numberless (see, e.g., Hubert and Mauss, 1898; Hertz, 1907–9; Durkheim, 1912; Wilson, 1951; Evans-Pritchard, 1956; Wallace, 1966; Douglas, 1968; Rigby, 1968). The products of an innovating force (see Chapter 7 below), they embrace all manner of magical beliefs and practices, religious prayers, and all sorts of substances and techniques, including fire. Among them, water has perhaps been most common. Water has at all times been used to remove material stains. At all times it has been taken for granted that it removes moral impurities as well. So deep-rooted is the feeling that water cleanses moral stains that at one time in the Judeo-Christian world it was generally believed that the biblical Flood had been a great catharsis sent by God to cleanse the impure morals of earthly inhabitants.

In conclusion, the interplay of biological and sociocultural forces appears to be visible enough in the manifestations of the need for self-purification and in the fact that these, in turn, intensify the need—as when, for instance, a born-again Christian converts by pious example another person to his own beliefs. Not so clear is the problem of adaptation. Understandably, this issue becomes more clouded as we approach the context of genuine altruism. Still, it seems reasonable to say that to the extent that a defiled person or a transgressor of some moral rule, real or imaginary, can, through a self-purifying act, achieve peace of mind and full reinstitution in the group context, even if only subjectively felt, self-purification must be viewed as an adaptive mechanism. It restores the afflicted individual to the group support and to the fitness-enhancing resources at its disposal. Or self-purification may permit an individual to perform with impunity an activity that is otherwise defiling but essential to group survival.

The following is a case in point. Among the Todas of India, who held the buffalo to be sacred, dairymen were subject to an ordination ceremony not unlike, in its essentials, the ordination ceremonies of priests in many religions. "The essential feature of all the ordination ceremonies is a process of purification by drinking and washing with the water of a stream or spring used for sacred purposes only. . . . In every case the water is drunk out of certain leaves, and the body is rubbed with water mixed with the juice of young shoots of bark" (Rivers, 1906: 144). Remaining at the proximate level justified by the highly symbolic nature of the problem at hand, Rivers (1906: 231) hypothesizes that this elaborate ritual has most probably arisen as a means of counteracting the dangers likely to be incurred by a profanation: making dairy products from the milk of the sacred buffalo, "in other words, as a means of removing a taboo which prohibits the general use of the substance." While anything said about the origin of the specific taboo itself would be pure speculation, it is fairly clear that this ceremony of purification itself is adaptive at the group level. Having for unknown, but possibly biologically relevant, reasons prohibited the use of the nourishing products of the buffalo, the Todas turned and harnessed the interdiction through a purification ceremony. The behavior is manifestly adaptive, for it adds to good health by restoring a vital source of self-sustenance to the Todas diet.

Other cases of ritual purification defy attempts at labeling them adaptive. Consider the manner in which the Nyakyusa men of Tanganyika coped with cuckolding, according to a study carried out in 1934–8. The author (M. H. Wilson, 1957: 78) quotes an informant as follows: "if my wife has run off with another man and then comes back to me with his semen in her, then I take some powdered medicine and eat some and throw a little on the fire, and if I do not do this then my legs will hurt, I shall be unable to walk properly, and my body will become weak." To be sure, to the extent that the ceremony does prevent a weakening of a cuckolded husband's body, it may be declared adaptive behavior. Just the same, it would seem that a vigorous maximization drive would have led to a ceremony wherein concern was less with the legs and more with the semen. Clearly, we are approaching types of nonadaptive behavior.

6 Sociality, II: Ascetic Altruism

Chapter 5 conceptualized the predispositions of sociality as being largely at the service of self-enhancement forces, but it did not assert total dependency. In fact, we caught a glimpse of a nonadaptive dimension of the need for recognition, as when a scholar is so busy attaining immortality of name that he has no time to tend to his genetic immortality through reproductive activities. Likewise, the need for self-purification suggested a strong tendency on the part of the individual to live in a state of recurring anxiety lest the sanctity of the group be violated—and therefore a certain inclination to subordinate personal well-being to group well-being, as attested, for example, by various rituals of expiation.

The latter propensity, to the extent that it is real, reinforces my inclination to argue that group selection is both an evolutionary fact and a theoretical necessity of biocultural science. This position, however, does not entail the proposition, typically held in sociobiology, that true altruism (behavior so socially oriented that it violates the maximization principle) must be explained in terms of group selection. Group selection and genuine altruism may sometimes go hand in hand, but there is no relationship of necessity between the two, and this chapter will propose an alternative route to the explanation of true altruism.

My intention has been to stress that, as the work of Emile Durkheim and W. I. Thomas, among others, has shown, living in a group context entails a certain dialectical tension between the pursuit of the individual advantage and the pursuit of some collective good. We need not decide whether, at close examination, the collective good often turns out to be to the benefit of a subsocietal group such as a dominant class, as Karl Marx among others argued, or of the more inclusive aggregate that both Thomas and Durkheim seemed to envision under the label of "society." In either case, group life entails the natural selection of individuals predisposed to play roles that yield differential life chances. It is in view of this perspective that we can now introduce the problem of altruism. Note again, however, that to tie differential life chances to group life does not necessarily entail looking for true altruism from the perspective of group selection.

The present chapter pursues the hypothesis that sociality has evolved in the direction of the selection of altruistic behavior on the part of at least some members of a given population, whether sporadically or for a lifetime, whether partially or *in toto*. The aim, specifically, is to attempt a theory of what may be termed genuine or *ascetic altruism*. The effort will be tied to the current argument on altruism in sociobiology and will elaborate on a previous report on the subject (Lopreato, 1981). I shall, therefore, continue to operate within the context of natural selection. By gradual steps, however, the emphasis will shift to the influence of culture on biology. At certain junctures, the argument may even seem critical of sociobiology. But there should be no need to reiterate here my commitment to the proposition that there is an interplay between biology and culture; or my adherence to the principle that systems of knowledge thrive best when they are linked by open lines of communication.

6.1 On the Conceptualization of Altruism

Do human beings act only for their own good and that of their kin, or do they also behave willingly and without expectation of gain for the good of others? The question is ancient. Certainly, it has been debated by moral philosophers for millennia; and of course it has been crucial, in one form or another, to social science at least ever since Auguste Comte coined the term "altruism" a century and a half ago (e.g. Kropotkin, 1902; Durkheim, 1912; Sorokin, 1950a, 1950b; Friedrichs, 1960; Sawyer, 1966; Nagel, 1970; Gouldner, 1973; Phelps, 1975; Wispé, 1978). Most recently, according to E. O. Wilson (1975b), the natural selection of altruism has become the central theoretical problem of sociobiology.

The sociobiological participation in the debate is a welcome development because, as we shall presently see, it introduces concepts and perspectives that, while restricting somewhat the scope of the inquiry, promise a partial extrication from the confusion that so far has surrounded the problem. With few exceptions, social scientists and philosophers have found human beings to be selfish or egoistic. But, because they have approached the issue largely in ethical terms, they have run into various problems that tend to complicate the problem rather than to solve it. Accordingly, the problem of altruism in the socio-philosophical tradition remains an "intuitive issue" (T. Nagel, 1970). In general, philosophical inquiry has moved in the direction of precepts that would reconcile the principle of selfishness with the principle of altruism. Consider Jeremy Bentham's approach. A complex discussion, in which the selfish tendency is time and again asserted, concludes with the normative argument that social disapproval is harmful to individuals, and, consequently, it is to our advantage to avoid it by shunning selfish acts. The solution, however, is flimsy, and Bentham himself was not unaware of the fact. Reprehensible acts are not always detected, for example, and one bent on self-interest would be tempted to take advantage of that circumstance. Bentham never did solve the problem, but in the process he devised the celebrated formula of "the greatest happiness for the greatest number," which was intended as the proper object of governments in their activities of reconciling the various private and public utilities.

A related answer had been furnished by Spinoza who proposed that "nothing is more useful to man than man." Therefore, "men who seek their own welfare under guidance of Reason desire nothing for themselves that they do not desire for other men, and so are just, honest and of good faith." Other philosophers, for example, Hobbes and Locke, partaking of the basic logic of social contract theory, argued that society and the individual relationships to it are founded on certain natural laws; those who violate such laws do harm to society and consequently to themselves.

On the whole, all these theories tend to resolve the problematic relationship between the selfish principle and the altruistic principle by invoking the postulate that in working for the community we in effect work for our own good. The solution is faulty for a variety of reasons. It does not account for undetected cheating. It neglects the effect of competition over highly desirable and scarce goods. It fails to account for the fact that there are indirect as well as direct effects of actions that may be differently valued; so, for example, a thief may suffer an indirect harm of little personal consequence by robbing the community of which he himself is a member but at the same time gain from his action a large direct advantage.

The socio-philosophical literature on altruism is enormous, and I have barely touched on it. From a more detailed review we could derive a number of notable lessons. First, it would seem that altruism and its opposite (termed sometimes "selfishness" and others "egoism," though a distinction between the two terms is at times made) are interdependent phenomena or, perhaps better, inseparable sides of the same coin. The point is made concisely, even if somewhat hyperbolically, by Herbert Spencer (1895–8) as follows: "If we define altruism as being all action, which in the normal course of things benefits others instead of benefiting self, then from the dawn of life, altruism has been no less essential than egoism. Though primarily it is dependent on egoism, yet secondarily egoism is dependent on it."

Second, in the last century or so, many philosophers of ethics, taking notice of evolutionary theory, have had to conclude that there are neither logical bases nor good reasons of any sort for the general acceptance of any given set of moral standards to apply cross-culturally. For our purposes, that amounts to saying that the question whether or not human beings are altruistic to any degree may not lend itself to a definitive solution in moral philosophy (see, e.g., Quillian, 1945; Flew, 1967a, 1967b).

Third, there is no one socio-philosophical definition of altruism on which there is a substantial degree of consensus. Conversely, definitions abound. Dividing them into sets according to internal affinity we find that in one of them altruism refers very broadly to the willingness to behave, without ulterior motives, in ways which in some sense benefit others (e.g. T. Nagel, 1970). Another set seems to view altruism as behavior that reduces one's own personal welfare, the latter notion being more or less explicitly conceived in terms of life expectancy. As we shall see, this second class of definitions is rather closely related to the one current in sociobiology. But to my knowledge, no calculus has been devised to assess the units of life expectancy that are sacrificed when acts of altruism are in fact performed.

In the absence of a proper criterion of measurement, the debate between those who assert altruism and those who hypothesize selfishness is either inconclusive or is won by the latter merely through the sophistic technique of *reductio ad absurdum*. For instance, any behavior, *b*, that X would define as altruistic is subject to a selfish redefinition by Y on the basis that *b* fetches the actor such rewards as social approval, historical or scholarly glory, the avoidance of censure and material punishment, or even spiritual salvation. If you, for example, term me an altruist because I did you a good turn without expectation of reward, I may answer that, no, I was really being selfish because, after all, doing good for you made me feel good—it increased my pleasure—or set me straight with God. We should note that this technique is often employed in real life by persons whom we term "modest," "self-effacing," "humble," "unassuming," and the like. Yet they often define themselves as selfish, and there is little or no hard ground on which to prove them wrong.

For these reasons, and because sociobiology has recently posed a challenge to the moral and social disciplines on the issue of altruism, I propose to accept, with some necessary variations, the sociobiological definition of altruism. That may afford me an opportunity to provide an interdisciplinary theoretical input while operating on the sociobiologists' own conceptual grounds.

Representing fairly wide consensus, Wilson (1975b: 578) defines altruism as "self-destructive behavior performed for the benefit of others." Self-destructive behavior must be understood by reference to the maximization principle, or the tendency of organisms to behave so as to maximize their inclusive fitness in the degree to which they are under the influence of natural selection. Inclusive fitness refers to the sum of one's contributions of genes to future generations through one's own offspring and/or through the offspring of blood relatives. It follows, therefore, that altruism goes counter to the logic of the maximization principle: it tends to minimize one's inclusive fitness.

Unfortunately, however, what sociobiologists define as altruism and what they usually label as altruism are two entirely different things, and we need to look at this complication in order to better confront the task at hand. In general, when sociobiologists, indeed behavioral evolutionists in general, are keen to the rigid demands imposed by the definition (your gain is my loss), the tendency is to deny altruism in any species. We shall return to this problem. When, conversely, the definition loosens its rigid hold, they set out to explain its presence. The result is a concept that is formally one thing but practically another, namely, a synonym for the less exacting concept of beneficence—beneficence, however, that redounds to the benefactor.

The multitude of statements that weave together this discrepancy is naturally a source of unnecessary confusion, in biology as well as in social science. Accordingly, on the way to my main goal, it will be useful to review briefly the major contexts within which sociobiologists study "altruism." For clarity's sake, the concept will be placed in quotation marks when in fact it is a misnomer.

6.1.1 THE SOCIOBIOLOGICAL CONTEXTS OF "ALTRUISM"

6.1.1.1 Kin Selection There is, first, the kin selection context surrounding W. D. Hamilton's (1964, 1972) path-breaking work on the social insects. To understand its importance we must recall that the struggle for existence which Darwin, like Wallace and Spencer, derived from Malthus's theory of population led him to a principle of natural selection in which organisms were viewed as competing to safeguard their biological fitness through the reproduction of viable offspring. This focus on offspring soon ran into some embarrassing facts. Some birds, for example, are perfectly capable of laying a four-egg clutch, and yet lay one of three. Clearly, four are in principle more promising of fitness than three. Various sorts of animals give danger signals, at the sight of a predator, that sometimes put them in jeopardy. Worker bees sting suicidally in defense of the hive. There is no paucity of like examples. Perhaps the most clamorous refers to the huge caste of workers in social insects. With few exceptions they are nonreproductive; however, they are exceedingly diligent in catering to the needs of their mother, the queen, and her prodigious production of brood. It will be helpful to look briefly into this Darwinian puzzle, for it is its solution that at one and the same time broadens the Darwinian concept of fitness with the concept of inclusive fitness and puts some teeth into the maximization principle.

The problem is evident: if organisms are programmed to maximize their own reproduction, how do these seemingly foolish insects fit in the evolutionary scheme? Quite well, as it turned out, and Hamilton showed how. These ancient insects have a rather odd system of reproduction and sex determination known as haplodiploidy. The queen lays two kinds of eggs: one sort, the majority, is diploid, or fertilized; they become (female) workers. The other is haploid, or unfertilized, and yields what are termed drones or kings. Thus, these carry only the chromosomes inherited from the mother, namely, half the usual complement. It is also they, of course, who mate with queens.

Now, let us assume the common, though not always true, case that a given queen is fertilized only by one king. What is the result in terms of the genetic relationship existing among her daughters, the nonreproductive workers? This is an important question because one of them will one day become a queen and represent, with her reproduction, the inclusive fitness of her sisters. According to Mendel's First Law, each has inherited 50 percent of her mother's genes, and on the average she shares one-half of that (25 percent) with each of her sisters. The other 50 percent has been inherited from the father's side. But since he is haploid—deriving from an unfertilized egg, he has only half the normal complement of chromosomes—that means that all his daughters have received *all* his chromosomes and genes. His sex cells have not undergone the usual meiotic division because all his cells are meiotic to begin with, and thus behave (irregularly) according to the mitotic process of cell replication. In short, on the father's side, the workers are genetically identical and share thereby 50 percent of their genes with one another. Add this 50 percent to the 25 percent that they have in common on the mother's side, and note Hamilton's discovery: if the workers had their own offspring, they would share 50 percent of their genes with them; if, on the other hand, they assist their mother and remain themselves nonreproductive, they share on the average 75 percent of their genes with one another. In short, in these haplodiploid species it is more genetically fit for the females (who are the majority) to be nonreproductive than to be reproductive. The reason, of course, lies in the fact that, as noted, the next queen will be a sister rather than a daughter, and her brood will be more closely related to their aunts than to their grandmother. Specifically, workers share, on the average, 37.5 percent of their genes with their nieces, whereas queens share only 25 percent of their genes with those same individuals, who are their granddaughters. Here, then, is the solution to an old Darwinian puzzle, a solution that corroborates dramatically the maximization principle.

Note, too, that the old Darwinian concept of natural selection has been extended to include what Maynard Smith (1964) has called "kin selection," a term that is sometimes used synonymously with Hamilton's (1964) concept of inclusive fitness. Consequently, natural selection is now best viewed as the differential contribution of genes, rather than offspring, to future generations. In short, one can achieve fitness through all one's blood relatives, not just one's children—hence, Hamilton's concept of *inclusive* fitness.

Hamilton's work led him to state the following inequality on "altruism": $k > \frac{1}{r}$. According to this formulation, the probability of "altruism" increases (1) as the coefficient of relationship (r) between dispenser and recipient increases, and (2) as the benefit accruing to the beneficiary exceeds the cost incurred by the benefactor (the k ratio). The label of altruism is, *stricto sensu*, misplaced within this context, and some scholars would consider "nepotism" a more felicitous choice of terms (e.g. Alexander, 1974; Barash, 1977; van den Berghe and Barash, 1977; Wilson, 1978).

Whether Hamilton's inequality is applicable to *Homo sapiens* as much as to other species remains to be seen. Nevertheless, its clearly quantitative and probabilistic approach to altruism is happily practical. It is not the case that, as the literature sometimes seems to imply, altruism either exists or does not exist. It is rather the case that *degrees* of altruism exist or fail to exist. Thus, it can and does happen that acts of nepotistic beneficence often fetch the dispenser benefits, in terms of units of inclusive fitness, that fall short of her/his costs (for a brief but excellent discussion of kin selection in general, see Irons, 1979b; also West Eberhard, 1975; for a recent discussion of common misunderstandings of Hamilton's work, see Dawkins, 1979).

6.1.1.2 Reciprocal "Altruism" Kin selection theory is basically a theory of reciprocal favoritism, and sociobiologists, like other social scientists, are developing a healthy awareness of the fact that, as the discussion of reciprocation pointed out, human relations may in part be viewed as a network of interpersonal exchanges that transcend the boundary of the kin group. E. O. Wilson's (1978) previous distinction between soft-core "altruism" and hard-core "altruism" is a case in point. The latter is associated with kin selection and engenders provincialism and blood feuds. By contrast, through soft-core "altruism" we manage to oblige one another beyond family boundaries—to render real the old saying, so keenly appreciated by social exchange theorists, "You scratch my back and I'll scratch yours" (Homans, 1974). Accordingly, Wilson argues that "reciprocation among distantly related or unrelated individuals is the key to human society. The perfection of the social contract has broken the ancient vertebrate constraints imposed by rigid kin selection."

As noted, the basic sociobiological work on reciprocal behavior has been done by Trivers (1971, also 1974). Recall that Trivers's model of reciprocal "altruism" shows how "altruistic" behavior can be selected for without invoking any assumptions about genetic kinship. Indeed, the formulation accounts for "altruism" even in interspecific contacts, for example, in cleaning symbioses among fish. The basic conditions are: (1) a high probability that the recipient of a beneficent act will in the future reciprocate toward the original benefactor; (2) that, in keeping with Hamilton's aforementioned inequality, the benefit of an "altruistic" act to the recipient be greater, in terms of inclusive fitness, than the cost of it to the dispenser.

There is one catch to reciprocal "altruism." It is ultimately self-serving behavior, and hence it does not fit the technical definition of altruism. As Trivers himself (1971: 35) puts it, under certain conditions natural selection favors reciprocally "altruistic" behaviors "because in the long run they benefit the organism performing them." Nevertheless, I shall argue later that the selection of reciprocal "altruism" may have constituted a stepping stone for the evolution of an altruism shorn of any *quid pro quo* provision.

6.1.1.3 Group Selection To the extent that sociobiologists visualize true altruism at all, the tendency is to associate it with group selection (see Wynne-Edwards, 1962, for a classical statement on this). The idea of group selection in evolutionary biology is deceptively simple. We must envision a population divided into a number of distinct groupings or demes, each of which is characterized by random mating and very little or no migratory contact with the outside—that is, the demes are diverse genetic units. They are also in competition with each other; hence, the concept of extinction is crucial to the idea of group selection. If we assume that the probability of a given deme's extinction decreases if certain genotypes multiply, then the crucial idea is that such genotypes are "favorably selected as a species characteristic although disadvantageous to their individual bearers" (Boorman and Levitt, 1980: 7; see also 1972, 1973; also Levins, 1970).

The early reactions to Wynne-Edwards's thesis tended to be critical (see Williams, 1966, for a review). They, however, have been followed by a considerable number of statements showing that, albeit under conditions that are difficult to satisfy, selection at the group level is indeed possible (e.g. Levins, 1970; Williams, 1971; Boorman and Levitt, 1973, 1980; Gadgil, 1975; E. O. Wilson, 1978; Alexander, 1979; D. S. Wilson, 1980). To some extent, I suspect, the debate over individual versus group selection reflects the differential intensities of predispositions of self-enhancement and sociality among the scholars partaking of it. Not just philosophers, but also scientists often see the world through the hypothalamic-limbic system.

For social scientists, and in this volume, group selection is an important and useful concept. The basic reason is that, as we shall better see in the discussion of Durkheim's theory of religion in Chapter 8, ours is a highly symbolical species, and as such it has evolved moral orders whereby societies undergo a sort of apotheosis. That means not only that, at least under some circumstances and at special times, the concept of society is coterminous with the concept of divinity. It also means that the society becomes somehow detached from the numerical and demographic conception of the membership. It achieves an emergence; it becomes a being in its own right; and in its name the individualistic, self-oriented, interest sometimes breaks down. Even the maximization drive may cease to operate. We shall see the point better later in this chapter.

But while the concept of group selection is valuable for sociocultural science, it cannot be borrowed unmodified from evolutionary biology. Two changes are especially important. First, group extinction must be understood broadly enough to encompass partial as well as total disappearance. It often happens, for example, that in the process of wiping out another society, the vanquishing society absorbs at least a few of the defeated members. Such was the case, for example, in the encounters between the European settlers of the American continent and a number of the pre-Columbian American societies.

The second change rejects the notion that the genotypes that are selected as a group characteristic are disadvantageous to their individual bearers. At least, I do not see a relation of necessity in the problem in question. Consider this scenario, from which for obvious reasons we must exclude kin selection. Two demes, A and B, are in competition for resources in a given space. The former consists of individuals who are weak in public spiritedness; the latter is composed of persons ready to sacrifice their lives for one another. War breaks out. The members of A join the battlefield in disarray; those in B, as a well-organized army. The latter group has a selective advantage. But so do its members, for they have an average lower chance of dying when compared to their counterparts in group A. So conceptualized, group selection does not unequivocally represent the individual in self-sacrificial behavior (see also Durham, 1976). Indeed, it might be argued that group selection is really an extension of Trivers's reciprocity model. The members of a deme may be viewed as randomly taking chances for one another "in order to" increase all their chances of survival (for a brief discussion of individual versus group selection, see Irons, 1979b).

Nothing I have said, of course, can gainsay the possibility of genuinely altruistic behavior when this is conceptualized in relative and variable terms. Just the same, insofar as the selection of altruism is concerned, these considerations about group selection constitute further incentive for pursuing alternative avenues to the solution of the problem.

We may conclude this section on group selection by defining this phenomenon as *the differential contribution of genotypes, by diverse populations, to the world population of future generations*. When we wish to extend the concept to comprise sociocultural evolution as well, we must think of it as *the differential contribution of genotypes and/or sociocultural variations, by diverse populations, to the world population and/or sociocultural repertory of future generations*.

The psychologist Donald T. Campbell, one of the first social scientists to enter the group selection debate within a sociobiological context, has argued that sociocultural evolution and biological evolution have taken place concomitantly throughout human history, and indeed have often supported each other. But at some point in time, the two became separate and largely autonomous processes. Biological evolution is alleged to have optimized individual fitness while sociocultural evolution has optimized social system functioning. Altruism, therefore, is a cultural phenomenon, and, given the selfish tendency inherent in biological evolution, "social [read: sociocultural] evolution has had to counter selfish individualistic and familistic tendencies" with moral injunctions to achieve the development of "self-sacrificial altruism" (Campbell, 1975; see also 1972, 1979).

Campbell's provocative work has not specified the selective mechanisms that allegedly have more or less successfully pitted an altruistic morality properly speaking against the imperious selfish undercurrents. Moreover, Campbell has associated self-sacrificial altruism with group selection. Still, he has evocatively proposed that if such forces as moralistic aggression and ethnocentrism do exist as innate tendencies, then they "support a social equivalent of group selection and predispose social customs enforcing self-sacrificial altruism on reluctant individuals, just as would a purely social evolution of beliefs in transcendent social purposes, rewards and punishments after death and social organizational features optimizing organizational survival" (Campbell, 1979: 43). This is an important idea intimately related to the principal aim of this chapter.

6.1.2 FROM NEPOTISTIC FAVORITISM TO ASCETIC ALTRUISM

I shall turn to that goal presently. In the meantime, we can fairly agree with Trivers's (1971) blunt verdict that available "models that attempt to explain altruistic behavior in terms of natural selection are models designed to take the altruism out of altruism." Under the circumstances, however, why the continued use of the false label of "altruism"? Why allow it to engender confusion? For example, not everyone can satisfactorily extricate oneself out of Wilson's (1978: 22, 149) reasoning when he states both that the workers of insect societies are "more altruistic than people" and that human soldiers sometimes throw themselves on top of grenades to shield comrades who, we might add, are total strangers to them and, given the finality of the act, will not be in the position of the reciprocator in the future.

With few if any exceptions, all behaviors labeled altruistic by sociobiologists may be comprised under the general rubric of self-serving *favoritism*. There are various forms of such behavior. Until better times, I propose the following classification (see also Lopreato, 1981):

(1) *Nepotistic favoritism* encompasses acts of beneficence toward one's relatives and, thus, concerns kin selection.
(2) *Reciprocal favoritism* comprises acts of beneficence between unrelated individuals who have a written or unwritten rule that one good deed

deserves another, and thus accounts for the facts of "reciprocal altruism."

(3) *Phenotypical favoritism* may be conveniently employed to accommodate certain related facts of growing interest to social scientists and behavior geneticists. The ancient Romans had a saying, *similis simili gaudet*—like takes pleasure in like. Studies of homogamy and assortative mating provide considerable support for the thesis that individuals, short of incestuous relations, tend to favor their own kind as marriage partners (e.g. Beckman, 1962; Eckland, 1968, 1972; Vandenberg, 1972; Murstein, 1976; Karlin, 1978; Thiessen and Gregg, 1980). The latter team, for example, has reported a strong inclination toward positive assortative mating that tends to increase parent–offspring relatedness above the 50 percent rate accounted for by Mendel's First Law.

(4) *Ascetic altruism*. My chief aim is the search for, and the explanation of, altruism properly speaking: what, as we have seen, is defined as behavior that reduces the inclusive fitness of the benefactor while increasing that of the beneficiary. I propose the label of *ascetic altruism* for this type of behavior, rather than simply altruism, because of the just-noted ambiguity associated with the current use of "altruism" and because of the notion of self-denial that is inherent in the concept of asceticism.

The prevailing definition of ascetic altruism cannot, however, be accepted without tinkering with some difficulties associated with it. One concerns the postulation that one individual's loss is necessarily another's gain, and vice versa. Consider the following case, stated in a form to avoid the context of reciprocal favoritism. You are a tourist at the beach in a foreign country and go to the rescue of a total stranger and native who is drowning. At considerable danger to yourself, you save his life. Your act prevents the *reduction* of his fitness, but it cannot be said *stricto sensu* that it increases it. More important still, your own Darwinian fitness is not reduced, although *it could easily have been*. Or suppose that the poor fellow drowns despite all your efforts. Did you not put your life on the line in his favor anyway? Consider, indeed, the possibility that you both drown. Your own fitness is certainly reduced, but you have not done a blessed thing for the other man's fitness. The only combination of events that the current definition of altruism seems to take into account is one in which the stranger's life is saved and yours is either lost or at least disabled to some degree.

Surely this is a problematic case. The difficulty is tied to the widespread tendency to define ascetic altruism solely in terms of consequences (e.g. Barash, 1977: 77). This position is not acceptable for several reasons. First, a definition by consequences violates the theoretical thrust of behavioral biology, with its emphasis on *directed* behavior and the theoretical import of the biogram, whose discovery is allegedly a major task of evolutionary science (Wilson, 1975b: 548). My rationale is based on the plain fact that consequences can be incidental and contrary to behavioral *predispositions*. A man, for example, sets out to steal another man's wife, thereby acting quite selfishly. The other man reacts by killing him and turning the tables around, thereby ending up with two wives *and*, according to the definition of altruism by consequences, transforming the selfishness of the dead man into altruism. We need a way out of this conceptual pickle.

A definition of ascetic altruism in terms of consequences is in part a defense against the excessively rationalistic assumption of the environmentalists. In a sense, then, the error is a result of a polemical stance. But the case for such a definition is also presented on the basis that natural selection operates on consequences. Still, it is a misguided application of natural selection. To argue against a definition of ascetic altruism in terms of consequences is not to deny that natural selection acts on consequences. The latter is a mechanism through which we grasp modifications in gene frequency, and it could care the proverbial dry fig for the difference existing between what organisms "want" and what they get. The concept of altruism, by contrast, is unavoidably both relational and *dispositional* in character. It directs attention to the quality of interorganismic relations and the manner in which organisms are *inclined* to act toward one another. Natural selection records particular consequences of those actions, but no one can deny the importance of behavioral inclinations or predispositions for the scientific purposes of grasping the quality of social relations revealed by actions. In short, a definition of altruism in terms of consequences would disallow the *social-behavioral* dimension that the New Synthesis would introduce into the Modern Synthesis.

But the problem is even more serious. If we insist on a definition of ascetic altruism in terms of consequences and at the same time deny the presence of ascetic altruism, we end by implicitly and absurdly denying evolution by natural selection. For, if we negate behavior that by consequence reduces the inclusive fitness of the dispenser while increasing that of the recipient, we deny by implication that natural selection favors certain genotypes at the expense of others. By *consequence*, in a relational context, the unfavorably selected *are* in fact ascetic altruists. Consider. Where a dominance order reigns—for example, in North American wild turkeys, among whom brother–brother competition yields competitive groups whereby one dominant male does most of the mating (Watts and Stokes, 1971)—a large evolutionary payoff accrues to a few males. Thus, those who by consequence are ascetic altruists in varying degrees constitute the majority, although, to be sure, not all subordinates will remain such all their life (see, e.g., Guhl *et al.*, 1945; Errington, 1963; DeFries and McClearn, 1970; Geist, 1971; Wiley, 1973; LeBoeuf, 1974; Barash, 1977).

Certainly one may doubt that the majority is by natural selection predisposed toward a result that clearly suggests ascetic altruism. Thus, a definition of ascetic altruism by consequences is theoretically self-defeating.

All of this is not to say, however, that I intend to promote a definition of altruism in terms of intentionality as this term is normally understood. Rather, I aim at the utilization of a widespread scientific strategy whose value is implicitly, if not always explicitly, quite widely recognized in sociobiological science. It is an aspect of system analysis. Systems, even physical ones, commonly behave as if they were programmed to pursue a sequence of preferred goal states. Therefore, they are sometimes termed goal-oriented or directed systems and logically treated *as if* they featured intentionality. A case in point is the target-seeking missile, a systemic mechanism that is treated mathematically as if it were a purposive entity. Natural selection has produced organisms that are goal-oriented (dispositional) as a minimum in this special sense suggested by systems theory. More directly to our point, individuals may be viewed as being goal-oriented by virtue of being energized by their behavioral predispositions.

In turning now to my definition of ascetic altruism, I am pleased to note in passing that at least one sociobiologist has avoided the consequences trap. According to Wilson (1978: 213), altruism "may be entirely rational or automatic and unconscious or conscious but guided by innate emotional responses." I shall amend the prevailing definition of true altruism as follows: *Ascetic altruism is behavior, conscious or unconscious, which, guided at most partially by innate predispositions, potentially reduces the inclusive fitness of the actor and potentially increases the fitness of other(s)*. Obviously, however, the problem of altruism is still laden with various conceptual difficulties that require a great deal of interdisciplinary attention.

6.2 Is 'Homo sapiens' Ascetically Altruistic?

Evidence against anything except minor cases of ascetic altruism, and in favor of the maximization principle, is overwhelming insofar as non-human animals are concerned, though some baffling cases are not lacking. A review, even the scantiest, is beyond the scope of this book (but see, e.g., Lack, 1954, 1968; Bruce, 1960; Perrins, 1965; Schaller, 1972; Power, 1975; Wilson, 1975b; Dawkins, 1976; Barash, 1977; Hardin, 1977; and of course, recall Hamilton's stunning findings on nonreproductive social insects).

Consider family planning as the basis of a general statement on the point at issue. Why, for example, should the individuals of a given bird species vary in the number of eggs they lay? If some birds lay a clutch of four, why have they not been favored by natural selection over those who lay a clutch of three? David Lack (1954, 1968) suggests that there is an optimal clutch size insofar as the probability of successful offspring is concerned. Indeed,

Perrins (1965) has shown that fledglings in smaller clutches have a greater probability of reaching maturity than fledglings in larger clutches. By and large, the larger the clutch, the scarcer the food and the greater the danger from predators, for the parents must go farther from the nest and leave it longer without protection. Findings such as these have led one scholar to put the matter bluntly: "Scratch an 'altruist' and watch a 'hypocrite' bleed" (Ghiselin, 1974: 247).

We should emphasize, however, that altruism should properly be viewed as a variable, and its manifestation is likely to be related to properties of both endogenous factors (behavioral predispositions) and exogenous or environmental ones. Thus, we must in principle consider the following possibilities: (1) some individuals, however few, are always and fully altruistic in our (ascetic) sense of the word; (2) others are periodically driven to self-enhancement at such a low level of intensity (perhaps because of low rank in the social hierarchy, for example) that they manifest behavior that *temporarily* casts doubt on the maximization principle; (3) still others are in the throes of so intense a need for self-enhancement (perhaps because they are dominant, or actively seek dominance) that they clearly fit the maximization principle.

But what about *Homo sapiens*? I think we must say about human beings at least what we have just said about our fellow creatures. I intend to say more, of course. But for the time being the focus is on what sociobiologists, or evolutionary behavioral biologists in general, have to say on the topic. The typical answer to our query is provided by Alexander (1975: 90), according to whom "all organisms should have logically evolved to avoid every instance of beneficence or altruism unlikely to bring returns greater than the expenditure it entails." True altruism is maladaptive behavior, and thus it should occur only adventitiously if at all.

Some ambivalence, indeed more, about the problem is not lacking, however. Lewontin (1979), for example, has convincingly argued that "there are a number of evolutionary forces that are clearly nonadaptive." As Wright (1931) showed, there are multiple "selective peaks" when a given trait is influenced by multiple genes. That means in practice that natural selection can follow alternative paths of evolution, with the "choice" followed depending on chance events. More generally, in discussing the sociobiology of human behavior, Barash (1977: 313–14) shows some flexibility in favor of Campbell's (1975) argument on the role of moral injunctions, and rather freely entertains an emergentist hypothesis to the effect that human society may have "an existence and interest independent of its constituent members."

Certainly an occasional statement may be found that appears to attribute genuinely altruistic behavior to human beings. A case in point is afforded by Wilson's (1978: 149–67) discussion, already encountered, of hard-core altruism and soft-core altruism. In the former case, we may add here, "the bestower expresses no desire for equal return and performs no unconscious actions leading to the same end." Such behavior, it may be recalled, has been molded by kin selection or natural selection at the level of tribal units. In the soft-core variety, conversely, the benefactor is "calculating," and expects reciprocation. This type of behavior is the result of individual selection and is "ultimately selfish." Note the implication that the hard-core type is not selfish. One unaware that, by and large, the two types of behavior correspond, respectively, to nepotistic favoritism and reciprocal favoritism, both of which are comprised by the maximization principle, would be led to conclude that hard-core "altruism" is true, ascetic altruism.

Actually, Wilson is somewhat predisposed against finding ascetic altruism, and the above position is rather an insidious result of our indiscriminate and undisciplined use of the altruism concept. Referring to a conversation with Malcolm Muggeridge (1971), he (1978: 164–5) asks: "What about Mother Theresa?" Mother Theresa is a member of the Missionaries of Charity, and therefore nonreproductive. In Calcutta and elsewhere she cares for the poor, the forsaken, the sick, the foundlings. She has received international honor and recognition, including the 1979 Nobel Prize for peace. Yet, as Wilson notes, she lives a life of poverty and grinding hard work. Is she the "perfect" altruist?, he asks rhetorically. The answer is negative, and he explains by resort to Jesus according to St Mark: in a word, even Mother Theresa is selfish, for she seeks the salvation of her soul.

If the issue is stated this way, there are by fiat no altruists, for we shall find that in one fashion or another we all seek for ourselves "good" as contrasted to "evil." We thereby slip into the metaphysics or the intuitionism of altruism. The problem at hand requires strict adherence to the genetic definition of altruism. The sociobiology of altruism concerns the salvation or perdition of genes, not of souls, although later I shall argue that the evolution of the soul concept did play a crucial part in the retention, if not the evolution, of ascetic altruism. As matters stand now, it is hard not to entertain the hypothesis of ascetic altruism in Mother Theresa. After all, she has sacrificed her reproductive fitness in the service of countless individuals who are as unrelated to her as they possibly can be.

One of the least equivocating statements about the alleged selfishness of human beings is boldly offered by a sociologist who has certainly done more than his share to help open lines of communication between social and biological sciences. According to van den Berghe (1978a), "we are programmed to care only about ourselves and our relatives." Such a statement, however, is unwarranted even in view of the theory of reciprocal favoritism.

The issue of selfishness versus altruism has, of course, a long tradition in social science, although we have had no general principle in which to anchor our arguments. Indeed, our approaches have been so different that in terms of the Darwinian logic we may be said to have sometimes mistaken altruism for selfishness, and vice versa. What Emile Durkheim (1897) terms egoistic suicide, for example, strongly suggests altruistic behavior, inasmuch as such conduct may reduce one's fitness either through the failure to reproduce or by leaving one's dependents impoverished and vulnerable to various forms of predation from neighbors and the law. Still, many social scholars have held a conception of selfishness that approximates the Darwinian cast. Consider, for example, B. Malinowski (1926b), Lévi-Strauss (1949), and indeed nearly all great scholars to a remarkable extent—from Marx, Pareto, Durkheim, Weber, and W. I. Thomas to the present-day theorists of exchange theory. According to Blau (1964), for example, "the tendency to help others is frequently motivated by the expectation that doing so will bring social rewards." Likewise, many social scientists have, with varying degrees of correspondence to the Darwinian context, paid close attention to the problem of altruism (e.g. Durkheim, 1897; Kropotkin, 1902; Pareto, 1916; Sorokin, 1950a, 1950b, 1954a, 1954b; Friedrichs, 1960; Gouldner, 1973; and Wispé, 1978; among many others). Thus, social science has been less precise in the delineation of the issues in question, but future developments may well show that it has also been more balanced than sociobiology—the tendency to find both selfishness and altruism in *Homo sapiens* will most likely be vindicated.

Clearly, what is needed is a cooperative effort between the two traditions. Surely there is warrant for supposing that the very indefiniteness characterizing socio-philosophical treatments of the altruism–selfishness problem is rooted in our failure to tie the problem, at least for an initial thrust, to concepts that are relatively easy to operationalize, and lend themselves to application in propositions that are subject to refutation. The isolation until recently imposed upon social science by the reaction to Social Darwinism has robbed it of the opportunity to probe the possible utility of concepts and perspectives that, through the Modern Synthesis, are currently flowering in the New Synthesis. Conversely, it is probably equally true that, however much the New Synthesis seeks both to guide social and moral disciplines toward the Modern Synthesis and to help bring their potential input to full fruition, sociobiologists have an inordinate amount of trouble conceptualizing the influence of culture on biology (but see Lumsden and Wilson, 1981). Their failure to discover true altruism in *Homo sapiens* is, I believe, at least in part a result of this theoretical involution of their own.

6.3 Asceticism

The next section will illustrate cases of ascetic altruism, and then the chapter will set out to explain the selection and retention of this controversial but intriguing phenomenon. Before we go any further, however, it is necessary to introduce briefly a behavioral predisposition that is closely associated with true altruism and whose label has been used to qualify the type of altruism under examination. I refer to *Asceticism*, which accounts for acts of self-denial and self-punishment. Ascetic acts are so widespread in time and place that we may again hypothesize that they are anchored in a biological force. My general hunch is that within a relational context, such as that represented by reciprocity, asceticism has combined with several other biological and possibly sociocultural selections to yield one or more classes of ascetic altruism.

Social science offers a number of hints at asceticism as a behavioral predisposition. One scholar stated the case explicitly as follows:

> Observable in human beings is a special group of sentiments that has no counterpart in animals. They are sentiments that prompt the human being to seek sufferings or abstain from pleasures without design of personal advantage, to go counter to the instinct that impels living creatures to seek pleasurable things and avoid painful things. They constitute the controlling nucleus in the phenomena known as asceticism (Pareto, 1916: 1163; see also Weber, 1904–5; Durkheim, 1912).

Asceticism, or something very similar to it—for example, the requirement to "control emotions, needs, desires, or instincts"—is also a central element in a theory of hominid evolution originally suggested by Chance (1962) and later elaborated by Fox (1972). Both scholars emphasize extreme sexual competition among breeding males as the chief factor shaping mental evolution. Fox proposes that cortical control of drives and the breeding structure of hominid populations developed in correlation with the rapid expansion of the hominid brain. According to Fox (1972: 291),

> The whole process of enlarging the neo-cortex to take-off point was based on competition between the dominant and sub-dominant males in which those who survived were those best able to control and inhibit, and hence time, their responses. Here then are the beginnings of deferred gratification, conscience and guilt, spontaneous inhibition of drives, and many more other features of a truly human state.

These comments hint at some possible causes of the selection of asceticism, and the referent, in classical Darwinian style, is adaptiveness.

Briefly, those individuals who were most able to make investments of time and resources with a view to a greater probability of success at a future time had a selective advantage over others, especially many of the younger pretenders to positions of dominance who were impatient to achieve the coveted goal.

It would seem, therefore, that asceticism is based on the utility of moderation, saving, and deferred gratification when such traits are crucial components of success in competition (for a brief discussion of the related phenomenon of self-punishment from the viewpoint of learning theory, see Bandura, 1977). Once the ascetic tendency arises, however, it can have various repercussions, and even become hypertrophic. A strictly restrained sex drive, for example, can save a man much trouble, if not his freedom or even his life. Again, where there is a shortage of food, fasting may be advantageous to individual as well as to group life. The ascetic predisposition does not, therefore, deny the maximization thrust; it merely attenuates it.

My interest just here, however, is less in explaining the selection of asceticism than in illustrating varying but universal manifestations of it, that is, in pinpointing the universal and variants possibly associated with asceticism as a biological force. I shall proceed in what may seem a polemical key because the present problem affords us another particularly good chance to grasp the approach to human nature in the absence of an anthropological genetics.

Suppose we begin with a Christian man who does penance of some sort because he believes that in so doing he makes amends for his sins and thereby pleases his God. We call him an ascetic. Shall we as behavioral scientists conclude that the cause of his penance is his belief in a God who accepts self-inflicted pain as a means of self-redemption? If our focus is on the manner in which beliefs are associated with actions, the conclusion is certainly justified. Indeed, the forthcoming theory of ascetic altruism will find beliefs quite useful to explanation and verification. But the focus on beliefs, at least in the present context, freezes us at a level of scientific explanation that is utterly *ad hoc*. Another man at another time and place will give another explanation for behavior that must also be labeled "penance." That makes our previous explanation uncomfortably fickle, for it is adequate at best to cast light on specific acts in specific times and places.

If we are serious about the search for scientific laws, we must, therefore, widen the circle and see whether the penance of the Christian falls in a larger class of phenomena. We find, for example, that the Greek Cynics were famous for their ascetic ways and yet they had no religious conception of such behavior. The Spartans were superb at suffering pain as a means of maintaining strict military discipline. The Buddhists have practiced asceticism by stultifying their vital energies. Certain tribes of

American Indians chopped off their fingers and suffered various other wounds for a variety of "reasons" (e.g. Kroeber, 1908; Flannery, 1953).

Indeed, if we do not insist on focusing on the unique aspect of sociocultural facts, we discover that similar phenomena take place every day under our very nose. From time to time we read, for example, that a fraternity pledge in an American college drowns in a campus pond or asphyxiates in a self-made sand tomb on a beach. College hazing sometimes entails many forms of suffering, some of which end in tragedy. Then there are those who avoid entirely the consumption of any alcoholic beverages, who condemn sexual activity, or who make war on erotic literature and the like. In the volume discussed in Chapter 3, Max Weber (1904–5) pointed to the stern, severe privations of certain seventeenth-century Protestants whom he aptly labeled ascetic.

Illustrations are literally numberless. Among the extreme cases of asceticism, first preference would probably go to the Stylites, individuals of uncanny Christian faith who in penance spent portions of their lives on top of columns. St Simeon, called the Stylite, may have been the originator of this peculiar form of asceticism back in the fifth century. The practice among Christians lasted at least four centuries. What is interesting is that identical cases are observable in history in places where there was no trace whatever of Christianity. The author of the *Syrian Goddess*, for example, discussed at some length the behavior of people who were Stylites in all but the name.

Flagellation, another form of asceticism, is of all times and places. The Spartans practiced it. So did Christian penitents in the Middle Ages. Muslims in some parts of the world still practice self-flogging. On 1 December 1979 the public media reported that ambulances in Nabatiyeh, Lebanon, had to rush to the hospital Shiite Moslem admirers of Ayatollah Khomeini because they were bleeding and unconscious from self-inflicted beatings with swords. As to Khomeini's fellow citizens, during the highly publicized days of the revolutionary period, they were often seen whipping themselves with chains on Moslem holy days.

A common form of asceticism is masochism (see Reik, 1941; Menaker, 1979). It has lately come out of the closet, again. Throughout the United States of America, a sizable class of people, especially from the well-heeled business and professional ranks, flock to one "chateau" or another where they find dungeons full of racks, pillories, cages, whipping posts, shackles, wooden crosses for spread-eagle positions, and any number of sundry other implements which, in the hands of a trained staff, inflict what is apparently expensive pain. *Playboy* once offered tips on the right paddle to use for sexual spanking. Ankle and wrist restraints are much in demand across the country. Handcuffs and whips are big sellers; so are male body harnesses.

In recent times, a widespread form of asceticism was prohibitionism.

Fundamentally, it entailed abstinence from sex, or at least a holy horror of the sexual urge. Presumably, it reached its peak in the Victorian age, but has recently been vigorously challenged by the "sex revolution." Several decades ago, prohibitionism in the United States of America focused on alcoholic beverages of all sorts and managed to occasion a constitutional amendment. It was spearheaded by Christian fundamentalists, who somehow were unimpressed by the common knowledge that Christ freely enjoyed a glass of wine. That goes to emphasize the tyranny that non-rational forces often exercise upon us, despite apparent cultural counter-measures.

Perhaps the most common, and durable, form of asceticism concerns fasting. For one alleged reason or another, fasting has been a cultural feature of many peoples on all continents. In extreme cases, people starve themselves to death. Witness the young IRA members who in 1981 fasted to the last breath in the jails of Northern Ireland. Self-starvation is also fairly common among Jain monks. Among Judeo-Christian-Islamic peoples, milder forms of fasting, the avoidance of certain foods, and the like, have been matters of biblical injunction. Until only a few years ago, Catholics were forbidden to eat meat on Fridays and on certain other days of the Christian calendar.

In recent years, a hungrier-than-thou way of life has swept the United States of America. There is an industry of lose-weight techniques that probably surpasses in volume the distribution of all sociology books. Television idols host programs on weight-losing. Millions upon millions are dieting, and many of them bear bodily weights below what modern medicine recommends.

Shall we accept uncritically the reasons given for dieting, which propose health and beauty? To some extent we have no choice. But chances are that today's dieting is at least in part the manifestation of the same ancient force that accounted for the fasting and related ascetic acts of times long gone by. Certainly, the fact is that for one alleged reason or another people have fasted with faithful recurrence. This universal fact demands scientific recourse to a universally applicable force. The point is that without that force, without the ascetic predisposition, there would probably be no fasting even for aesthetic reasons.

The ethnographic literature shows that initiation and other rites of passage are very often marked by such things as severe fasting, temporary isolation, enforced silence, and privation of water and sometimes of sleep and movement (e.g. Durkheim, 1912, bk 3: ch. 1). Durkheim further refers to "systematic asceticism" as one of the "essential elements of religion." In his discussion of initiation rites, van Gennep (1908: especially ch. VI) writes of a wide diffusion of mutilation, including dismembering of the sex organs in both sexes, of physical and mental abuse harsh enough to deprive the neophyte of his senses, and of flagellation, intoxication, and severe beatings.

Sociality, II: Ascetic Altruism 215

I have said little or nothing directly about the sex drive, which is central to evolutionary theory. The subject is extremely complex, and I have deliberately chosen to deal with it indirectly at several junctures. At this point, however, a direct though brief confrontation of the topic can no longer be avoided. For my purposes it is necessary to show that if asceticism is a force associated with genuine altruism, then it must be found to play a mitigating role in the reproductive expression of the sex drive, at least among a portion of a given population. The mitigation entails a potential reduction of the Darwinian fitness.

E. O. Wilson (1975b) has argued that sex is to a large extent an antisocial force in evolution. Social bonds are created in spite of it and not because of it. Marriages are formed on sex, but they are also frequently broken because of it. Sex often interferes with friendship, with relations of kin, even with brotherhoods and sisterhoods. The potency of the sex urge in principle produces an egocentrism whose inevitable result is a conflict of interests. After all, in sexual species, sex is the first servant of the struggle for existence.

In view of the enormous vigor of the drive, group life would, therefore, seem to be evolutionarily precarious. History, however, shows, and scholarship records (see, e.g., Freud, 1938), that sexual appetite becomes conditioned or checked by the force of asceticism through an at least circumstantial and partial sex taboo as well as the mortification and self-mutilations associated with the taboo. Many scholars (e.g. Durkheim, 1912; Pareto, 1916; Freud, 1938) have noted that a sexual asceticism is associated with religious "virtue." Little wonder that today's "sexual revolution" is one of the deepest worries of much organized religion, which attacks premarital sex, fornication, promiscuity, or just plain "immorality." Two thousand years ago, sex morality was called in support of Christianity during its war on paganism. A fundamental idea of Augustine and other Fathers of the Church was that paganism was false because it was obscene. Later, Protestants and philosophers of the eighteenth century used that same weapon relentlessly against the Catholic Church.

A strong but blocked sex impulse often leads to loathing for the sexual act. In some Christian saints it led to misogyny. At times the stultification of the urge became so violent as to produce hallucinations, with the Devil tempting one to sins of impurity. It also led to a glorification of virginity and to zeal, on the part of priests and moralists, in safeguarding women from "temptations."

But the fundamental drive remains. Thus, if in some individuals the restricted sexual energy is redirected toward morally acceptable activities, in others it breaks through with a vengeance. Whence arises the paradox that immorality is greatest precisely where it is most severely condemned by morality and by law (Freud, 1930). The fable of the forbidden fruit is a universal classic. The paradox lies in the fact that a hindered force that breaks through is often a force that has had to pick up extraordinary momentum to achieve that feat. Hence, its manifestation excels the levels that are observable under normal circumstances. If the recent American prohibition of alcohol produced, in greater proportion, the very licentious and criminal behavior that it intended to eliminate, much the same may be said of the sex–religion alliance everywhere and at all times. "Whenever the worship of Cythera is banned, the rites of Sodom and Lesbos come into vogue" (Pareto, 1916). The drive will have its due. If its normal activities are interfered with, the drive assumes other forms. In the process it may, in turn, interfere with reproductive activity.

6.4 Ascetic Altruism

Ascetic altruism is possible in direct proportion to the mitigating, or redirecting, action of asceticism on the sex drive. To the extent, moreover, that sex, religious behavior, and asceticism are associated, we may expect to find that at least one class of ascetically altruistic behavior takes place within a religious context. In fact, there are several classes of behavior that with varying degrees of confidence may be considered ascetically altruistic. The present section points briefly to a few. It is difficult to estimate the percentage that together they incorporate in a typical national population. The problem is compounded by the fact that, as we have noted, ascetic altruism is a matter of degree. Furthermore, it is a variable properly speaking, and hence may also be expected to fluctuate in the course of time in keeping with changes in the factors that modify behavioral predispositions and cultural phenomena.

6.4.1 AVOIDANCE OF PARENTHOOD

We have known for some time that contemporary societies are experiencing a considerable decline in the birth rate (e.g. Tsui and Bogue, 1978) and that many feature a *completed* rate of childlessness (ever-married childless women aged 45 or above) falling in the 10–20 percent range. Little or nothing, however, was known until recently about the *deliberate* avoidance of parenthood, which is a less uncertain indicator of ascetic altruism. A study by the demographer Dudley Poston (1976) distinguishes between voluntary and involuntary childlessness. The data are from the US 1965 National Fertility Study based on an area probability national sample of 5,600 married women under 55 years of age. Blacks, constituting roughly 12 percent of the total, were excluded from Poston's calculations. Thus, my remarks concern only a sample of roughly 4,928 women (including

377 individuals who were pregnant at the time of the interview and are herein treated as mothers).

The childless women were 440 in all. Of those, 138 were defined voluntarily childless by virtue of the fact that, while they were biologically capable of motherhood, they reported a preference against children, coupled with the use of contraceptive measures. It should be noted that humans as well as animals have long been known to delay reproduction until survival contingencies have improved. But here we are faced with the fact that 50 percent of the voluntarily childless women fell in the 45–54 age bracket; 86 percent were thirty years or older. In 1965 the probability that a childless woman in the 30–34 age category would bear a child during that same five-year period was only 6.7 percent. If she was still childless at the age of 40, the probability of her reproducing before age 44 fell to $\frac{5}{1000}$ (Shryock and Siegel, 1971; for related sources, see Rossi, 1968; Bumpass and Westoff, 1970; Presser, 1971; Mason, 1974). We may, therefore, estimate that, if Poston's measure of voluntary childlessness is valid, approximately $\frac{130}{4928}$ or 2.6 percent of the sampled women, would be voluntarily childless at the end of their reproductive cycle (for an analogous estimate for Canada, see Veevers, 1972). How many voluntarily childless husbands are represented by this figure, we cannot tell. What we do probably know is that, according to these data, in some individuals the maximization principle seems to operate at an extremely low level of intensity, if at all.

The data, of course, say nothing about kin selection and inclusive fitness, a consideration that comes immediately to mind. It can be easily argued, however, that a direct reproductive strategy is likely to confer greater fitness on the individual than the indirect route through kin selection (see also Barash, 1977: 90), although to be sure there are exceptions to the rule (e.g. Hamilton, 1964; Wolfendon, 1975).

We must guard against the temptation to consider the avoidance of parenthood as a selfish act. It may, to be sure, be so defined from a socio-philosophical viewpoint. But we are following an evolutionary perspective, and from that viewpoint, the phenomenon entails the reduction of one's genetic fitness, which by definition refers to altruism.

Should we wish to venture beyond the realm of reasonably hard data, evidence of ascetic altruism would prove to be both ample and rich of the cultural variants that may act as proximate causes of the phenomenon. For example, humans may develop so intense a feeling of evil about society that their very reserve of affection for offspring may prevent them from having any. So, one of Charles Dickens's characters in *A Tale of Two Cities* states: "I say, we were so robbed, and hunted, and were made so poor, that our father told us it was a dreadful thing to bring a child into the world, and that what we should most pray for, was, that our women might be barren and our miserable race die out!"

Before turning to the next class of ascetically altruistic behavior, we must return to the definition of ascetic altruism and raise a query that is unavoidable in view of what was said at that point. The definition, it will be recalled, makes a comparison between benefactor and beneficiary in terms of relative fitness. Who, we must now ask, is the beneficiary in relation to one who deliberately avoids parenthood? The question can be answered in at least two ways. One, the narrow type, would require specifying one or more individuals who would derive a direct benefit from the nonreproductiveness of a given altruist. This specification is not always possible, and I do not believe it is necessary. Indeed, I reject the convention in sociobiology to employ such terms as "dispenser" and "recipient" in the definition of altruism, except in a loose, metaphorical sense, because it suggests an excessive concern with social encounters in small groups and, thus, avoids the problems of the macrosocial context.

Fortunately, the principle of natural selection itself comes to our rescue with a clue for a second, and broader, answer to our query. If we think in population terms, as really we should in evolutionary science, natural selection, which favors some organisms over others in terms of relative fitness, suggests that the "others" of those who avoid having children are in principle all those in a given breeding population who conversely do have offspring. The former lose fitness relative to the latter, and in comparison to these they are ascetically altruistic.

6.4.2 HOMOSEXUALITY

This less narrow sense of ascetic altruism makes it possible tentatively to subsume various classes of behavior under the label in question. For example, Paul H. Gebhard (1979), director of the Institute for Sex Research in Bloomington, Indiana, reports that 4 percent of US American men and 2 percent of women are "predominantly homosexual at any time." Most of these, especially the males, are nonreproductive. E. O. Wilson (1978) hypothesizes that "homosexuals may be the genetic carriers of some of mankind's rare altruistic impulses," although an afterthought inclines him toward explaining the recurrence of these nonreproductive individuals by recourse to kin selection. Again, however, it is hard to see how homosexuals can match direct reproductive fitness through that indirect route. The current sociobiological emphasis on kin selection and inclusive fitness may turn out to be excessive. Certainly, it is colored by the sociobiological emphasis on social species, which provide a ready context for kin selection and inclusive fitness but at the same time represent a minority of the living species. As a minimum, it would thus be valuable to distinguish systematically between minimal, maximal, and optimal reproductive strategies.

6.4.3 HEROISM

A fairly obvious class of ascetically altruistic behavior refers to acts of bravery and self-sacrifice for the good of strangers we shall never see again. The phenomenon has frequently been remarked upon and, in part, has even inspired a famous theory of "altruistic suicide" (Durkheim, 1897). The young soldier who saves the lives of his buddies, typically total strangers before the common military experience, by offering his own life to an exploding grenade, commits an act of genuine altruism, and he may expect nothing in return. At other times, he volunteers to be the first to run the risk of tripping a mine, or he lays his life on the line to rescue the wounded. Similar cases, though proportionately perhaps less frequent, are observable also away from the battlefield. As Wallace (1966), among others, notes, there is a widespread phenomenon in human society whereby the individual is prepared by various rituals to die for God and country. Indeed, while it is true that we value greatly our life, and therefore are cautious about dangers to it, human beings commit numberless acts of beneficence for each other in contexts that recall neither kin selection nor reciprocal favoritism, and can be subsumed under the label of ascetic altruism. A common type, though fluctuating in time and place, concerns the defense of women and children. The type is exemplified by the rescue preference given to old as well as young women in cases of disaster, such as wrecks at sea. Of what possible genetic value is it to save women long past their reproductive period at the expense of men who are at the peak of their productive and protective as well as reproductive years? Yet morality does demand the sacrifice.

The treatment of heroic behavior requires that we return to a prior context with a comment of clarification. Previously, I sought to cast doubt on the validity of the association between altruism and group selection on the grounds that a public-spirited, heroic behavior on the part of the members of one deme would give them an *average* selective advantage in relation to the members of a less solidary group with whom they might be in competition. The present discussion would seem to contradict that argument. I must, therefore, explain that I see a major difference between the previous context and the present. In the former case, the warring group was conceived as being largely coterminous with the society; hence, the dead could count on the survivors to aid their surviving relatives. In the present case, a soldier who singly absorbs an exploding grenade sacrifices his life in favor of strangers who, typically, will be strangers to the hero's survivors. In short, reciprocity is more clearly relevant in the previous context than in the present.

6.4.4 RITUALISTIC CHASTITY

Certain forms of behavior within a religious context probably provide the most unequivocal examples of ascetic altruism. Particularly striking is the phenomenon of priestly or monastic chastity that has been common to many religions, including Catholicism and Buddhism among today's great faiths. This type of behavior also takes into partial account voluntary avoidance of parenting through avoidance of marriage and sexual activity altogether, a phenomenon that was not considered in the section on avoidance of parenthood. In 1979 there were approximately 200,000 Catholic nuns, monks, and priests in the United States of America alone. In the world as a whole, they added up to at least 1,500,000 individuals. In some countries, they represented very substantial numbers. Italy, for example, included 210,000 such people. Nearly three-quarters were nuns. When compared to women of comparable age, they represented roughly 2 percent of the total (*National Catholic Almanac*, 1980; *Annuario Pontifico*, 1980).

It should further be noted that, while Protestant men and women of the cloth have been subject to no institutional rule of chastity, many nevertheless have been driven to beat the bush, often in the most recondite corners of the earth, in search of souls to save. In this pursuit, not a few have been unlucky enough to suffer great deprivations and run into dangers that have taken their lives or effectively limited their fitness by demanding defensive if not ritualistic chastity.

6.4.5 ET CETERA

Sometimes we succor the poor and strangers we shall never see again, and without expectation of gain we give refuge to our conspecifics at mortal risk to ourselves. Philip Halley (1979) writes of the people of a French village who, during the years of the Holocaust, outfoxed a Nazi SS division stationed nearby and turned Le Chambon, their town, into a sanctuary for thousands of refugees.

Despite the lack of systematic data, there are also those, few to be sure, who prefer to adopt and raise strangers' children to begetting and nurturing their own.* There may be complex proximate causes behind such

---

* At least in the United States of America, the Census Bureau does not distinguish between natural children and adopted ones. According to *Current Population Reports* and their Special Studies Series, for example, "related" children in a family "include own children and all other children in the household who are related to the householder by birth, marriage, or adoption." A few states, for example, Colorado, Idaho, and Alaska, do report the number of adoptions in relation to certain other demographic variables. To my knowledge, none, however, sheds light on the problem of ascetic altruism. Whatever data on adoption are available, furthermore, appear to be based on voluntary reports by state welfare departments (see, e.g., US National Center for Social Statistics, 1980, Report E-10; for a general statement on "who adopts," see Bonham, 1977).

behavior. One may be of an aesthetic nature: giving birth sometimes spoils the beauty of the body. It is revealing that in a recent and informal survey in two of my large lecture classes, two-thirds of the young ladies considered a hysterectomy less objectionable than a radical mastectomy for women of their age.

Considering the adoption of strangers, Barash (1977: 312–13) is inclined to view it as a vestige, likely to eventually disappear, of a practice viable at a time when human societies were small bands of a few individuals, and orphans were, therefore, related to those who adopted them. That is, he invokes the lingering effect of kin selection in a non-kin context. Likewise, he rightly points to the reciprocity advantages of adoption, in the sense, for example, that the adopted child (and the society) may repay the adopting parents and/or their other relatives with assistance at a later date. The fact, however, that this sort of adoption may be a vestige of a previously self-serving type of behavior does not gainsay its ascetically altruistic character today. Rather, it emphasizes the reciprocal dependence of biology and culture. Further, while the vestige hypothesis is reasonable, it is less reasonable to assume that the vestige is likely to disappear through eventual accommodation to culture's rapid development. As to the recourse to reciprocal favoritism, the argument makes some sense in cases where adoption is additional to personal reproductiveness, but none in those, however few, where it may be a deliberate substitute for it.

Many examples could be included in the present residual rubric, beginning with Freud's (1930) cogent discussion of the "death instinct." Charity comes immediately to mind. There is a legend in some parts of the Catholic world according to which on a cold day Christ appeared to St Martin on a country road in the guise of a beggar. St Martin, then a nobleman, took pity on the poor wretch, took his mantle off, severed it into halves with his sword, and gave one-half to the beggar. The legend, of course, does not confirm an act of true altruism. But it does sensitize us to a widespread phenomenon whereby people often give scarce resources to the needy without expectation of return, typically under the influence of deeply entrenched moral injunctions (Yang, 1945: 96–7).

In conclusion of this section, I wish to stress two points. First, my intention has been to show that ascetic altruism is in all likelihood a feature of human society, not to suggest that it is the rule—that we are all and always ascetic altruists. On the contrary, the facts at our disposal have already indicated that most of us, most of the time, are plainly nasty, brutish, and self-seeking. Moreover, a good turn to A often entails the most vicious attacks against B.

Second, the study of altruism is a complex enterprise requiring intense knowledge of many disciplines and rich bodies of data. I am, therefore, aware of my own temerity in attempting an explanation of the subject, and have a lively appreciation of the pitfalls to which I am susceptible. Some of my arguments and facts will no doubt be found wanting. But if I succeed in establishing sufficient plausibility, and directing greater interdisciplinary attention to the possibility of ascetic altruism, I shall have accomplished my goal.

A final remark: To speak of different classes of ascetic altruism is very possibly to speak of different causes of the phenomenon. That complicates the problem beyond the outer limits of temerity. Accordingly, in what follows I propose to endeavor a more or less systematic explanation of only one class of ascetic altruism, namely, the one isolated under the label of ritualistic chastity. Within this context, Mother Theresa represents the pure type, wherein she has considerable company. My discussion of social approval will also help to explain other classes, for example, heroism; but the focus is on the religious context.

6.5 An Explanation of Ascetic Altruism

I return to Trivers's theory of reciprocal favoritism, a valuable bridge between biology and social science as well as a useful step toward the solution of the problem at hand (for theoretical applications of reciprocity as well as kin selection theory to ethnographic data, see also Essock-Vitale and McGuire, 1980). My aim is to conceptualize the selection of a cultural component, the idea of the soul, that in a certain portion of a population, presently unknown with any exactness, appears to check the maximization tendency and produce behavior that violates the maximization principle. How such behavior may be retained in a population will become clearer as we go along.

6.5.1 RECIPROCAL FAVORITISM, CONFORMITY, HIERARCHY, AND SOCIAL APPROVAL

Acts of mutual beneficence take various forms and carry varying degrees of danger to the benefactor. They run the gamut from a mere renunciation of certain simple enjoyments to the ultimate sacrifice of self-immolation. Most carry little danger to the actor; in general, the more costly the act of beneficence, the less likely it is to take place. Moreover, reciprocal favoritists may be viewed as evolving *pari passu* with mutual aid organization and, thus, as having the power to kill, ostracize, wound—in short, as being able to defend themselves in various ways against cheating, which, as noted, is a possible hindrance to the evolution of reciprocal favoritism. On the average, then, there are evolutionary payoffs accruing to reciprocators from their mutual support, although, as we have seen, the more marked the dominance order is, the easier it is to exploit the conformists' resources. The evolutionary advantage of reciprocity, we might add, was fairly well understood in early behavioral studies (e.g. T. H. Huxley, 1863, 1892; Kropotkin, 1902). Mutual aid and aggression against cheaters selected in favor of a high degree of conformity. At an ontogenetic, though not at a phylogenetic, level, social scientists are now reasonably familiar with the development and functions of conformity in social groups (e.g. Festinger, 1957; Milgram, 1974; Anderson, 1980).

We have noted two sides to conformity, corresponding to the enforced and the voluntary varieties, respectively. They may represent two different evolutionary stages. But they could not have been selected far apart from each other. As one type was being selected, the other should have followed closely, for those who featured the double-edged predisposition were in most instances at a selective advantage over the others—they could both give aid and demand it in return. I am not suggesting that the selection of double-edged conformity eliminated cheating from social relations altogether (see also Trivers, 1971). Both types of conformity tend to vary in intensity as a likely result of a dynamic tension existing between the two. In a given population they, thus, manifest themselves in time along a fluctuating line. In a sense, that is what historians and philosophers have meant with their theories of alternating periods of "faith" and "skepticism." Faith stands in part for high solidarity and low rates of cheating combined with an intense sense of guilt amongst at least some cheaters (Freud, 1930). The opposite is true of skepticism.

I rather suspect that faith has been the more faithful companion of the masses. Human beings are indeed absurdly easy to indoctrinate. The pages of history betoken the readiness with which people everywhere have embraced, often without the slightest comprehension, the mythology, the theology, or the sheer madness of one demagogue after another. Such extreme forms of indoctrination respond in part to the tendency of behavioral predispositions to radiate far afield from their original functions. Who among our primeval ancestors would have thought, for example, that their eat-or-die hunt would eventually give rise to the British fox hunt? In part, however, they reflect the fact that systems in dynamic tension often feature behavior that is located far from equilibrium (e.g. Perry *et al.*, 1963). The phenomenon of hypertrophy (Wilson, 1978) is related to this tendency. It is probably true that just as isolated systems are subject to entropy, so open systems in dynamic tension—or elements therein—are susceptible to hypertrophy, for they are highly dissipative of energy (see, e.g., Prigogine *et al.*, 1972).

It is at this point that we catch a glimpse of ascetic altruism. In an hierarchical society, such as human society has probably always been, submissive individuals are more exposed to cheating than dominant individuals; in the system of reciprocities they get the short end of the stick. Trivers (1971) was well advised to postulate a weak dominance hierarchy as a facilitator in the selection of reciprocal favoritism. How did the cheated cope with their disadvantage? One possible strategy was to remonstrate and seek distributive justice. There is reason to suspect, however, that, unless this behavioral form has undergone radical change in the course of human evolution, it was in the distant past, as it is in the present, a strong trait in a minority of the human population but a weak one in the majority, at least most of the time. We have seen one cause for this circumstance: under adverse conditions it pays to be submissive and to defer gratification. Further proof of sorts lies in the observation that rarely are political revolutions truly popular uprisings. Rather, they are typically executed by minorities, usually during periods of crises of authority and moral disorder (Pareto, 1916; Brinton, 1938; Eisenstadt, 1978; Skocpol, 1979). The leadership of the revolutionary class, if successful, usually replaces, in substance if not in form, the overthrown rulers. In retrospect, as Chapter 4 argued, revolutionary behavior turns out to be self-seeking behavior. This first reaction to cheating, therefore, is likely to be a mechanism of mobility into positions of dominance on the part of a small number of individuals, those endowed with an intense climbing maneuver.

Another means whereby to cope with a personally disadvantageous system of exchanges is self-ingratiation. It may be the chief response of the majority. Subordinates tend to ingratiate themselves with dominant individuals for whatever little favors they may receive. To better understand this, we may recall some fundamentals of human dominance. In a broad sense, dominant individuals constitute what in political sociology is severally known as a ruling class, a governing class, a power elite (e.g. Mosca, 1896; Pareto, 1916; Aron, 1953; Mills, 1956). Such a grouping consists of a small minority of people who tend to monopolize group decisions and under normal circumstances have the means with which to maintain internal order, *inter alia*. Hence, they oversee the struggle for meager resources taking place in the subject class, and are in a position to help or hinder individuals below. Evidence in support of this argument is available from animal studies as well (e.g. Alexander, 1961; Wynne-Edwards, 1962; Eisenberg and Kuehn, 1966; Guhl, 1968; Wilson, 1975b). To be in the good grace of powerful individuals, therefore, may bring considerable advantages to submissive individuals. Natural selection probably favored those in the subject class who, if they had little or no chance to rise to the top, were most diligent in catering to the will of dominant individuals.

The important point, however, is that there are times when leaders can do nothing of material value for their subjects. For example, resources are too scarce. Or two ingratiators ask for a favor that can be granted only to one. Under the circumstances, subservience and ingratiation must settle for substitute rewards. One of capital importance is *social approval*, a type of reward that, as we have seen, reinforces behavior without necessarily paying for it with a tangible coin. Social approval is a sign that we are doing right and perhaps a *promise of future reward*. This latter property—the virtual invitation to defer gratification—may have facilitated the evolution and retention of the need for approval.

Like conformity and many other forces, however, the need for approbation is a variable that is subject to considerable oscillation. At one extreme, its presence may be imperceptible. At the other end, it may reach such a high level of intensity that the individual is capable of anything—including, in extreme circumstances, self-immolation—in order to attain it. To the extent that approval becomes an end worthy of any cost, it becomes shorn of Darwinian adaptiveness. The individual so endowed places himself at the service of those whose approbation he covets, including "society" as a whole, and exchanges units of inclusive fitness for intangibles that satisfy his highly moralized emotions rather than the immortal thrust of his genes. The evolution of the need for social approval to an hypertrophic state probably reinforced asceticism and marked the dawn of ascetic altruism. It helps to explain the acts of heroism mentioned earlier and such other cases of altruism as those listed above in the residual category. Fundamentally, ascetic altruism is the result of an hypertrophic need for social approval expressing itself within a context of hierarchically arranged relations, although a glimpse of ascetic altruism may also be caught, as we have seen, from the perspective of a sublimated need for social recognition.

I might add that we are dealing with a large portion of the total society, that is, very probably with a goodly size of the subject class, which constantly replenishes itself by virtue of the scarcity of positions in the higher reaches of the dominance order. Moreover, I am conceptualizing the trait in variable terms. Hence, the problem of the retention of ascetic altruism is not crucial, but we shall return to this problem.

6.5.2 SELF-DECEPTION, THE IDEA OF THE SOUL, AND ASCETIC ALTRUISM

A unique characteristic of human beings is that they live simultaneously in two domains. We have been dealing with this world. But there is the transcendental-spiritual, too. It is from this latter perspective that we can probably descry the ultimate in ascetic altruism. Such behavior is epitomized by ritualistic (religious) chastity or nonreproductiveness—what we may term the Mother Theresa complex. Away from the ideal type, however, may be observed numerous cases of partially ascetic altruism.

As we have seen, E. O. Wilson (1978: ch. 7) has implied, and D. T. Campbell has stated, that "a purely social evolution of beliefs in transcendent social purposes [and] rewards and punishments after death" "predispose social customs enforcing self-sacrificial altruism" (Campbell, 1979: 43). How could such beliefs have evolved? The answer, I think, lies in the biocultural evolution of the *soul*. But such a selection depended, in turn, on other evolutionary events. In particular, the evolution of the soul seems to have been a by-product of a series of variations *through natural selection* that resulted in the phenomenon of *self-deception*.

Self-deception is a salient feature of human behavior. "Since the mind recreates reality from the abstractions of sense impressions, it can equally well simulate reality by recall and fantasy. The brain invents stories and runs imagined and remembered events back and forth through time: destroying enemies, embracing lovers, carving tools from blocks of steel, travelling easily into the realms of myth and perfection" (Wilson, 1978: 75). Social scientists are familiar with deception, of self as well as others, under such labels as false consciousness, ideology, sophistry, and rationalization. Litterateurs and philosophers are extremely keen to it (e.g. Girard, 1965; Trilling, 1972). Charles Dickens has his highly moralized Mr Pip declare that "all other swindlers upon earth are nothing to self-swindlers." With the most marvelous sincerity we con ourselves into believing that we are better than others because we are Protestants, or Catholics, Jews, Americans, communists. We are the superior race or the chosen tribe. Countless societies have appropriated the label "human" as their very name and excluded other peoples from it. Millions upon millions throughout the globe have believed that this life is but a prelude to a more lasting one in a world beyond. The list is endless. Our mind is like a valley animated by self-mocking echoes.

I cannot hypothesize with any great confidence about the evolutionary circumstances that led to self-deception, although in keeping with Wilson's theory of soft-core altruism, it may have evolved *pari passu* with the increasing complexity of the reciprocal system and as a means of coping with the flagrant cheating of an evolving dominance order. It is reasonable to argue in any case that, whatever the origin of self-fraud, once it was evolved, it developed a rich repertory of mischievous tricks. One of these may have had the effect of leading the gene itself absurdly astray. As consciousness reached higher levels of complexity, the curiosity instinct or explanatory urge that is a broadly adaptive trait of many species (Hardy, 1964) became thoroughly restless. Having pried into the nature of external things, the mind gradually turned on itself to inquire about the nature of thoughts, desires, fears, self-centeredness. Could the "selfish" thrust that was the motivating force of the immortal replicator have become itself the object of self-intuition and introspection? Could introspection, under the influence of the need for social approval, have threatened to reveal that the "evil" of others was in fact lodged in one's own being?

But where? Searching for a pure product of the mind, our ancestors could only find a fiction. The fiction, moreover, would be more likely to evolve if it mimicked at the mental level the fundamental characteristic of the gene: the search for immortality. What goes by the name of soul fits the requirement to perfection. Thus, the emergence of the soul was probably a sort of moral echo of the gene, a sublimation of a portion of pure genetic activity into moral, altruistic activity. I say a portion because for most individuals that is all that is involved. They may be observed sacrificing bits of fitness (through the "good works") in the pursuit of a metaphysical fitness: the salvation of the soul.

But how did the idea of the soul come about? I have already speculated that, under the influence of the need for social approval and the curiosity instinct, the mind turned on itself and caught a glimpse of evil within. This inward action was most probably in a relation of interdependence with many other forces. It was this interaction that yielded the soul concept. Let us hypothesize about the involvement of one such force: the "social self" (Cooley, 1922; Mead, 1934; also Freud, 1938). Following a Meadian perspective, the self is conceptualizable as a social process executed by two major mental forces in constant interaction. One, the "I," represents memory, awareness, personal interest; and manifests itself as "the response of the organism to the attitudes of the other." The second, the "me," is "the organized set of attitudes of others which one himself assumes." It thus represents the extent to which individuals have incorporated the interests and expectations of others into their own selves. To the extent that individuals behave in me-fashion, therefore, they take "the role of the other." The more completely social and moral an individual is, the greater the influence of the me within the self. For the me contains images, or moral clones as it were, of a number of others, especially the "significant others" of the kin and peer groups. From another perspective, the me is a sort of synthesis of bits and pieces of other persons or social selves.

It follows that my self-image and my very sense of being are intimately tied to the social self and existence of multiple others. The existential question now arises: what happens to me when they die? If they thereby cease to exist, I am a moral, emotional, and cognitive cripple. For they are no longer available to constitute and attest to the wholeness of my social and moral existence. Thus, the tendency toward the maintenance of personal integrity—of a coherent self-image—may well encourage the denial that the persistence of an individual is coterminous with the life of her/his body. From another perspective, the fact that the memory of the I persists in relation to the departed others may be an indication that their own persistence is assured in the presence of the living.

I tentatively submit that, as the social self was developing in the course of human evolution, a related tendency also developed to conceive of death as a purely organic event that was outlasted by the psychological and social dimensions of the individual. These came to be known as the "soul," the indestructible phase of the person. It is this sort of process, I believe, that lies behind what has been termed "the persistence of relations between the living and the dead," namely, behavior manifesting itself in such phenomena as funerals, memorials, and commemorations (Pareto, 1916). The concept of the survival of the soul is at bottom an extension of the feeling inherent in the self that the individuality of a person is a unit over the course of time. We shall return to this context in Chapter 8 and the discussion therein of predispositions of *Relational Self-Identification* and *Denial of Death*.

I now state *in nuce* my overall argument: *The evolution of self-deception consolidated the ascetic altruism selected through an hypertrophic need for social approval within an hierarchical organization by inventing the soul concept and redirecting a portion, sometimes overwhelming, of the gene's quest for immortality toward the quest for the eternal salvation of the soul. The emergence of the soul was a sort of moral echo of the selfish gene.*

The subject is complex; and my argument is necessarily speculative though, I hope, plausible. An empirical test of sorts does come to mind; and perhaps there are other, more ingenious, ones that other scholars can devise. Basically, my test is predicated on the assumption that if the theorized connection between ascetic altruism and the search for eternal grace is valid, we must find (1) a fairly universal belief in the immortality of the soul, and (2) a causal association between "doing good" for others and the intention to achieve the salvation of the soul. The remainder of the chapter will be devoted to this task after a brief definition of the soul concept.

First, however, a note of warning. I am keenly aware of the fact that religion can in no way be interpreted exclusively as a source of beneficence and goodwill toward others. Religious tenets and beliefs have often provided the rationale for meting out numberless forms of cruel punishment to countless numbers of humans both within and without the group. My aim is merely to establish plausibility for the proposition that the quest for salvation of the soul has also provided moral support for the demands of the social contract and that in some instances—of which the Mother Theresa phenomenon is the ideal type—the quest has entailed ascetic altruism.

6.6 Definition of the Soul

The subject is not without controversy, but we can steer fairly clear of dispute by focusing on facts and arguments about which form is more controversial than substance.

In his magisterial work on the fundamental principles of religion, Emile Durkheim (1912: 273) stated: "Just as there is no known society without a religion, so there exists none, howsoever crudely organized they may be, where we do not find a whole system of collective representations concerning the soul, its origin and its destiny." Thirty years later, on the basis of an extensive ethnographic compilation, G. P. Murdock (1945) included "soul concepts" among the traits common to all known societies. More importantly, many properties associated with the soul concept suggest that human consciousness in this area of behavior has evolved to mimic genetic properties and to reflect a latent genetic influence. Thus, keeping to such characteristics as are most common and fundamental, we may build the following composite portrait of the soul: (1) It is the "animating principle" of the individual—the very cause "of life, of thinking, willing, and knowing." (2) Hence, it refers also to the individual's "characteristic skills, motives, and capacities," that is, to "the resources which he brings to his relations with the environment and through which his behavior toward that environment is developed" (Swanson, 1960). (3) In many societies the soul "has generally been regarded as existing prior to the formation of the body and as surviving its decomposition at death" (Brandon, 1962). (4) Again, in many societies, whatever the soul's abode after its separation from the body, it does not remain there forever; after a varying period of time, and through diverse series of events, the soul is reincarnated (e.g. Hertz, 1907–9; Swanson, 1960), with the result that many peoples have believed the conception of a child to be the return to life of an ancestor (e.g. Durkheim, 1912). (5) Finally, a central feature of virtually all religious systems is activity to save, or facilitate the proper survival of, the soul (see Hertz, 1907–9; Durkheim, 1912; Brandon, 1962, among many other such reports).

6.7 Saving the Soul: Supporting Facts

"And when, closer to us, the Christian Church guarantees 'the resurrection and the life' to all those who have fully entered it, it only expresses, in a rejuvenated form, the promise that every religious society implicitly makes to its members" (Hertz, 1907–9). This generalization is not beyond dispute. The crucial question for my purposes, however, does not concern so much its universal validity as it does the obligations that must be fulfilled before the promise, wherever it is found, can be "realized." With this clarification in mind, we may enter a brief discussion of several major religions.

I cannot make an effort at a comprehensive examination of subject matter; it has been estimated that humankind has produced about 100,000 religions (Wallace, 1966). My treatment will nevertheless comprise a large portion of the world's population. Nor is it necessary to enter a detailed review of the relevant literature, which is enormous. Fortunately, the several dozen sources consulted are fairly agreed on the main points to be presented below. To a high degree, those on the great historical religions are even summarized in the excellent text by S. G. F. Brandon (1962), which brings together this scholar's Wilde Lectures in Natural and Comparative Religion delivered at Oxford University in 1954–7, and has the virtue of viewing religion as an institution that embodies the interpretation of human nature and destiny. For these religions I shall, therefore, rely primarily on this volume (but note also these helpful sources: Swanson, 1960; Bellah, 1964; Flew, 1972; Lessa and Vogt, 1972, among others).

6.7.1 TOTEMISM

We may begin with totemism, a particularly problematic subject. In his famed discussion, Lévi-Strauss (1962) termed totemism "a false category." Indeed, it is fairly generally agreed that at best there have been heterogeneous forms of totemism (Elkin, 1954). The disputation, however, does not concern the soul concept in any way that is significant for our purposes.

According to E. B. Tylor (1871), one of the first great anthropologists to study religion among preagrarian peoples, most such societies saw no connection between morality and the supernatural. "Retribution theory," rare in primitive society, might belong to an "intermediate" stage of development. Later scholars with fieldwork experience, however, took issue with this position. Reo Fortune (1935), for example, defined the religion of the Manus as "a concentration on setting man right with man as the way of setting man right with the supernatural." Malinowski (1935, 1948) viewed the Manus case as fairly typical. Swanson (1960) found that whatever controversy existed was due to largely implicit disagreements about such terms as ethics and morality, and showed cogent data from a sample of societies in favor of Malinowski's hypothesis on the association between supernatural sanctions and moral behavior.

The strictest link between religious behavior and morality was made by Durkheim (1912), according to whom the totemic god is guardian of the morals of the tribe and treats the people severely when they violate them. After death, he separates "the good from the bad" according to their relative faithfulness to the community ceremonies, to their relative piety toward the elders and one another in general, and according to whether or not in life they excelled in war and in other service to the community.

There is a widespread and diversified phenomenon in human society known synthetically as "the sacrifice." The topic is vast (see, e.g., Sahagún, 1530; Hubert and Mauss, 1898; Evans-Pritchard, 1956; Wallace, 1966), but my grasp of the subject finds credibility in the following statement by Hubert and Mauss (1898: 63–4): "In the course of religious evolution the notion of sacrifice has been linked to ideas concerning the immortality of the soul." Further: "There is no sacrifice into which some idea of redemption does not enter." Nevertheless, as we have already seen, sometimes the sacrifice is clearly selfish; the person performing the sacrifice endears self with the deity at the expense of the property or life of others. Other times, however, it is to a degree altruistic: individuals often sacrifice their own property; less often, their kin; less frequently still, themselves.

6.7.2 THE GREAT HISTORICAL RELIGIONS

According to Bellah (1964: 366), religious action in the historic religions is "above all action necessary for salvation." In ancient Egypt one's actions and moral duties were definitely associated with the concept of the immortality of the soul and the accompanying idea that its health in the eternal life was a function of just and upright behavior in this life. The fact transpires from a variety of practices, for example, the "negative confessions" or affirmations of sinlessness that were to be recited by the dead upon entering the "judgement hall." A confession might include such statements as: "I have not committed evil against men"; "I have not done violence to a poor man"; and "I have not taken milk from the mouths of children."

The ancient Hebrews had a God who rewarded for "righteousness before him." Sectarian differences did exist. Both the early Christian documents and Josephus, the highly reputable Jewish historian of the first century AD, depict two leading religious parties, the Pharisees and the Sadducees, as being respectively characterized by belief in a resurrection of the dead and by the rejection of that belief. In general, however, the twin ideas of resurrection and judgement became firmly established in Jewish thought, and a person's eternal future was believed to be decided by one's moral conduct in this world. This is evident, for example, in Daniel 12:2 and 3.

About Greek religion, Brandon writes of two major currents. One refers to the older, Aegean culture, which taught that eternal salvation could be obtained through divine grace and the performance of morally appropriate acts and rituals. The other view, which is alleged to have been a consciously constructed interpretation of life on the part of the poet or school known to us as Homer, held that life was essentially a tragedy, at the end of which lay virtual extinction. The great dramaturgists—for example, Aeschylus, Sophocles, Euripides—fall under this doctrine, too. The teachings of Socrates and Plato, however, are much closer to my thesis. The *Phaedrus* contains a doctrine entailing the concept of a *postmortem* judgement of the living's moral qualities. Moreover, two of the major cults of Greek religion, Eleusis and Orphism, believed in the immortality of the soul and provided the moral means, including many acts of self-abnegation, for its salvation.

The case of Christianity is fairly clear despite the famous dispute between Paul and the original Mother Church of Jerusalem, or the later debate between Augustine of Hippo and Pelagius on the question whether individual moral effort without the intervention of divine grace was sufficient for salvation. Christianity has emerged through the centuries as a cult of personal salvation through humility and self-denial. The fact that most Christians pay only lip service to the moral injunctions of the faith does not gainsay my basic argument. What matters is that the moral principles are there and that many individuals have to some extent been guided by them. Ascetic altruism within Christianity is dramatically suggested by the story of God's sacrifice of "His only begotten Son." Matthew's Gospel puts it in a nutshell: "For what is a man profited, if he shall gain the whole world, and lose his own soul?" Not a few have believed so, and to save their souls they have practiced monastic chastity, severe privations, self-castration, and many and sundry other ordeals.

Islam, the religion established in the seventh century AD by the prophet Muhammad, is much influenced by both Judaism and Christianity. Its concept of God is dominated by Muhammad's obsession with the thought of divine judgement. Allah's revelation to the prophet conveyed the assurance of eternal salvation provided that the individual submit in obedience to the will of Allah. Submission takes the form of faith and various requirements of an ethical and ritual character: piety, probity, self-privation, even self-flagellation.

India has seen numerous cults. I shall focus briefly on the Brahmanic period, when certain individuals practiced austerity of great severity. The practice continues to this day among the Hindu holy men. The Upanishads, some of the oldest documents of Indian religious thought, contain statements to the effect that "one becomes good by good action, bad by bad action." And there is, of course, the matter of the transmigration of the soul. There are some speculation and debate about the meaning of this concept. But it is fairly generally understood that a pious and virtuous life, an asceticism of both body and mind, has a positive bearing not only on the several stages of reincarnation but also on the eventual release from the cycle of ever-recurring birth and death.

The corpus of at least one major branch of Buddhism was originally inspired by a profound concern about the suffering of the living. Gradually it came to systematically advocate conduct designed to obtain deliverance from suffering by explaining the cause of the unhappy lot and the means whereby it could be avoided. The liberation from the life's suffering entails piety toward others, a thoroughly ascetic mode of life, and a stultification of the senses and the natural bodily needs. The reward runs the gamut from a mere destination to a heaven, from which there is rebirth, to a lot wherein the individual is not only no longer liable to rebirth but has also lost the desire for existence. From a biocultural perspective, this is the *non plus ultra* of ascetic altruism, the very and ultimate negation of the commandment of natural selection. Genes may indeed use us as their "survival machines," as a scholar has cogently argued (Dawkins, 1976). But in some machines the genes are subjects of a brain so specialized in self-deceit that they come to a grievous and mortal end. There is probably no truer statement than Marx's famous verdict that religion is the opium of the people, although the great revolutionary does not seem to have fully appreciated the complex role played by religion in human evolution.

Again it will be useful to lean on a caveat: I have sought to demonstrate a link, possibly causal in nature, between attempts to save the soul and acts that to varying degrees may be termed ascetically altruistic. But it has not been my intention to argue that such religious behavior has been highly efficient in producing ascetic altruism. Most of the time, and for the generality of people, the religious influence on the "good works" resembles fairly closely Huckleberry Finn's Darwinian philosophy in the following passage:

> I set down, one time, back in the woods, and had a long think about it. I says to myself, if a body can get anything they pray for, why don't Deacon Winn get back the money he lost on pork? Why can't the widow get back her silver snuff-box that was stole? Why can't Miss Watson fat up? No, says I to myself, there ain't nothing in it. I went and told the widow about it, and she said the thing a body could get by praying for it was "spiritual gifts." This was too many for me, but she told me what she meant—I must help other people, and do everything I could for other people, and look out for them all the time, and never think about myself. This was including Miss Watson, as I took it. I went out in the woods and turned it over in my mind a long time, but I couldn't see no advantage about it—except for other people—so at last I reckoned I wouldn't worry about it any more, but just let it go. (Twain, 1885)

6.8 Conclusion

J. Maynard Smith (1978) has inquired in a recent article: "How can a gene that makes suicide more likely become established?" This is in all likelihood a misleading question for the study of genuine altruism. If there is validity to the argument of this chapter, especially to the functions assigned to self-deception and the soul, *then it is entirely possible to have genuinely altruistic behavior WITHOUT altruistic genes*. This is my overall conclusion. One important point associated with this position is that the question of the retention of ascetic altruists in a population is no longer a problem. As long as self-deception and the idea of the soul exist, a certain percentage of individuals, very probably varying in time and place, are likely to fall victims to them. And that may be considered a recurrent fact.

The question naturally arises as to why self-deception should have evolved in such a way as to contravene the maximization principle. The answer is in principle simple. Indeed, I have already provided an answer suggesting, along with Wilson's (1978) theory of soft-core altruism and civilization, that in an increasingly complex society it satisfied the need to cope with the cheating that is inherent in dominance hierarchies. More important, science shows that a single cause can produce more than one effect. Genetic material, as we have noted, has pleiotropic effects. This is one principle behind the multiplier effect. We have to assume, for example, that the evolution of haplodiploidy favored inclusive fitness. But it is also true that this intriguing phenomenon offers differential evolutionary payoffs to queens, workers, and drones. Ascetic altruism, in short, may be a by-product of a self-deception whose main effect was, and may still be, the maximization of inclusive fitness—for example, by favoring reciprocal favoritism, as the hypothesis of soft-core altruism and civilization suggests.

A further question arises as to whether my argument may properly be regarded as falling under the logic of natural selection. I have no difficulty in answering affirmatively. But the chapter has tried to show that the biological and the cultural domains are in a state of strict evolutionary interdependence with each other. The evolution of self-deception is no doubt a genetic phenomenon, and its persistence attests to the continuing relevance of natural selection. But once self-deception was selected, it probably made possible conceptions of a cultural nature that at times thwarted the self-serving thrust of the gene. The soul, principal among them, may be thought of as the kernel of an internalized morality, reinforced by the will of fictitious forces, whereby some humans are in varying degrees led to subordinate their genetic fitness (and their self-interest in general) to the fitness and interests of others, even strangers who are in no position to reciprocate. Thus, the concept of the soul has to some extent modified the genetic action of natural selection.

What, then, does my reasoning do to the maximization principle? It leaves it in very good health indeed. At the same time, however, it proposes that a more flexible conceptualization would be appropriate, and might even grant the principle greater theoretical reach. As often happens with statements of theory in their days of ascendancy, the principle has at times been used not as an heuristic tool subject to an occasional surprise but as a source of preconceived assertions that for many critics have the ring of dogma.

It is useful to note that the practice of scientific theorizing frequently features a remarkable tendency to hedge theoretical bets. Thus, the laws of Newtonian physics—for example, Galileo's Law of Falling Bodies and Newton's own Laws of Motion—are typically stated as theoretical "idealizations." Their validity is predicated on the specification of ideal conditions under which the uniformities in question are applicable universally (Nagel, 1961; Lopreato and Alston, 1970). It is these conditions and the empirical efforts to cope with their "detractors" that enrich the laws and render them focal points of research and cumulative theoretical development.

A more flexible conceptualization of the maximization principle has the virtue of stimulating the search for cases that do not fit the maximization principle as well as for those that obey it. In practice, when we encounter examples that fall outside the principle, we have two theoretical options open to us: (1) We may continue to give the principle the benefit of the doubt and hope that at least occasionally we shall be rewarded with the sort of discovery made by Hamilton in his work on haplodiploidy—that the exception to the rule is only apparent. (2) Alternatively, we may tentatively accept them as genuine exceptions to the principle. If the latter decision proved valid, it should lead to qualifications, auxiliary propositions, and the like resulting in a more nearly complete theory of altruism, of which the maximization principle would be the focal point. From one perspective, the present chapter is, thus, an exercise in finding one or more classes of exceptions to the maximization principle and explaining them within the logical limits of natural selection and the biocultural hypothesis, just as a physicist might explain why, for example, two bodies that are not freely falling do not accelerate at uniform speed.

I conclude by suggesting the law that *organisms have evolved a tendency to maximize their inclusive fitness,.but the tendency is reducible by self-deception*. It is a tentative proposal, and I suspect that there are other qualifying conditions to be discovered. It is likely, too, that self-deception will be subsumed under a broader qualifier endowed with a more interspecific reach. For example, it is known that there are maladaptive genes in many species. According to V. A. McKusick (1975), *Homo sapiens* carries at least 2,336 variants that cause diseases and malformations ranging from mild anomalies to lethal organizations. What are the implications of this built-in self-destructiveness for the maximization principle?

7 A Model of Sociocultural Variation and Selective Retention

The preceding three chapters discussed a number of likely behavioral predispositions which, as products of natural selection in a state of interdependence with sociocultural evolution, account for certain classes of human behavior labeled cultural universals. From the viewpoint of natural selection, the discussion stressed the factors of competition and differential survival chances, for example, victimization and the quest for dominance. Still, we could not fail to note that competition is inextricably tied to cooperation, public spiritedness, even individual submission to group welfare. As the presentation moved toward the sociality end of the spectrum, the hypothesis of biocultural interplay, which is a cardinal feature of this volume, became increasingly salient. Thus, the explanation of ascetic altruism began with natural selection but drifted to the influence of a cultural selection, the idea of the soul, on the purely genetic thrust of natural selection.

Missing in the analysis so far is any systematic attempt to construct a theory of sociocultural variation and selection. From the perspective of the hypothesis of biocultural interplay, two crucial questions remain: (1) What biogrammatic forces are most directly responsible, at that ultimate level, for sociocultural transformation? (2) What other factors, including sociocultural ones, facilitate or impede their operation? The present chapter attempts to answer these questions in a general and largely abstract way. The next two chapters then will assay a specification and application of the model to be constructed here. I should add, however, that this last part of the volume should be considered a rough first approximation. At the present level of knowledge, any attempt at a theory of biocultural evolution is bound to be a bit premature if not altogether presumptuous. Success may require a theoretical synthesis of evolutionary biology and social science as well as richer and more systematic knowledge of the environmental, including cultural, conditions and historical vicissitudes of *Homo sapiens* and his society. Still, in the words of Sir Francis Bacon, better to dare than to stare. As we shall see later in this chapter, work in this intriguing area of science has started, but is still at the frontier.

7.1 Biocultural Evolution Defined

I begin with a conceptual clarification and a definition. The task facing us is an evolutionary theory of transformation in the sociocultural system; we are, thus, dealing with sociocultural evolution, and the chapter title underscores this fact. However, as I have just above implied, and as we shall presently see, sociocultural evolution appears to be linked to the action of certain behavioral predispositions, and in this sense we may properly speak also of biocultural evolution. Hence, for convenience sake, "sociocultural evolution" and "biocultural evolution" will be used interchangeably, but the latter term is for obvious reasons the broader of the two. Indeed, it is, *stricto sensu*, preferable because it underscores the mutual dependence of biology and culture at the same time that the *bio* portion of the label is inclusive of the term *social*. Often the latter concept is used by behavioral scientists as a synonym for "cultural," but that is a license that is best avoided. The reason is obvious: in other social species the social has little or nothing to do with the cultural, and a great deal to do with the biological (for example, the social insects). I have no pretension, of course, of defining biocultural evolution in so sweeping a fashion as to define at the same time biological evolution. The specific aim is to come up with a definition of evolution that is useful for the study of a species which is both cultural and bio-social—a definition, moreover, in which the biological dimension has only biogrammatic representation.

Biocultural evolution is accordingly intended to refer to a process of sociocultural transformation that is ultimately under the control of biological predispositions of innovation and selective retention, and proximately under the influence of various sociocultural and other environmental factors. I shall refer to the former as *forces (or predispositions) of variation and selection*, and to the latter as *criteria of selection*. The process is further understood to be in large part at the service of predispositions of self-enhancement and sociality and, more generally, of the drive toward the maximization of inclusive fitness.

It may be readily admitted that the biological emphasis is probably excessive. On the other hand, it is almost entirely lacking in social science, and may therefore be at least partly justified on that basis alone. It follows in any case that, at the level of this first approximation to a more nearly complete theory of biocultural evolution, I must eschew any considerable attention to the multiplicity of proximate factors that play a role in sociocultural evolution: for example, energy use, capital investment for research and development, the state of knowledge, freedom of thought or research from ideological impediments, accidental variations due to such other proximate factors as memory and language use, coordination among the various phases and competences of research or profit organizations, and cultural diffusion.

The latter is by general agreement a major source of behavioral variation in social systems in general and in sociocultural systems in particular, and hence it must sooner or later be dealt with systematically by the emerging theory of biocultural evolution. Some scholars have even considered the possibility that cultural diffusion is analogous to such biological processes as reproduction and recombination (e.g. Sahlins and Service, 1960; Blute, 1979). The concept of diffusion (cultural contagion, propagation, and so on) is also evocative of the concept of genetic drift (chance evolution) in evolutionary biology and of the dispersal of behavioral traits associated with genetic drift and migration (for a brief discussion of genetic drift, see Wilson, 1975b: 64–6; for general statements on diffusion, see Barnett, 1953; Kushner et al., 1962; Haggett, 1972; Ammerman and Cavalli-Sforza, 1973; Cavalli-Sforza and Feldman, 1973; Foster, 1973; Ruyle, 1973; Zaltman et al., 1973; Bandura, 1977; Hamblin et al., 1979; Lumsden and Wilson, 1981).

Another topic of major importance, but equally neglected, concerns the problem of creativity, which refers roughly to the differential capacity for innovation among individuals, and seems to be associated with various factors, including intelligence, knowledge, a strong need for recognition, and pure chance, among many others (for a review of literature on this poorly understood phenomenon, see Maslow, 1954, 1968; Kuhn, 1962; Kunkel, 1970; Stein, 1975; Taylor and Getzels, 1975; Arieti, 1976). To the extent that this and other more or less proximate factors of sociocultural variation enter my model at all, they will be treated more or less indirectly through the discussion of criteria of selection. The principal aim of this chapter is to propose a form of mental operation that is ultimately at the base of sociocultural evolution, and to discuss certain basic criteria which such operation tends to satisfy and which tend to facilitate it in turn.

7.2 Forces of Variation and Selection: a Conceptual Introduction

Plato long ago propounded the erroneous theory that human beings never invent or discover anything new. Rather, they come into the world equipped with *anamnesis*, namely, subconscious knowledge of all that is knowable, along with the capacity to retrieve such information through the development of a memory process. I mention this ancient notion because in substance if not in detail it remains a rather common belief. The Latin saying, *nihil novum sub sole* (there is nothing new under the sun) expresses a common bit of evolutionary skepticism that among modern people is sometimes rendered by the well-meaning mockery "you have rediscovered the wheel (or America, and so on)." Others say that "the more things change, the more they remain the same." In *The Organisation of Thought* (1917), Alfred North Whitehead put a related idea bluntly as follows: "Everything of importance has been said before by somebody who did not discover it."

To an extent, such a posture may be a function of perceptions typical of slow-moving times, or conversely of periods in which novelties arise so rapidly and in such a state of disarray that they are dismissed as aberrations by those who do not live long enough to adjust to them as institutional elements. But then again the Platonic error may be the result of the rather deceptive way in which the brain and the mind operate to introduce innovations in sociocultural systems.

The exact rules and genetic mechanisms leading to the processing and transformation of information in the nervous system are not yet known, although work on the brain and the neuronal basis of the mind represents one of the more lively activities of theoretical neurobiology and related sciences (see, e.g., Shepherd, 1974; Kandel, 1976; Kuffer and Nicholls, 1976; Bullock *et al.*, 1977; Edelman and Mountcastle, 1978; Gazzaniga and Ledoux, 1978; Grossberg, 1978; Jacobson, 1978; Cowan, 1979; Geschwind, 1979; Hubel, 1979; Simon, 1979; Stevens, 1979; Wickelgren, 1979; Crook, 1981). One thing, however, appears quite certain. The fundamental neurological activity of the brain consists of the control and coordination of the physiological and anatomical apparatus through sense organs or devices (eyes, ears, and so on) that translate external physical events, including symbols, into the pulse code of the neurons. If we are justified in the hypothesis that to a large extent we are genetically programmed to do "what is best for the genes"—to behave according to the maximization principle—neuronal activity must in the final analysis be oriented toward serving the quest for inclusive fitness. Sense organs must, therefore, be viewed as adaptive mechanisms.

If the maximization principle holds with respect to sense organs, the same may further be said of the mental operations of the brain: the mind. A brain which evolves to yield mental productions that interfere with reproductive success is a brain that seeks to be acted upon unfavorably by natural selection. With few exceptions, some of which were noted in the previous chapter, the mind must serve "the ultimate end" of life no less than do our eyes, ears, nose, and so forth. But how does it achieve this feat? No doubt, various mechanisms have been evolved to perform at its disposal. Here, however, I wish to argue the point that some have most likely been forged by analogy to organismic mechanisms and processes that from the beginning specialized precisely in serving the maximization drive. The chief hypothesis of this chapter is that *in part the brain has evolved to mimic in its mental operations the activities of some of the other adaptive mechanisms and processes of the body*. The hypothesis is suggested by the systemic or interdependent nature of our organs and by the assumption that evolution follows the line of least resistance—that it tends to economize by taking advantage of and building upon strategies that have already proven to be evolutionarily successful. These considerations lead to an interesting corollary. Consider for a moment the basic property of sexual organs. What do they do? What does sexual reproduction accomplish in an evolutionary sense? The answer is simply that it shuffles and reshuffles genes, typically by bringing together halves of two different genotypes. This process is known as recombination in biology.

My contention (the corollary) is that this process is simulated by the mind. Indeed, one of the mind's fundamental activities is to combine or associate together two or more objects, ideas, symbols, and so forth. In the process, just as sexual reproduction produces new genotypes through meiotic combinations of preexisting genotypes, so does the mind produce, through combinations or associations, new ideas, abstractions, symbols, sociocultural formations, and so on. This, I hold, is the fundamental source of variation in the sociocultural system, and one of the irreducible mechanisms of biocultural evolution.

Many genotypical combinations or organisms die out quickly. They are unfavorably selected. That is the picture we necessarily form of the entire world of living things. It applies to our own species, too, especially until only scores of years ago, when both birth and death rates were much higher than they are today. Many if not most mental combinations are likewise ephemeral. They never get past the individual level; or they are found harmful, repugnant, trivial by the public if not by the innovators themselves, and never become widely shared. Other combinations end by enduring in time. They enter and transform the existing institutional framework and the customs, norms, and beliefs that embody its behavioral repertory: tradition.

I shall refer to such enduring combinations as *variations*, or *selections*, terms that are comprehensive of what have previously been called cultural variants and universals. In keeping with the distinctions drawn in Chapter 2, however, universals may be thought of as complexes of highly persistent variations that emerge gradually in time and coalesce around one or more behavioral predispositions as their most direct behavioral manifestations. Universals, thus, are not only sociocultural formations that are general to human society. They also originate and change very slowly, and are consequently long-lasting features of sociocultural systems. One of the fundamental challenges of biocultural science will be to elucidate this extremely complex problem of the evolution of sociocultural universals. Variants, by contrast, are more like secondary appendages of the sociocultural system than like basic components of its structure. It follows that, while to a degree they endure in time, they have a relatively short life span and vary, moreover, greatly across societies according to a multitude of factors, for example, environmental differences, differences in political and socioeconomic history, linguistic diversity, and dissimilarities in technology and the state of knowledge.

The question now arises as to why certain combinations manage to become enduring variations or elements of the sociocultural system. The query requires an answer that appeals to both ultimate and proximate causes. We shall come to the latter in the next section and the discussion at that point of what will be termed criteria of selection, namely, factors that facilitate the selection of given combinations. Here I shall continue to argue that, just as there may be a behavioral predisposition, or a class of predispositions, which brings variations about, so there is a class of behavioral forces fixed by evolution in the mind that discards certain combinations and selects certain others for inclusion and persistence in the existing repertory. It is such selective agents that account for the endurance of combinations. Without them, all combinations would be ephemeral, or enduring variations would at best be the result of pure random occurrence and could not, therefore, be accounted for by the hypothesis that sociocultural systems are formed through fitness-enhancing processes.

The processes that are typically relied upon by cultural scientists to explain behavior—for example, socialization, internalization of norms, enculturation—are no substitute for this notion of selective retention. They are largely mechanisms of transmission that facilitate the institutionalizing function of the selective agents. They do not say enough about sociocultural change per se. Herein lies the fundamental "static bias" of cultural science. Curiously, there has been a tendency to attribute it only to certain theories, especially to some of the sociological theories of Talcott Parsons (e.g. Dahrendorf, 1959; Lopreato, 1971). In fact, the static bias is pervasive of our entire enterprise: it is an integral part of our fundamentally nonevolutionary perspective.

The forces of variation and selection correspond by analogy to the forces of variation and natural selection in biology and may be taken together to constitute the basic elements of variation and selective retention in biocultural evolution.* I conceive of them as being so closely tied together that for all practical purposes they are inseparable. For that

---

* A more complete treatment of this problem would have to include a discussion of selective extinction, or the process whereby previous variations are discarded from existing tradition. Selective extinction and selective retention are, however, closely related processes. The proposition is suggested by the adaptationist strategy that I have adopted. Thus, biocultural evolution entails not only the tendency to select combinations that are fitness-enhancing. The same tendency implies also the gradual replacement of less adaptive variations with more adaptive ones. To an extent, human beings and their groups may be viewed as being constantly engaged in a cost–benefit analysis in which old cultural items are allowed to atrophy and disappear, if they are not altogether banned, because the new alternatives prove to be more efficient, better suited partners of the universals, morally more comprehensive, and so forth (for some useful discussions of causes of extinction, see Barnett, 1953; Kushner et al., 1962; Price, 1975; Lenski and Lenski, 1978; Langton, 1979).

reason, they will not be dealt with in separate chapters. For convenience sake I shall refer to them henceforth separately as *Combiners* (or forces of combination, of variation; combination predispositions; and so on) and *Selectors* (or forces of selection, of institutionalization; selection predispositions; and so forth).

7.2.1 COMBINERS

Chapters 8 and 9 will turn to a discussion of particular predispositions of combination and selection and of the manner in which they cooperate in the evolution of certain sociocultural phenomena. Here I continue with a few further remarks about combiners and selectors in general. Charles Darwin (1871) grasped a variety of distinctions between *Homo sapiens* and the other higher mammals. One especially noteworthy was the former's "almost infinitely larger power of associating together [such things as] the most diversified sounds and ideas." Later Vilfredo Pareto (1916: 889), an economist and sociologist who wrote in an amazingly evolutionary key, argued: "Figuring as a sentiment in vast numbers of phenomena is an inclination to combine certain things with certain other things"; the tendency is intensely powerful in the human species and constitutes one of the important factors in civilization.

This theme, with minor variations, has been occasion of widespread consideration in behavioral science and bears a variety of closely related labels. Pareto (1916) referred to it as "the instinct for combinations." Auguste Comte (1896, III: 305–8) had written of "innovating instincts" which, under the pressure of demographic density, have the better part over the "conservative instincts" and help to determine the social movement or "rate of progress" in society. Later, W. I. Thomas (1923: ch. 1; also Thomas and Znaniecki, 1918) wrote rather extensively of a "desire for new experience" and made a reasonable effort to give it an evolutionary underpinning by tying it to "the pursuit, flight, capture, escape, death which characterized the earlier life of mankind." "Behavior," he continued, "is an adaptation to environment, and the nervous system itself is a developmental adaptation." It represents, among other things, behavioral inclinations that have developed in association with "a hunting pattern of interest." The social scientist who associates Thomas almost exclusively with primary group relations and symbolic interactionism must be surprised by this intense evolutionary and biocultural dimension. Indeed, Thomas discussed various behavioral manifestations of the need for new experience, for example, adventure, hunting trips, sports, courtship with its element of pursuit, newspaper sensationalism, and scientific or intellectual curiosity in general. Thus, the modern scientist allegedly uses the same mental mechanism as the hunter of early hominid evolution. "The so-called 'instinct for workmanship' and the 'creative impulse' are 'sublimations' of the hunting psychosis."

Other scholars have spoken of a curiosity instinct, or exploratory drive, that is a broadly adaptive trait of many species (Hardy, 1964; also Marris, 1974). Dobzhansky (1962: 214) has termed it "the restless and ostensibly idle and pointless curiosity that is expressed in the urge to explore, to pry into the nature of things, and to enjoy forms, sounds, colors, and thoughts and ideas."

All these conceptualizations entail combining or associating one thing with another, for example, the urge for "progress" with technology in the case of Auguste Comte. Thus, combination predispositions appear to be the source of the inventive faculty, the ingeniousness, creativity, imagination, experimental inclination, and curiosity of the species. Practical examples are not hard to find. We join a piece of wood with a piece of steel and get thereby a hammer. The death of Christ is associated with a particular day of the week, and the result is "black Friday." "Democracy" is combined with "poverty" and "justice," and we get all sorts of political maneuvers called "the war on poverty." The scientist combines two propositions in a given way, and out comes a theorem.

The strong combinatorial quality of innovation can perhaps best be appreciated by considering the nature of some of our most basic symbolic systems, for example, the decimal system, the alphabet, and musical notation. All of them function as bases for endless combinations and as stimulators of the very combinatorial predispositions that they are, in turn, major products of. Little wonder that even scientific geniuses sometimes feel obligated to confess that if they have seen further, it is because they have stood on the shoulders of giants. The new is typically the result of mental operations that sort out old ideas and sometimes transform them by perceiving them as problematic. Thus, scholars interested in the nature of scientific innovation rarely escape the conclusion that innovation entails two fundamental steps: (1) an initial perception of a problem, and (2) the search for a solution on the basis of knowledge already available. In other words, scientific and technological developments often if not always represent "recombinations" of what is already known (e.g. Price and Bass, 1969; an essentially similar view is presented by Kuhn, 1962).

In linguistics, Noam Chomsky (1965), among others, has argued that formal grammars are adequate for the description of language because they are "generative" models that account for the dynamic or "creative" property of language. "Thus an essential [creative] property of language is that it provides the means of expressing indefinitely many thoughts and for reacting appropriately in an indefinite range of new situations" (Chomsky, 1965: 6). This creativity is allegedly accommodated by the "universal language," a set of cross-linguistic rules that are embedded in the "deep structure," or the biological wiring, of human speech. More recently, some students of sociolinguistics have refined Chomsky's concept of creativity, which is formally treated by the mathematical property of recursiveness, by arguing that creativity consists in combining linguistic signs with an infinite number of contexts, for example, other linguistic signs, nonverbal signs, various types of available information, and possibly broad conceptions about the nature of reality (see, e.g., Douglas, 1975; Polanyi and Prosch, 1975; Leach, 1976; Lavandera, 1977).

In a broader key, the predispositions of combination appear to be part and parcel of the process of socialization itself. While I previously associated the emphasis on socialization with the cultural scientist's static bias, there is no denying that socialization is also responsible for innovation in various ways (e.g. Bandura, 1977). My special interest here is in the technique of social learning known as "modeling." It appears that as people learn from one another, the tendency to imitate particular individuals rather than others is rather pronounced. Conformity, or mere acquisition of received behavior, is problematic because, even with the preferential tendency to imitation, the acceptable models are multiple and to a degree heterogenous. This diversity in modeling fosters behavioral innovation through a combinatorial technique.

> When exposed to diverse models, observers rarely pattern their behavior exclusively after a single source, nor do they adopt all the attributes even of preferred models. Rather, observers *combine aspects of various models into new amalgams that differ from the individual sources*. [Moreover] *Different observers adopt different combinations of characteristics* (Bandura, 1977: 48—emphasis provided).

The manifestations of the combinatorial predispositions are legion. People everywhere and at all times have acted on the apparent belief that by reciting certain words, sometimes in a particular sequence; by performing certain acts; by joining together certain properties; by these and analogous practices, certain effects can be brought about. They may concern the reciprocation of unrequited love, the salvation of the soul, the completion of a safe voyage, communication with the dead, and so forth.

If such phenomena were not manifestations of specific predispositions, chances are that in modern, science-oriented society they would be absent, or at least on the decline. But in recent years many millions of people in the United States of America have bought ouija boards. Countless people read Tarot cards or faithfully follow the daily horoscope. According to a recent Gallup poll, over a quarter of the American people believe in astrology and think that their lives are governed by the position of the stars (reported in *Newsweek*, 1978: 32). Friday-the-thirteenth is considered a day of ill luck. A study of therapeutic practices associated with children's diseases revealed that even highly educated mothers sometimes try to treat their children with such strange concoctions as honey mixed with kerosene; the probability of such unorthodox behavior increases with the difficulty of the diagnosis and/or prognosis of the child's disease (Lewis and Lopreato, 1962). Millions play the lottery and other sorts of games, often addicted to particular numbers which they have seen in their dreams or which they associate with happenings that in some way have impressed them. On 7 July 1977 (7/7/77) the public media reported that at the horse tracks the number seven was extremely popular among bettors.

Legalized gambling has become widespread. Soccer lotteries are a worldwide phenomenon. In the United States of America, a majority of the states allow horse racing and bingo. Dog racing is legal in many states. Lotteries of all sorts are fairly common. Card rooms may be found in several states. Some states have legalized sports betting, and a few allow gambling on jai alai. The people who play the numbers are numberless. Gambling is much more widespread than legal activity suggests. Estimates of gambling turnover run as high as $100 billion a year. That places the practice among the biggest commercial activities in that nation. All across the country many thousands of individuals behave like Dostoyevsky's compulsive gambler who, cleaned down to the last coin reserved for food, inexorably returns to the gambling table only to lose that as well. Gamblers Anonymous chapters have sprung up all over the country. There are chapters, too, for gamblers' spouses, and there are even a few where gamblers' children seek emotional support rendered necessary by their parents' plight.

Why has gambling become such a common pastime? Why do some individuals behave like the millions of proles in George Orwell's Oceania for whom betting was the principal if not the only reason for remaining alive? There are many possible answers to the questions, of course. Where, for example, money has become the measure of all things, it is only natural that people will have great avidity for it and employ any means at their disposal to attain it. Little wonder then that crime and gambling are so intimately interconnected—crime big and small alike. A popular US magazine reported in 1976 the case of a chronic gambler who, by personal admission, engaged in digging up and selling coffins in order to get cash for a bet. Again, when the taste for work becomes purely instrumental in nature, at the same time that work offers little or no promise of self-aggrandizement, Fortune titillates the desire to risk. J. Cohen (1959) argued that the belief in luck is a social stabilizer; that is, it excuses our own failures, while the success of others motivates us to continue hoping for a lucky break.

Ultimately, however, the passion for gambling is the logical result of a mind gone berserk with the predispositions of combination, which entail the manipulation of the environment, *inter alia*. It is linked to the chief gods of rationalistic culture: Speed, Numbers, and Sports. Gambling within limits is what life is all about. We all gamble by taking chances in the goals that we pursue: we choose certain combinations, or strategies, over others. But gambling as an obsession is a sort of arrogant challenge to the tenacity of what Sumner (1906) termed the aleatory element, or the element of chance in nature.

There are a motor, active, aspect to combiners and a passive side. On the active side, an individual may, for example, be inclined to find ways of bringing about a moratorium on nuclear power plants "in order to" have a cleaner and safer environment. On the passive side, there may be merely the idea that the moratorium and a safe environment are necessarily conjoined, so that if the moratorium is instituted, an improvement of the environment will necessarily ensue. Similarly, in an active sense, politicians, entrepreneurs, scientists, and engineers may be at work on the combinations needed to solve the energy problem. In a passive sense, others may believe that those people have solved many problems before and they will solve this one, too. In this latter sense, people who are not themselves engaged in bringing about sociocultural change are nevertheless positioned to be swept along by selected combinations.

7.2.2 SELECTORS

So much for the forces of combination, for the time being. Selectors, too, have received a reasonable degree of recognition in behavioral science. We have already encountered, for example, Auguste Comte's (1896, III: 305–8) concept of conservative instincts, which he associated in part with a low demographic density and a relative paucity of social problems. The class of sentiments labeled the "persistence of aggregates" by Pareto (1916) contains items that may properly be termed selectors. Furthermore, this scholar (1916: 992) argued that after certain combinations have been brought about, "an instinct very often comes into play that tends with varying energy to prevent the things so combined from being disjoined." Others have written of such things as "the wish for security" (Thomas, 1923) and the "conservative impulse" (e.g. Marris, 1974: 4), namely, "the tendency of adaptive beings to assimilate reality to their existing structure, and so to avoid or reorganize parts of the environment which cannot be assimilated."

We need not be concerned about a comprehensive review of literature. One striking feature of all these notions is suggested by the recurring attribute of "conservative." It is in a sense a troublesome label because it tends to be uncritically associated with certain indefinite types of political attitudes and preferences. For me, however, the fundamental idea behind the conservative or selective forces is conveyed by the process of variation itself and the disarray that it tends to produce in established tradition. My reasoning goes as follows. Because we have thought of this process by analogy to biological recombination, which is a random mechanism of biological evolution, it is reasonable to think of variation also as a largely, though not exclusively, random mechanism of sociocultural transformation. By random variation I mean a process of change based on activity that may properly be labeled trial-error-and-retrial, or more simply trial-and-error.

This conception has been advanced or defended by various scholars (e.g. Darwin, 1859, 1871, 1872; Sumner, 1906; Keller, 1915; Campbell, 1960, 1965, 1974, 1975; Langton, 1979). It is also controversial, in part because it is linked to problems surrounding what in Chapter 2 was referred to as the rationalistic bias. The question, "do sociocultural events happen because we wilfully and rationally make them happen, or do they take place by chance?" is never far below the surface in models of sociocultural variation.

My own position is that the question is essentially misguided. Sociocultural variation is the result of both will and happenstance. It is wilful in the sense that human beings are goal-oriented and constantly engaged in the search for strategies or behaviors that are in some sense superior to old ones. In part such wilfulness may also be termed rational in the sense that the means employed in the search are efficiently related to the goals. Variation, however, is also a result of blind and random activity in the sense that what we aim at and what we in fact achieve are often two quite different things. Furthermore, while the incidental variations may sometimes be properly considered affinal alternatives to those envisioned in the wilful and rational strategy, all too frequently they are entirely unrelated to the aims in view—they are the results of serendipity. In this sense, it may be said that the process of sociocultural variation is partly random and blind, partly wilful-rational, and partly serendipitous.

In their attempt to summarize available knowledge about scientific discovery Price and Bass (1969) have reached certain conclusions that seem quite useful for understanding sociocultural variation in general. One conclusion is that innovation is a result of combinatorial behavior. According to another, the predictability of a discovery increases with its "rationality." Scientific innovation, moreover, typically calls into action a complex network of ideas and people arrayed along a multibranched and unpredictable path. Finally, scientific discovery is facilitated by communication between disciplines and interested parties.

Building on this work, Harold Schneider (1977) has proposed a useful distinction between linear and nonlinear innovation. In the former case, discovery involves combinations of what is already known. In the latter case variation takes place as a result of stepping outside of the existing repertoire and trying what is in effect an unusual combination—something that is not readily suggested by available knowledge.

The blind or random element in variation is probably more typical of nonlinear than of linear innovation. Inasmuch as systems of knowledge tend to become internally exhausted and thus to provoke extra-systemic activities (see Kuhn, 1962, for a discussion of this issue), it would seem that nonlinearity and the adventurous process of trial and error that it is typically associated with stand out as reasonably acceptable properties of the process of sociocultural variation. None of this is to imply, however, that nothing in sociocultural evolution is the manifest result of deliberate, purposive, and rational behavior. For example, the history of scientific developments leading to, and surrounding, attempts to attain *in vitro* conception in humans is too complex to be divested entirely of blind or random elements. The fact, however, is that for some time now such a mode of reproduction has been an avidly pursued goal of various branches of medical science, and has of late become cultural reality.

In all likelihood, therefore, there is a counterpart to the extreme rationalistic bias that is no less fallacious and restrictive of scientific development. A recent paper has reasonably well expressed the spirit of theoretical reconciliation that is required to achieve a synthesis of the purely mechanistic and the psychological perspectives on biocultural evolution. Boehm (1978), among others, has convincingly shown in a partial and constructive critique of the blind-variation-selective-retention model that, to a degree, sociocultural behavior follows a process of "rational preselection." That means that, even if evolution implies pure random or blind variation, culture has somehow managed to evolve in such a way as to "beat natural selection to the draw." Specifically, cultural variation moves, presciently as it were, in the direction of selections whose outcome fits reasonably well the predictions of the maximization principle. Rational, or seemingly rational, innovation converges with random variation on the maximization path.

It remains to be said that, while the scientist understandably wishes that all the biases went away, the fact remains that biases are an integral part of scientific and, therefore, sociocultural innovation. The bias toward the model of random variation and selective retention, for example, carries with it the hypothesis that our evolutionary course is strewn with wasteful and costly errors. Thus, countless lives were lost to numerous diseases and epidemics along the way leading from ancient medical practices to the medicine that has culminated in the modern miracle drugs.

Such an hypothesis, in turn, serves to cast light on the nature of the selective forces as they relate to the issue of conservatism. The point I wish to make, by reference to Sumner's (1906) renowned analysis of the mores, is that trial-and-error evolution implies, by virtue of its costliness, a strong attachment to, even a tendency to sanctify, those combinations that have been selected for retention. This is but one small step away from saying that a basic property of the selective forces is a *holy belligerence*. As constituents of tradition—of the institutional framework—selectors are what

Wilson (1975b: 168) terms "the ultimate refinement in environmental tracking." They thus tend to resist the transformation and disorganization of the established ways with agonistic behavior. It is in this sense that selectors may be considered "conservative" forces. More than other predispositions, they are the forces of tradition and group preservation. Moreover, inasmuch as transformations may often originate in other sociocultural systems, it follows that they tend to give rise to concepts of "us" and "them," "in-group" and "out-group," peoplehood and foreignness. It is in this sense that Pareto (1916: 2522) could state that the sentiments of persistence "constitute the foundations of society and stimulate the belligerent spirit that preserves it."

Little wonder that in the United States of America war is such a popular subject in literature, in the fine arts, and on celluloid. Nearly four decades after the conclusion of World War II, Herman Wouk's 1979 commemorative novel *War and Remembrance* threatened to remain on the bestseller list indefinitely. "Patton" was one of the most popular films of the 1970s. Television is cluttered with World War II documentaries and dramas. Many European societies feature analogous behavior. Indeed, the American fascination with the greatest war in history reflects the fascination with war by humankind in general. The underlying reason is that war is a fundamental mechanism of group as well as individual selection, and as such calls into action predispositions of selection. Thus, there is an element of sanctity, even transcendence, attaching to it. To be concerned with war, as members of groups, is to be concerned with survival viewed as a cosmic fact. Across the globe, and throughout history, one encounters one holy war after another.

There is a relationship of complementarity between combiners and selectors. Together they represent the basic constructors of biocultural evolution. From one perspective, combiners and selectors may be viewed as coupled adaptations reflecting the organismic needs to deal with both change and stability in the environment. They reveal both the ability to cope with the new, or to bring it about, and a tendency to be attached to relatively fixed behavioral forms. The capacity for the new enables individuals to absorb knowledge and to react to specific, idiosyncratic features of immediate contingencies arising during the unique course of individual life histories. Fixed behaviors are efficient in coping with environmental regularities that have exerted continuous selective pressure during the history of the species (Waddington, 1957; Dobzhansky, 1967; Lorenz and Leyhausen, 1973).

From another perspective, there is a relationship of dynamic tension between the forces of variation and those of selective retention. To an extent, that is due to the fact that innovation, in addition to adding to the existing sociocultural endowment, substitutes existing elements with new ones. That tends to clash with established ways of thinking and acting. It is, therefore, the function of the selectors to check the pace and the scope of the combiners. Their task is twofold and very complex. Konrad Lorenz (1974) has noted that all adaptedness of living systems is based on knowledge built into structure, which is static as contrasted to the dynamic process of selection. Selection, accordingly, presupposes some dismantling of old structures, for the new information inherent in variations inevitably demands some breaking down of existing knowledge. It follows that, metaphorically speaking, combinatorial forces tend to be profligate, spendthrift, wasteful, even destructive makers of noise in the sociocultural system. Conversely, the selective forces must not only be on the lookout, so to speak, for adaptive variations; but they must also be frugal, ever on the alert to guard carefully over the criteria of selection (see below) against a too ready acceptance of combinations that would overload and disorganize tradition.

It is the tension between the two sets of forces that accounts in part for the fact that history traces a rhythmic line along which the two sets alternate in intensity and predominance. In the Dark Ages, for example, forces of selection had the upper hand; the opposite has been true for some time in European society. Nevertheless, if we may continue with metaphors, combiners are to the pistons of a motor vehicle as selectors are to its brakes. The brakes vary in their efficacy, but both in principle and in practice they do have the power to stop a motor vehicle in its tracks.

In view of the above metaphorical reference to the combinatorial predispositions as profligate forces, and to the selection predispositions as frugal forces, it will be useful to return briefly to the property of intensity and hazard a few comments about the personality types that may be expected to be associated with combiners and selectors, respectively. Recall that in a given population the intensity of such predispositions varies from individual to individual, from social category to social category, from time to time. Accordingly, for simplicity's sake we may speak of two *ideal* personality *types* whose composition is at least vaguely suggested by the terms "frugality" and "profligacy." The two seem programmed to facilitate, respectively, the action of selectors and of combiners. People rich or strong in selectors tend to be cautious, tradition-loving, patriotic, religious, familistic, frugal in their economic habits, inclined toward the use of force and confrontation in political matters, adept at deferring gratification. Conversely, persons strong in combination forces are enamored of change; they are culture-relativists; they are hedonistic, rationalistic, individualistic, dedicated to spending and entrepreneurship; and inclined toward ruse, deception, and diplomacy in political matters.

A dramatic case of international politics, taken out of the mothballs during the 1980 US presidential campaign, may serve to illustrate the difference between the two ideal types. In the winter of 1948, three years after the end of World War II, East Germany set up a Soviet-inspired

A Model of Sociocultural Variation and Selective Retention 251

blockade of the Allied-occupied portion of Berlin. We are in the heart of the cold war. The United States of America and Britain responded to the challenge by dispatching to West Berlin cargo planes which over a period of fifteen months flew nearly 280,000 missions and delivered a total of 2 million tons of food, fuel, and clothing. Many hailed this response as a great success and feat for Western technology and ingenuity. Those were rich in the combining inclinations—shrewd but not assertive. Others, a minority, at least in appearance, would have preferred to respond to force with force. They argued that the Allies had by international agreement the right of free access to West Berlin. They pointed out, moreover, that Soviet Russia did not yet have the atomic bomb, and consequently a forceful response on the part of the West would have successfully called the Soviet bluff. These were rich in the selection predispositions—assertive but not shrewd.

Another example: It is said that after Alexander the Great's initial victory against Darius's army and his Athenian allies, a local sage led him to the Gordian knot and sought to test his right to rule by inviting him to unravel it. Alexander drew his sword and shredded the knot. That was a case of triumph of assertiveness, force, tradition, and steadfastness over shrewdness, cunning, inventiveness, and stratagem.

7.3 Criteria of Selection

Without the selectors, all combinations would tend to be ephemeral. By the same token, without the intervention of mechanisms to guide the process of selection, selective retention would in all likelihood be purely random. The present section will consider the question, how are certain combinations, among the plethora produced every moment in any given population, singled out by the selectors for retention and institutionalization? More specifically, what mechanisms or criteria come into play to steer the selectors in favor of certain combinations as against others? We are entering an enormously complex area, and I harbor no presumption of doing it justice. My aim is economy accompanied by plausibility.

7.3.1 THE ENHANCEMENT OF FITNESS

One answer to this query is required by the basic perspective followed in this volume. To the extent that we accept the maximization principle as a valid statement of theory, we must also subscribe to the hypothesis that combinations tend to be favorably or unfavorably selected depending on whether they are adaptive, that is, depending on whether they contribute to reproductive success. This statement, however, counsels reiteration of a prior clarification. Suppose we begin by defining the term adaptive in the sense in which it is here being used. Wilson (1975b: 577—emphasis provided) defines a trait (physiological, behavioral, and so on) adaptive if it "makes an organism more fit to survive and to reproduce *in comparison with other members of the same species.*" I have italicized a phrase that is crucial for a proper understanding of adaptation, if not altogether of natural selection. Compare Wilson's position, for example, with one taken by W. H. Durham (1978: 433; see also Barash, 1977: 33) in what is otherwise one of the most helpful endeavors in the emerging theory of biocultural evolution. This scholar hypothesizes that "cultural features of human phenotypes are commonly designed to promote the success of an individual human being in his or her natural and sociocultural environment," where success is best assessed "by the extent to which the attribute permits individuals to survive and reproduce and thereby contribute genes to later generations of the population of which they are members."

Statements such as this tend to encourage Lewontin's (1978) charge that sociobiologists follow an excessive adaptationist strategy. Durham and other students of sociobiology know better, despite their susceptibility to the human flaw of occasional carelessness. However, the terms "fit" and "adaptive," whether explicit or implicit, are of little value, if not altogether misleading, when they are not expressly employed in the relational, relative, and comparative context that is required by the principle of natural selection. Put otherwise, a cultural novelty need not be adaptive for all or even most individuals in a population in order for it to be favorably selected. Indeed, universal adaptiveness is contrary to the logic of natural selection (which, we recall, refers to the *differential* contribution of genes to future generations), and could accordingly be at best a nonevolutionary concept cloaked in evolutionary garb. It follows that to say that combination X is favorably selected because it enhances fitness is to say that *in principle* it enhances the fitness of all but in practice it increases the fitness of some individuals at the expense of others. Otherwise, we would end by denying natural selection and such things as the dominance order, among other sociocultural phenomena.

It will not be amiss now to note that with varying degrees of explicitness my Darwinian conception of biocultural evolution receives support even from the findings of scholars who are not normally associated with human sociobiology. There is, for example, some affinity between my hypothesis of fitness enhancement and a proposition of "exchange theory" which George Homans (1974: 43), at the microlevel of a means–end schema, terms (understandably but unfelicitously for us) "the rationality proposition." Writes Homans: "In choosing between alternative actions [read 'combinations'], a person will choose that one for which, as perceived by him at the time, the value, V, of the result, multiplied by the probability, p, of getting the result, is the greater." Value, of course, may be interpreted in various ways, including one that has relevance to reproductive fitness.

Likewise, some forty years ago, Ogburn (1942) wrote a most informative essay in which he showed how inventions affect the size of populations and, through this, influence human history. He noted, for example, that the population of France in the seventeenth century was probably four or five times larger than the population of England at that same time, whereas early in the twentieth century, as a result of the older Industrial Revolution in England, the two countries were about equal in population size. The new knowledge, technology, jobs, and means of sustenance introduced by the Industrial Revolution favored the English population in relation to the French. This, incidentally, is further proof of the fact of intergroup selection in human society.

Evidence in favor of the hypothesis here in question is really quite plentiful. It is well known, for example, that the selection of combinations is facilitated if they are introduced by individuals high in the social hierarchy (e.g. Barnett, 1953; Pulliam and Dunford, 1980). The circumstance is not due merely to the fact that social leaders may have special powers of persuasion and/or manipulation at their disposal. One would have to guess also that their "diffusive" task is greatly facilitated by the fact that, as leaders, they represent success stories. Their acceptance of variations, therefore, tends to bestow upon them the brand label of "adaptive." The topic is related to the previous finding that "model" learning or imitation tends to favor dominant individuals. We need not belabor the point here at issue. Basically it is unassailable. The proof that selections tend to be adaptive lies in the terrific population explosion that across the globe has accompanied, first, the agrarian revolution and, currently, the industrial revolution. The important point is the differential adaptiveness of sociocultural variations as among individuals and peoples.

There is no clear justification, furthermore, for the proposition that all cultural variations are adaptive for someone or other. The phenomenon of vestiges in anatomy and physiology, such as the human appendix, would suggest the presence of behavioral analogs. Ultimately it is the brain that must choose in favor of variation X as contrasted to Y, and there is no guarantee that the proper choice is made in view of possible alternative selections. The brain is quite economical in picking features of the conscious images transformed from cortical perception (e.g. Bullock *et al.*, 1977; Grossberg, 1978). In extracting select information from such images, moreover, the brain seems to follow the rules of speed, precision, and simplicity (e.g. Eibl-Eibesfeldt, 1975; Wickelgren, 1979; Lumsden and Wilson, 1981). Accordingly, we cannot exclude out of hand the hypothesis that the brain may sometimes act too hastily, and literally too simplemindedly, in favoring one course of action, one alternative variation, as against another.

In addition, the brain tends to reach decisions with a view that they may be properly fitted in its memory storage. The issue in question relates to a problem of human psychology first studied by the Gestalt psychologists. These scholars noted that perception is always organized into a *Gestalt*—that, contrary to what early learning theorists maintained, stimuli are combined into "total" and meaningful configurations (see, e.g., Combs *et al.*, 1976). Thus, a musical composition is grasped as something more than a mere series of notes. It is the problem of emergence appearing again in a new guise. But it is also a problem of cognitive logic or consistency. More specifically, the brain seeks consistency in the face of new information. But in view of the rules of speed and simplicity, the feat of reaching internal consistency may, for example, have to be a result of eliminating two strongly contrasting alternatives with a third choice whose simplicity is convenient but also indicative of adaptive neutrality.

There is an analogy to this sort of problem in the phenomenon of compromise that is a universal technique of social behavior. Some compromises are clearly adaptive. They represent the collective wisdom of various interests and orientations. Others, however, are more problematic —at least in the sense that they waste possibly precious time. A good case in point is the tendency in academic departments to avoid difficult but crucial problems through the apparently revered technique of perpetrating the problem into one committee after another.

Alternatively, it may happen that the brain's rule of speed leads away from mental coherence and introduces dissonant elements in its memory storage. It is possible, under these circumstances, that such conditions as mental illness, moral and social disorientation, mental confusion, and even poor memory itself are at least in part manifestations of a brain cluttered with excessively disharmonious information.

In conclusion of this section, the hypothesis that cultural variations are always selected on the criterion of their adaptiveness appears to be problematic, and certainly inadequate by itself. Still, it is reasonable and useful. For example, it helps to illuminate an important issue of human sociobiology. Consider, for example, the debate between sociobiologists and some cultural scientists over the notion that "genetic natural selection operates in such a way as to keep culture on a leash" (Lumsden and Wilson, 1981: 13; see also Wilson, 1975b, 1978). The metaphor may be a bit unfortunate in view of the fact that words are powerful and often treacherous tools. I am not surprised by so much revulsion to such expressions in cultural science. Nor does it help at this emotional level to read that, as Wilson and his collaborators correctly put it, the leash that ties culture to genes is quite long. Nevertheless, if there is any validity to the hypothesis that cultural variation follows an adaptive path, we can hardly gainsay the "leash principle." Moreover, we can devise a reasonable explanation for the hypothesized elasticity of the gene-culture link. If culture, like learning, is the handmaid of the maximization principle—not entirely passive or servile, as we have seen—and we, its bearers, are programmed to behave according to this arrangement, it makes very good evolutionary sense to have a long gene-culture leash. The strategy gives human beings, extremely intelligent animals, maximum latitude to explore, seek out, and adopt the most adaptive variations. In short, the gene-culture link is long precisely because a lengthy nexus allows for richness of variations and adaptive selections.

7.3.2 THE FACILITATION OF SOCIALITY AND SELF-ENHANCEMENT PREDISPOSITIONS

A second criterion for the selective retention of combinations is closely related to the first and to the interdependent nature of biocultural phenomena. It refers to the differential value that variations have for maintaining the evolutionary viability of the behavioral predispositions themselves. The more harmonious sociocultural combinations are with those behavioral forces and their sociocultural manifestations, the more likely they are to be favorably selected. This is another way of stating both the inextricable interplay of biology and culture and the systemic affinity existing between prevalent sociocultural phenomena and emerging ones. Thus, if the need to reciprocate and the systems of exchange and reciprocal favoritism that coalesce around it are adaptive mechanisms of the sociocultural existence, variations that grossly undermine them are bound to be at a selective disadvantage by comparison to those that reinforce them.

By the same token, variations that tend to maintain ascetic altruism at a low level of frequency in a population are likely to be favored over those that would yield a major increase of ascetic altruists; ascetic altruism is not harmonious with many other behavioral predispositions, and its evolutionary expressions must necessarily be viewed to an extent as being discontinuous with the overall cultural repertory. To use the Game Theory language of J. Maynard Smith and his associates, a large increase in ascetic altruists is not an "evolutionarily stable strategy" (ESS). The same may be said of a large increase in social cheaters. According to these scholars (Maynard Smith and Price, 1973; Maynard Smith, 1976; Maynard Smith and Parker, 1976), an ESS is a social relations strategy that is typical of the majority in a given population and is not easily bested by deviant individuals.

Consider, as an example, a hypothetical society consisting of only two types of individuals, reciprocators and non-reciprocators or cheaters, of whom the reciprocators constitute the majority. Assume, further, that a cultural variation favoring the cheaters becomes entrenched in the population. For a while, if we assume further that the reciprocators are not organized by sanctions applicable against the cheaters, the latter may have a survival advantage. They use the variation to receive or appropriate fitness-enhancing resources without the necessity to control their selfishness and share their advantages. But the resulting society of cheaters is a society in which each is at the mercy of one's own devices. The society comes to resemble the hypothetical one characterized by Thomas Hobbes as being in a state of war of all against all. In such a society, a mild illness, a little fatigue, the very state of sleep can be the ultimate disaster even for those who are normally strong, healthy, and alert. Indeed, uncontrolled conflict ends by destroying the social nature of the population, for its members cease to share their learning and other useful skills as well as their goodwill. Above all, as was argued in the previous chapter, their disorganization places them at a disadvantage in relation to a group of reciprocators with whom they might come in competition for the resources existing in a given space.

Under these circumstances, we are justified in hypothesizing that natural selection favors those people who are driven by behavioral predispositions to generate cultural variations that support reciprocal favoritism and maintain cheating under reasonable control. We may further guess that, given the properties of distribution and intensity attaching to behavioral predispositions, societies oscillate in time in their ratio of reciprocators to cheaters. But fluctuations are relatively mild, for the wild ones are likely to produce unstable and maladaptive results. That is another way of saying that societies that persist and thrive pursue an evolutionarily stable strategy in which the ratio of reciprocators to cheaters is relatively stable. Analogous arguments are possible in connection with each of the predispositions discussed so far.

7.3.3 THE ENHANCEMENT OF SATISFACTION OR CREATURE COMFORTS

If it is doubtful that culture has severed the umbilical cord tying it to inclusive fitness, it is even more certain that sociocultural evolution has accentuated the search for what has often been termed "creature comforts." All learning animals, as far as we know, tend to avoid pain and discomfort and to seek pleasure and comfort (Pulliam and Dunford, 1980). Bird nests, for example, are not useful only for reproduction; they are also comfortable to rest in. Chimpanzees build nests that seem to be unrelated to reproduction. The quest for comfort appears, however, to have reached a high pitch in human evolution, and there is every reason to think that such drift has played a major role in the selective retention of cultural combinations.

The circumstance has led some recent evolutionists in social science to theorize by placing notions of "struggle for satisfaction" (Ruyle, 1973) and "struggle for reinforcement" (Langton, 1979) in a conceptual framework analogous to Darwin's struggle for existence. Langton (1979), for example, stresses that people must struggle with their physical and social environments to obtain the things that satisfy their daily needs and wants.

G. P. Murdock (1940: 367) argued cogently that "elements of culture . . . can continue to exist only when they yield to the individuals of a society a margin of satisfaction, a favorable balance of pleasure over pain." By the same logic, new elements of culture are accepted into the existing system on the basis of the amount of bodily and mental pleasure that they yield. An assured food supply derived from a cultivated patch of land is more comfortable than the more desultory pattern of hunting and gathering. Thus, the horticultural revolution brought pleasure as well as a population explosion. It is even quite likely that pleasure, which is filtered through our senses, is intuited as a signal for the fitness-enhancing value of possible selections. A more or less permanent hut built on the cultivated land is a variation from the ramshackle accommodations of a nomadic life that provides both greater security and greater comfort in relation to an ever-dangerous environment. Gradually, it leads to items of furniture, such as a table and a bed, which further add to the pleasures of living and leisure time.

Again, joining a wheel or two to a cart-like formation and harnessing an animal for the transportation of merchandise are variations that may have been positively selected with great delight. The same may be said of many, if not most or all, other technological innovations. Consider the airplane, the end result and synthesis of an enormous series of combinations. There is no intrinsic reason, for example, why this invention should not have been relegated exclusively to military uses or purely commercial applications. But, to give a striking example, the airplane came in handy for the millions of Europeans who in the wake of World War II flocked as migrants to faraway Australia. The first to go typically travelled thirty or more days by sea, and many of them learned to enjoy the very thought of death in the process, as anyone who has suffered severe and sustained seasickness can attest. Again, unless one seeks adventure, natural beauty, and the like, flying from New York to San Francisco beats slugging it out in a week of cross-country automobile driving.

In a related key, living with chronic toothaches must have been a dreadful affliction for endless numbers of our ancestors. Having a tooth pulled out in the raw was an improvement, but not a great one. Losing a gangrenous leg to a dull saw in the absence of anesthesia is probably the nearest experience to crucifixion. In short, the inventions of painkillers, of expeditious and pain-free means of transportation, of ways to transfer one's most burdensome toil to others—these innovations and numberless others analogous to them have met with great favor in the continuing reshuffling, reorganization, and enrichment of the sociocultural system.

I have made no attempt to systematically define creature comforts, and none may be possible. But I should think that the label comprises at least those variations that reduce hunger and pain, facilitate learning and the storage of information, and encourage leisure, expediency, and economy of action expenditure.

7.3.4 SYSTEMIC IMMANENCE

A fourth criterion of selection concerns a crucial property of systems that has been termed "the principle of immanent change" (e.g. Sorokin, 1957: ch. 38). Immanence refers to the fact that a system such as a human organism or a human culture is in a constant state of flux due to its own inherent or determinate dynamics. These internal dynamics may be thought of as a sort of inner program orienting the system toward "preferred goal states," and thus toward certain changes rather than toward others. The notion of the evolutionarily stable strategy discussed above is related to what is being said here, for the ESS suggests that any variations that tend to meld with the existing ESS are favored over those that would disturb it. More broadly, as Aristotle long ago pointed out, an acorn can only become an oak, if it becomes anything at all. Immanence thus favors what previously were termed linear variations, namely, those changes that are theorems, as it were, of existing selections. The selection of combinations under the influence of systemic immanence is the selection of built-in potentials. It may thus be said that once systems are formed, they contain in themselves a considerable part of their future, what we may term the normal or predetermined part of their transformation.

This is not to deny that other sorts of change take place. It is to underscore the fact that certain changes are *ex initio* "in the cards" by virtue of the thrust that inheres in the dynamics of systems. So, for instance, a sociocultural system that has incorporated the invention of the wheel is likely to continue to favor variations that entail using the principle of the wheel. Indeed, human society since the horticultural revolution has been to a large extent energized by wheel-related variations: from the oxcart to the wheelbarrow to the chariot to the spinning jenny to the automobile and to the computer tape, to mention but a handful of millions of wheel-related innovations that have endured for more than fleeting moments.

There is, however, what may be termed an entropic or self-exhausting quality to immanent innovation. The point may best be made by reference to the process of scientific discovery. Much if not most of it in any given science represents a steady extension of existing knowledge obtained by the additive or combinatorial manipulation of available statements of theory. Discovery, in short, takes place within the context of a more or less close, fertile, and orienting paradigm (Kuhn, 1962). In this sense it is immanent in nature because it merely reaps the inherent productivity of the paradigm. Sooner or later, however, the paradigm is confronted with surprises, unsuspected phenomena, that cannot be explained in terms of existing theory. Kuhn terms them anomalies. It is the awareness and recognition of anomalies that gradually lead to new theories and methods that transform the paradigm from without. Discovery has now become nonlinear. The immanence of the preexisting paradigm grinds to a halt.

Combination becomes truly creative, for an anomaly requires associations of phenomena and ideas that transcend established theories and procedures. A case in point is the history of the theory of light and color. Pre-Newtonian theories were anomalous because, as Newton discovered, they did not account for the length of the spectrum. Newton's own theory became, in turn, anomalous and required a nonlinear, non-immanent, correction by wave theory because it could not account adequately for diffraction and polarization effects (Kuhn, 1962: chs VI and VII).

The self-depleting nature of systemic immanence is probably related to the phenomenon of cultural diffusion. Offhand it would seem that the more internally creative a sociocultural system is, the less prone it is to cultural contagion or diffusion from the outside. Conversely, the more impoverished is the system in its systemic immanence, the more susceptible it is to novelties deriving from elsewhere, provided of course that the system is geographically and politically open to external influences. Cultural diffusion, thus, may to an extent be viewed as a measure of cultural vitality and imperialism for those sociocultural systems that transmit, and conversely as a measure of cultural ossification for those systems that receive.

Diffusion so viewed addresses the problem at the intersocietal level. If we, on the other hand, focus on a given society and view it as a collection of interdependent subsystems (political, ideological, economic, religious, artistic, and so on), we may observe the waning and waxing of immanent change as a function of the changing links existing between the various subsystems. This is a form of diffusion, too. A small example from the history of music will help to make the point. The religious connections or orientations of medieval painters, architects, and sculptors inspired a style of music, the Baroque, that employed a complex harmonic language, rich polyphonic procedures, highly organized forms. Then came the Age of Reason with its devotion to rationality and simplicity, its disdain for theological contrivances of all sorts, and ironically, its sanction of the prevailing hedonism of the eighteenth-century aristocratic courts and salons. The Baroque style—that remnant of "barbarism" and cousin to the "ridiculous portals of our Gothic cathedrals," as Rousseau put it in the *Ecrits sur la Musique*—thus gave way to the classical style of a Mozart, a Haydn, and partly, a Beethoven.

But the Enlightenment, too, came to an end, roughly with the demise of the Napoleonic era and its vision of the Prince-Savior. And with the passing of the Enlightenment, musicians had to find new sources of inspiration. Beethoven, for example, was caught in such a period of transformation, with the result that his compositions go from the Classical through his special brand of it, the Heroic, and to a variation of the Baroque through a concentrated exploration of counterpoint and polyphonic textures, an intense interest in the music of Bach and Handel, and a new awareness of church themes, *inter alia* (Solomon, 1977).

7.3.5 THE MULTIPLIER EFFECT

The principle of systemic immanence basically says something about the quantity of variations that are endemic or intrinsic to a given sociocultural system. The multiplier effect, already encountered, is closely related to it, and states roughly that, up to a point, the rate of sociocultural variation is a positive function of preexisting sociocultural information and change. It thus concerns the speed at which variation takes place. I say "up to a point" because, as was noted on a previous occasion, without this qualification we run into an absurdity. It would be absurd, for instance, to hold that the greater the rate of sociocultural change, the greater the future acceleration of change. Up to a point, the proposition is historically valid. There are periods when change begets change at an extremely rapid rate. That is the source of what some writers have termed "future shock" (Toffler, 1970; more formally, see Ogburn, 1922). The reason for the absurdity lies in the fact that at that exponential rate of change, everything at any given moment would eventually be new, and selective forces would cease to operate. Thus, there must be a degree of entropy to cultural and biological phenomena. Change cannot, and does not, always follow a positive slope. Our present problem is related to what scholars in chemistry and biology term autocatalytic reactions, namely, processes that, once initiated, accelerate in direct proportion to the effects that they produce. As Wilson (1978) notes, however, such reactions never expand to infinity and the very processes (chemical, biological, sociocultural) acted upon by catalytic reactions normally change through time and eventually bring autocatalysis to a halt.

In evolutionary science, such phenomena have been termed reversals. Thus, human history reveals dramatic evidence of "dead" civilizations. The arts and sciences of the European Dark Ages, for example, were retrograde by comparison to the arts and sciences of Greco-Roman civilization (Sorokin, 1957). Likewise, common sense notwithstanding, it is quite unlikely that the extremely rapid pace of current sociocultural change will continue uninterrupted.

We know little or nothing about the specific mechanisms that come into play to slow down or interrupt altogether autocatalytic reactions, a concept analogous to the multiplier effect. At a very high level of abstraction, however, we are dealing with something that is somehow related to the second law of thermodynamics, according to which isolated systems tend toward the degradation of energy. Consequently, if it can be argued that all systems are more or less isolated, it is also possible to reason that they are subject to a loss of energy. To the extent that autocatalytic reactions are guided by an immanent principle, it is useful to assume that the repertory of possible variations inherent in an immanent "program" will sooner or later have to become exhausted.

From another perspective, perhaps related and in any case equally abstract, there appears to be a certain similarity between the forces of variation and the phenomenon known as charisma (Weber, 1946: 224–9). Both tend to effervesce and then gradually fizzle to "routinization," bureaucratization, and the like. The public media have recently reported, for example, that the US Patent Office is literally swamped by a massive backlog of paper work, with the result that it now takes, on the average, two years to clear a patent through that office. The circumstance encourages the theft of inventions, especially those of independent thinkers unconnected with large research and development firms. Such a state of affairs, in turn, provokes litigation, which eats up time, money, and innovative energy.

The suffocating action of ossified patent offices is further aided by a variety of factors that may be synthetically considered under the label "the state of the economy." By and large, vigorous, expanding economies stimulate the predispositions of combination by providing means for research and development and by offering promise of handsome profits. The opposite is true of sluggish, relatively stagnant economies. Some fifteen years ago, in the United States of America expenditure for research and development peaked at 3 percent of GNP. By 1980 it had slumped, with a slump in the economy, to about two-thirds that amount. Concomitantly, the yearly number of patents granted in this country has been decreasing in recent years by an annual average of about 20 percent.

Still, Herbert Spencer (1857b) singled out one of the chief factors of sociocultural evolution when he in effect stated the multiplier effect with the following explanation of his law of evolution: "Every active force produces more than one change—every cause produces more than one effect." Consequently, many if not all cultural selections may be expected to have a number, sometimes large, of repercussions. Imagine, for example, the reverberations of the variations that led to the discovery of the Americas, of the laying of the railroad across the North American continent (Mazlish, 1965), of the inventions of the steam engine and the automobile (Flink, 1970), and long before that of the invention of the wheel and of the bow and arrow.*

---

* In conclusion of this section, it should be noted that a more nearly complete theory of biocultural evolution will have to account for a number of other criteria of selection. One that may be of special importance concerns aesthetics. Just as some combinations are selected because they contribute essentially to bodily comfort, so others may be favored because they satisfy what Immanuel Kant called the feeling of the beautiful and the sublime. The possibility of a distinction between the adaptive and the nonadaptive within this context should represent a particularly challenging theoretical venture (for some especially useful statements on aesthetics, see Kant, 1763; Hegel, 1835; Tolstoy, 1898; Bell, 1916; Knox, 1936; van der Leeuw, 1937; and, in a psychobiological key, Berlyne, 1971).

7.4 Intra-Societal and Intersocietal Selection

This volume has stressed the point that in the human species natural selection operates at both the individual and the societal levels. In a biocultural key, we may now define individual, or intra-societal, selection as a process of variation and selective retention wherein individuals contribute differentially to the future sociocultural and genetic repertories of a given society. Group, or intersocietal selection, by contrast, refers to a process of biocultural evolution wherein, as we have already noted, societies make a differential contribution of both genotypes and phenotypes (both organisms and sociocultural patterns) to the future generations of world population and culture.

In putting the matter this way, I intend to expand the concept of fitness to embrace some such concept as "relative autonomy of sociocultural systems." Thus, some societies disappear in the sense that the individuals composing them die out without leaving descendants. The major technological revolutions of the past probably produced many examples of this process. Other societies disappear in the sense that they are more or less culturally absorbed by other societies. A number of pre-Columbian American societies are cases in point. As a third, perhaps more common, possibility, societies become extinct in part as biological entities and in part as autonomous sociocultural systems.

Glancing back at our evolutionary history, it may be reasonably conjectured that selection favored those societies that were first to invent or adopt such variations as the spear-thrower, the bow and arrow, the use of metal in work tools and weapons, the plow, the catapult, and so forth. To an extent the same logic applies to intra-societal selection; thus, biocultural selection in agricultural societies most likely favored those farmers who were alert to the economic, and thus fitness, benefits of crop rotation and such modern implements as the tractor and the threshing machine. My emphasis in this section, however, is on intersocietal selection, and from this perspective we may say that biocultural selection favors those societies that either compete successfully with other societies in producing and selecting combinations that are efficient for competition or at least feature a readiness to borrow such selections from their competitors through the process of cultural diffusion.

Another crucial variable operating in group selection has already been encountered under such titles as reciprocal behavior and ascetic altruism. Together, reciprocity and a pinch of ascetic altruism tend to reinforce social solidarity, mutual organization, pride of one's own traditions, and therefore a readiness to engage in concerted action in defense of one another and those traditions. While this tendency is no guarantee of success in a world of probabilities and multiple unknowns, on the average it is a good safeguard against extinction.

The adaptiveness of variations at the level of intersocietal competition is perhaps most obvious with respect to those selections that may be termed technological. We have already been reminded by Ogburn of the population advantage accruing to England, as compared to France, from the Industrial Revolution. We may add that the English achieved a major advantage over the Germans with the invention of radar during World War II. More poignantly, during the same war, the United States of America not only gave us a glimpse of doomsday by massive destruction of Japanese lives with atomic explosions over Hiroshima and Nagasaki; it also won the war as if by the swing of a magic wand. Thus, the more adept societies are at technological innovation and application, the greater the probability, *ceteris paribus*, of their success in case of intersocietal struggle. This statement is a special case of Lotka's principle (known also as Volterra's Law) according to which natural selection favors those populations that, on the proviso of the availability of energy, are most adept at converting and controlling energy sources and forms (Lotka, 1922, 1925; for an excellent discussion of energy and evolution, see Adams, 1975).

What about "spiritual," as contrasted to technological, selections? Do they, too, tend to bestow evolutionary payoffs on societies? To an extent, what was said of technological breakthroughs may be said of these as well. For example, it has very often been written that Christianity destroyed the Roman Empire. The validity of this proposition is dubious. What is true about the relationship between Christianity and the fortunes of the Roman Empire is that Christianity established itself in the Empire at a rate roughly equal to the speed at which the Empire was becoming decadent. Indeed, one might argue that the rise of Christianity would have reinvigorated the decadent Roman Empire, had not the latter become simultaneously the object of the Barbarian invasions. It is not surprising, therefore, to find, almost everywhere in the records of medieval history, that whatever flicker of past civilization could eventually ignite what came to be called the Renaissance was kept alive precisely by monks and scribes of the Christian monasteries. Moreover, if Christianity may be credited with the Renaissance, this, in turn, may be held responsible in large part for the birth of modern science and technology, the discovery and settlement of the New World, the Industrial Revolution, and the Protestant Reformation within Christianity itself. Weberian excesses or inconclusiveness notwithstanding, aspects of the Reformation, in turn, gave a boost to modern capitalism, the making of the British Empire, the emergence of the United States of America as a global power, and the modern scientific revolution. The centuries that experienced those selections were fraught with warfare and wholesale destruction of mostly non-Christian peoples and cultures, especially among the preagrarian societies of the African, Australian, and American continents. In any case, if religion is capable of destroying an empire, it is also capable of facilitating the rise of another.

7.5 Conclusion

In conclusion, a word is in order about the relationship existing between the theoretical sketch presented in this chapter and other contributions to the emerging theory of evolution in the sociocultural system. My goal is neither an exhaustive review of literature nor a detailed account of the similarities and differences existing between my work and other endeavors. It is rather to highlight several features that are unique to my work (for some useful discussions or reviews, see Harris, 1968; Blute, 1979; Lumsden and Wilson, 1981).

Most existing theories are descriptive in nature and tend to emphasize the course or the stages followed by societal development (Blute, 1979: 47). Such a focus will be an important ingredient of a full theory of biocultural evolution. Shorn of the adaptationist and biogrammatic phases, however, it is grossly incomplete. Most existing theories, further, are not theories of *biocultural* evolution; with varying degrees of emphasis and explicitness they argue that the mechanisms of sociocultural evolution operate independently of natural selection (e.g. Childe, 1951; Steward, 1955; Greenberg, 1957; White, 1959; Sahlins and Service, 1960; Tax, 1960; Cohen, 1962; Parsons, 1964, 1966, 1971; Campbell, 1965, 1975; Dawkins, 1976; Lenski and Lenski, 1978).*

My own approach falls within the so-called model of variation and selective retention. Some of the main proponents here are Keller (1915), Campbell (1965), Ruyle (1973), Durham (1978), and Langton (1979). The principal feature of their theories is that they attempt to describe the mechanisms that produce sociocultural evolution by analogy to the Darwinian model of random variation and selection. Analogy aside, however, with the partial exception of Durham's work (1978), which relates sociocultural evolution to inclusive fitness (see also Cloak, 1975; Emlen, 1976, 1980; Richerson and Boyd, 1978), these studies stress processes that are sociocultural and/or psychological in nature, and neglect both the importance of a biogrammatic input and the possible relevance of the maximization principle in evolution. In short, they lack the genetic dimension.

A case in point is John Langton's stimulating paper. This scholar's (1979: 288) goal is to show that "the synthesis of an abstract, Darwinian model of systemic adaptation and the behavioral principles of social learning produces a theory of sociocultural evolution having the logical structure and explanatory power of the theory of natural selection presented in *On the Origin of Species*." Langton's argument, however, does not include

---

* For other, more general, statements on sociocultural change, which typically include "evolutionist" attempts, negative or positive, see Nisbet (1969), Lenski's (1976) critique of Nisbet's anti-evolutionist argument, Nordskog (1960), and Eisenstadt (1970), among many others. For a useful critique of nonbiologically oriented theories, see Granovetter (1979).

consideration of biogrammatic forces and genetic adaptation. Thus, after defining sociocultural selection by analogy to Darwin's principle of natural selection, he takes variations to be "favorable" or "injurious" depending on "whether they aid or handicap individuals in the struggle for satisfaction or reinforcement" (Langton, 1979: 302). In short, the theory stresses roughly what I have called the creature-comfort criterion of sociocultural selection and the pertinent principles of learning theory.

Nevertheless, Langton does introduce a dimension that, along with the previously mentioned emphasis on sociocultural stages, should eventually help to produce a more complete theory of biocultural evolution. As Marion Blute (1976) suggested in her judicious and informative critique of E. O. Wilson's (1975b) volume, a full theory of biocultural evolution will have to add facts and propositions from the epigenetic sciences—for example, embryology and the psychology of learning—to a sociobiological perspective. Langton's (1979: 297–9) work has the virtue of guiding attention precisely in this direction by pointing out, among other things, that the Darwinian "struggle for existence" is related to "the law of effect," which is an important principle of such psychological behaviorists as Thorndike, Watson, Skinner, and Bandura (see, e.g., Skinner, 1953, 1974; Bandura, 1969, 1977).

E. O. Wilson appears to have been himself quite keen about the necessity of introducing an epigenetic dimension into the theory of biocultural evolution, and his more recent work with C. J. Lumsden emphasizes precisely this aspect of the problem (Lumsden and Wilson, 1981). It is that work that, when all the strengths and weaknesses have been summed up, represents the first major attempt at a theory of biocultural evolution (for related but less developed efforts, see also Feldman and Cavalli-Sforza, 1976, 1979). As previously noted, the present volume was essentially complete when the Lumsden–Wilson book came off the press. A brief note of comparison between the two efforts may, however, be of some utility.

Aside from questions of technique—for example, the Lumsden–Wilson volume is highly mathematical, while the present volume is discursive—both are works on coevolution or the evolutionary connection between biology and culture. The fundamental difference of substance concerns the respective ways in which biology and culture are linked together. According to Lumsden and Wilson (1981: 343, 370), the "pivot of gene-culture theory is epigenesis," namely, the "processes of interaction between genes and environment that ultimately result in the distinctive anatomical, physiological, cognitive, and behavioral traits of the organism." There can be no disagreement with this point, and my broad concept of adaptation as the functional interdependence of organism and environment, along with my hypothesis of biocultural interplay, espouses, if it does not pursue, this perspective.

The search for "epigenetic rules," a concept similar to my "behavioral predispositions," leads Lumsden and Wilson to a very useful examination of the brain and the mind, where they argue that laws governing culture must be derived from the principles governing the mind (Lumsden and Wilson, 1981: 177). This point, I believe, is unassailable: however much culture may be anchored in genetic structures and processes, the influence of those is necessarily filtered through the brain and mind.

The problem arises when we consider that the exact rules of pattern formation in the human brain are still unknown, and consequently an epigenetic approach must for the time being rely largely on the possibility of inferring the general features of such rules from the results of epigenesis studies in other organisms (Lumsden and Wilson, 1981: 303). Again, I find nothing wrong with that position, and nothing in the present volume could suggest otherwise. In the absence, however, of more precise information on human epigenetic rules, an alternative course may also be partially useful. The present volume has in effect proceeded on the assumption that the human epigenetic rules, or behavioral predispositions, may be inferred from knowledge of sociocultural universals. For the moment, such an approach is bound to be as fruitful as the search for interspecifically relevant epigenetic rules. Eventually, however, both approaches will have to be checked against whatever direct knowledge we shall develop on human epigenetic rules.

Two more differences may be worthy of mention. First, the present volume is more daring, and doubtless more reckless, in extending the scope and the variety of behavioral predispositions: while incomplete, and in part possibly superfluous, the biogram discussed herein is fairly extensive by comparison to Lumsden and Wilson's reliance on epigenetic rules associated with such behavioral categories as color naming, mode of infant carrying, phobias, facial expressions, and incest avoidance. Second, in the search for a theory of biocultural evolution, the present volume has attempted to isolate specific behavioral predispositions that are ultimately responsible for biocultural evolution, namely, the combiners and the selectors, which will be specified further in the next two chapters.

One last comment in conclusion of this chapter. Analogies for me are tools with which to attempt a theoretical excursion from what is known to what is unknown. They are heuristic devices pure and simple whose basic function is to suggest hypotheses in relatively uncharted areas. As tools of demonstration, however, they have not the slightest value. I have, therefore, earnest keenness regarding the imperfect, at best partial, and tentative nature of the model of evolution presented above in part by analogy to Darwinian theory.

8 Behavioral Predispositions and Religious Behavior

This chapter initiates discussion of specific predispositions of variation and selection that ultimately give rise to, sustain, and transform the institutions within which the behavioral forces of self-enhancement and sociality work themselves out. Working at the level of a first approximation, my intent is not to isolate a comprehensive set of such predispositions nor to deal with all major social institutions in all their evolutionary, behavioral, and organizational complexities. My aim is rather to underscore the joint action of a few combining and selecting forces within certain institutional settings that have of late been of special interest to human sociobiologists. Certain aspects of religion and the family are thus major candidates for our attention. The latter institution and its underlying predispositions, moreover, will in part be viewed as the nucleus of such evolutionary offshoots as ethnicity and ethnocentrism. These topics will be approached in the next chapter. The present chapter will tend to religion; to a degree, however, it will relate also to knowledge in general, or *scientia*, which is intimately connected to religion. For reasons that may concern my latent prejudices or lacunae in my theoretical reach, the maximization principle will continue to play a part, but to an attenuated degree.

8.1 Explanatory Urge; Relational Self-Identification

In approaching matters of religion it will be useful to return briefly to Chapter 6 and the discussion therein of the evolution of the soul; in part, the present chapter is an extension of that earlier one. In that connection it was argued that: (1) the soul was a cultural by-product of the natural selection of self-deception; (2) the "exploratory drive" (a term employed in Chapter 7 to describe the combinatorial forces in general) aided the work of self-deception by turning the mind upon itself; and (3) the emergence of the concept of the soul was associated with a process of self-identification that incorporated the social being of significant others in one's own social self. We can now retrieve some of those notions and elaborate them for inclusion among the combiners and selectors.

The first of the combinatorial forces to be considered is the *Explanatory Urge*. The "quest for certainty," a term employed by some scholars (e.g. Dewey, 1929), or the "need for logical developments," a closely related notion (e.g. Pareto, 1916), might be equally acceptable labels. They are, in any case, affinal concepts. They all relate to a tendency to find explanations for the conditions, observations, and experiences of the human existence. The tendency toward self-identification by reference to others will be subsumed under the selective forces with the label *Relational Self-Identification*. Starting with these two behavioral forces, I aim to strengthen the previous argument on the origin and evolutionary import of the soul and to achieve a biocultural grasp of religion, at an elementary though fundamental level, in part through the gradual introduction of additional predispositions of variation and selective retention.

One of the most unique traits of *Homo sapiens* is that we not only live and accept the fact of our existence but we are also profoundly conscious of existence as a *problem* demanding special attention and explanation. The cosmic forces, the recurring mystery of success and failure, the awesome cycle of birth, life, and death, the formidable challenges of the environment, the tenacious authority of traditions, the tremendous complexity of the creation—these and endless other facts seize the imagination and meet with a curiosity that in the long run increases *pari passu* with the volume of explanations already produced. The explanatory urge, thus, represents the evolutionary context within which what Weber termed "the problem of meaning" arises and is in a fashion resolved. The urge is inexhaustible. It is always looking for causes: the causes behind the Creation, the purposes for which human life was ordained, the causes of good and evil, the causes of tides, of capitalism and communism, of happiness and misery. When it does not find causes, it invents them. Hence, the force here in question is the origin of all systems of knowledge—scientific, literary, artistic, and philosophical as well as religious, theological, and folkloric. Indeed, one could assert that, were it not for theology, experimental science would not even exist, or vice versa. For those kinds of mental activity are manifestations of one or more closely related predispositions in the absence of which they would both vanish simultaneously.

The explanatory urge underlies religious (and magic) behavior as well as scientific conduct not only in the intrinsic sense that it provides the very same thrust to inquire into the unknown but also in the indirect sense that the scientific enterprise creates skepticism about sacred principles at the same time that it enfolds itself inevitably in a paradox: the more we invent, discover, explain, the more enormous becomes the ever-expanding periphery of the scientific domain. Thus, the more we know, the less we know, in the sense that we know that there is ever so much more to learn about. This is another way of expressing Kuhn's (1962) previously encountered notion of the anomaly in scientific discovery. As a consequence, the sphere of the unknown and the uncertain grows in direct proportion to that of the known and the verifiable. In a sense, then, science tends to undermine its seeming dominance by fatiguing the human need for certitude.

Most striking perhaps is the fact that, as noted above, human beings have evolved to have consciousness of the fact that we, too, are part of the existence, and that, moreover, we have the actual capacity to explain and to an extent control the existence, ourselves included. Herein lies the tendency, again previously alluded to, to find explanations of events and observations not only in natural forces and transcendental fictions but also, and perhaps especially, in the human will itself. Thus, while extremely useful in giving social scientists clues to the multifarious causes of human behavior, the explanatory urge is often also our most crippling nemesis. For, at bottom what people want is to think. It would almost seem that thinking, explaining, has become an end in itself. Unavoidably, many explanations are fabrications pure and simple. People are specialists in the art of covering behavior that is rooted in nonlogical, phylogenetic, forces with layers upon layers of sophistic logic. There is a clue here to one possible origin of self-deception: this unique human trait may very well be a by-product of thinking become an end in itself; hence, it may be a fundamental property of the explanatory urge.

Rightly then are "social illusion, superstition and socially conditioned errors and forms of deception" (Scheler, 1926) considered important components of the subject matter of the sociology of knowledge. Scheler divided knowledge into two major categories on the basis of their degree of artificiality and susceptibility to change. One consisted of "natural *Weltanschauungen*," or worldviews, namely, the cultural axioms of a group which are the (ontogenetic) bases of all knowledge and, like our cultural universals, change very slowly over time. The other comprises the much more unstable and "artificial" worldviews. He singled out various subtypes of this latter category, hypothesizing that as artificiality increases, so does the facility of change in time. Myth and legend are the least subject to change, whereas technological knowledge is the most susceptible to alteration over time. Between these two extremes may be found, in ascending order of variability, natural folk language, religious knowledge, mystic knowledge, philosophic-metaphysical knowledge, and positive or scientific knowledge.

Despite the seeming emphasis on social illusion and socially conditioned errors, Scheler came to the conclusion that all types of knowledge derive from certain natural urges or drives: for example, the urge to relate to an overpowering reality, which constitutes the search for religious knowledge, and the striving for dominance over nature and over men, which allegedly underlies the search for science. Note that in an area of sociology which would seem to be furthest removed from any evolutionary theorizing we find one of the founders, Max Scheler, in effect equating changeability with artificiality, and placing myth and legend among the least artificial forms of knowledge. That would seem to suggest that if we wish to grasp the foundations of human knowledge in ways that span large arcs of space and time, we need to examine those forms of knowledge which, being perhaps most directly linked to biological urges, are relatively unchangeable, like myth and legend, *even if it is precisely here that illusion or self-deception is at its highest, and the work of sweeping away the camouflage the most difficult.*

The great end of life, Thomas Huxley remarked, is not knowledge but action. That applies to humans as well as other forms of life to a greater degree than is normally recognized. Men first act and then think, Sumner argued. Still, with *Homo sapiens* the issue in question is not an either-or proposition. Knowledge, like action, *is* one of the great ends of human life. In a recent article Isaac Asimov (1980) asks, "What's fueling the popular science explosion?," and his answer stresses our desire to know whether technology and technological knowledge are about to overwhelm us. There may be an element of truth in that, but there are various sorts of fuels in this as in other flare-ups. One of them, the explanatory urge, is currently in a frenzy under the heightening action of systemic immanence and the multiplier effect. The more we know, the more we wish to know. That we might also wish to know specifically what technology will do to us is largely incidental, and itself part of the rationalistic varnish that the explanatory urge produces in great quantity. The proof lies in the fact that scientists themselves sometimes seem to act as if they had been bitten by the tarantula. Some scientists, for example, take a guess at the total number of protons and electrons existing in the universe. Others play with the decimal system to come up with numbers containing googols of zeroes.

I do not wish to suggest that such an extremely active explanatory urge is devoid of adaptive value. Up to a point at least—short of random thinking, which is not implied here—a vigorous need to think and explain reveals a mind that is diligently at work in seeking useful solutions to existing problems. To put it otherwise, a hyperactive explanatory urge generates a wealth of hypotheses about environmental challenges, and the probability of adaptive solutions is probably a direct function of the wealth of hypotheses generated. One would have to guess, for example, that the difference between underdeveloped societies and economically more advanced societies in their respective proficiency at producing economic solutions is in part a function of the greater intensity of the explanatory urge among the more affluent societies.

Still, while the probability of finding solutions to problems may be a function of the number of hypotheses or combinations proposed by the explanatory urge, it may be equally true that waste, "noise," pure random and useless behavior are functions of the same predisposition. Thus, the recurrent crises of industrial economies are not unrelated to a rate of theoretical production that verges on chaos. In the arts and sciences, and all the more so without, an excited explanatory urge expresses itself in endless garrulity. As the need to explain grows, the tongue wags insistently. Outside the sciences, witness the prattle, the banality, the "coachly" patois of television sportscasters in the art of "explaining" what is only too obvious to any spectator. Consider the plethora of happy-talk shows on radio and television where hosts and their guests are authorities on subjects A-to-Z. And one must not neglect to mention the spectators or listeners who call in to register with great relish their own opinions. Ironically, at close examination it is all too often rather clear that no one is listening, and few are saying much of anything in the first place. To talk, to explain, verges on being an end in itself. As Peter Farb (1973) puts it, evolution has created "Man the Talker."

Man the Scanner of Divine Portents, too. Everywhere we turn nowadays, we see or hear men and women of God who dip into the Bible with the greatest of ease, and with the help of a few words or a parable, explain everything under the sun—anything from headaches, the atomic revolution, homosexuality, infidelity, the ill behavior of children, and the assumed connection of temple stones in Jerusalem to the allegedly imminent Second Coming. Such people abhor, and yet are kindred souls of, the likes of the pious woman who in a New Mexico town will point to a tortilla and proclaim: "I believe this sign means that Christ will return to earth." The "sign" is a configuration burned on a tortilla cooked several years ago. One can make anything of the burned spot, including the mournful face of Jesus Christ crowned with a wreath of thorns. At the latest popular report, two or three years ago, the pilgrims flocked by the thousands to see "Jesus on a Tortilla."

8.1.1 ON THE SOUL AGAIN

Returning now more directly to the previous discussion of the soul, it is not unreasonable to suppose that, under the pressure of the need for social approval, the explanatory urge could operate in such a way as to sublimate, to an extent, the immortal tendency of the gene into the moral fiction that has been termed the soul. The scenario in principle is simple. All that is needed is the development of the concept of immortality.

I have already suggested two complementary ways in which this conception could have come about. First, there is every reason to believe that the mind has evolved to mimic many of the very genetic processes that it coordinates. Thus, as consciousness and the explanatory urge grew intenser with the development of the neocortex, the immortal tendency of the gene could have echoed in the mind as a sense of *personal* immortality.

Increasing consciousness promoted introspection, and this in turn may have led to what might be termed a primeval psychology of human biology. Second, the social-relational nature of the self would have supported this possible development by providing the notion that the deceased survived in a noncorporeal sense after the death of the body. The noncorporeal entity came to be conceived as the soul. This is another way of saying that the explanatory urge generated the idea of immortality, by combining genetic with mental processes, while the relational self-identification selected it for retention. At least one of the criteria of selection aided in this selection, namely, the compatibility of the idea of the soul with the forces of sociality, especially the need for social approval.

The connection between relational self-identification and the concept of the soul may be pursued a little further. With few if any exceptions, death has been viewed as but a transition to a different state of being in an after-life world. "The brute fact of physical death," Robert Hertz (1907–9: 81–2) argued, "is not enough to consummate death in people's mind." The image of the deceased continues to be part of the order of things of this world. For some, depending in part on age and degree of relationship, the image persists only a short time and then gradually disappears. For others—and we are all one and the other in relation to different deceased—the persistence is a lifelong affair. We have put too much of ourselves into the deceased, and too much of them by necessity remains in us. Such ties are not easily severed. "The 'factual evidence,' " Hertz adds, "is assailed by a contrary flood of memories and images, of desires and hopes."

Peter Marris (1974) has cogently suggested that the persistent attachment of the living to the dead is a part of a larger basic "need to sustain the familiar attachments and understandings [that] make life meaningful." An examination of such varied subjects as widowhood, political revolutions, the effects of urban renewal projects on the sense of neighborhood, and the activities of tribal associations in Nigeria leads Marris to argue that unwanted but radical change of any kind brings a devastating and long-lasting sense of loss conveyed in feelings of grief and bereavement. We cope with it in part through a tendency to deny the loss.

The loss of someone whose social being is intricately intertwined with our own threatens to rattle and disorient our very sense of being—our social self. A marvelous character in Graham Greene's *The End of the Affair* (1951) reminisces: "A week ago I had only to say to her 'Do you remember that first time and how I hadn't got a shilling for the meter?' and the scene was there for both of us. Now it was there for me only. She had lost all our memories for ever, and it was as though by dying she had robbed me of part of myself. I was losing my individuality. It was the first stage of my own death, the memories dropping off like gangrened limbs."

This is grief at its peak, and in this state of the emotion death is experienced as truly contagious. The living self has only recently been wounded, and the wound seems fateful. But there is medicine for the disease of grief, too, and its name is *mourning*, a universal phenomenon. In his classical discussion of funerary rites, van Gennep (1908) argues that mourning is a transitional period which the survivors enter through rites of partial separation from society and quit through rites of reintegration into society. The duration of the mourning depends on the closeness of the preexisting relationship between the living and the departed: in general, the more intense the relation, the lengthier the period of mourning. In some cases, reintegration of the living coincides with the alleged incorporation of the deceased into the world of the dead. During mourning, the deceased and survivors both are suspended between the two worlds. Neither is socially whole.

Mourning may accordingly be viewed as a period of social-psychological convalescence during which the injured part of the "me" is allowed full freedom of grieved expression. The process amounts to a degree of resocialization and thereby helps toward a complete return to the significant circle of the survivors. Any variations that facilitate this behavior tend to be favored by relational self-identification with the aid of the criterion of creature comforts; for there is no greater feeling of well-being than the relative sense of comfort experienced on the way back from a state of morbidity.

It should perhaps be added that whether mourning is highly ritualized social behavior, as is the case in many societies past and present, or constitutes merely personal experience is to an extent a function of the varying complexity of societies and of the size of one's reference groups. Thus, mourning is most ritualized, and perhaps most intense as personal experience, in traditional societies. In general, the greater the intensity of kin selection, the greater the intensity of mourning, for the social self is deeply rooted in one's relatives, and their death is thus a major jolting experience.

Conversely, in highly complex societies, the high rate of movement across space and social boundaries is associated with a certain estrangement from family ties. As a result, mourning is frequently an experience that would seem to vanish with the conclusion of the funeral service if not as quickly as the completion of a telephone conversation. If, however, mourning has the therapeutic value attributed to it just above, then the seemingly cavalier fashion with which nowadays we often receive the news of death is likely to be one basis of the loneliness, the despair, and the personal disorientation that are the hallmark of today's civilized people. Relative estrangement due to structural impediments does not necessarily undermine the emotive ties of kinship that early formed a major part of our social self. In the words of a psychoanalyst, "as long as the early libidinal or aggressive attachments persist, the painful affect continues to flourish, and vice versa, the attachments are unresolved as long as the affective process of mourning has not been accomplished" (Deutsch, 1965: 235).

What we have been saying helps to better grasp a number of other facts. In his survey of experiences with the spirits of the dead, Swanson (1960) finds that, with the exception of the great leaders and medicine men, the dead return, whether through dreams or apparitions, only to their kin and to those who have intimate relations with these; most are also recently deceased. That is to say that they return along the networks of relationships that constitute the core of living selves. Furthermore, ancestral spirits continue to relate actively to their surviving relatives in the sense that, as Swanson (1960: 98) notes, it is "the purposes and influence of the dead in their role as relatives which persist among the living." Finally, Swanson finds that a belief that ancestral spirits are active in human affairs is strongest where kinship groups are larger than the nuclear family, for example, in clans. In such extended kin groups, kinship is typically established by virtue of the fact that the members of the group have a single name or totem representing them all (e.g. Durkheim, 1912; Sahlins, 1976). Thus, kin selection is closely related to reciprocity selection, and such "spiritist" beliefs would seem to be adaptive by reinforcing the reciprocal basis of kindredness.

We can thus conclude that the selection of the soul and such related variations as rites of mourning by the predisposition of relational self-identification is supported by nepotistic and reciprocal favoritism. That is another way of saying that the dead influence positively inclusive fitness through their support for networks of nepotism and reciprocity. To the extent, however, that, as Swanson showed, the soul of the dead is most active where kinship is a social as well as (or rather than) a biological fact, the selection of the soul is not only harmonious with the maximization principle. It also performs a function in favor of the evolutionary tendency in human societies to develop through stages wherein networks of relationships and moral horizons become increasingly complex.

8.2 The Denial of Death

Relational self-identification is the primary selector of the idea of the soul. But other selectors have come to its aid. Of special importance is what has severally been termed "the quest for immortality," "the denial of death," and so forth. Chapter 6 has already established that the belief in and the quest for spiritual immortality are universal phenomena. They are proof enough that people countenance neither the idea of mortality nor the idea of an unhappy existence beyond the tomb. This is what is meant here by the *Denial of Death*.

The denial of death has been applied to good theoretical use in social science. It may be recalled, for example, that it is a crucial element in Weber's theory of the relationship that developed between the Protestant ethic and modern capitalism beginning in the first half of the seventeenth century. As Weber saw it, the uncontrollable need to know whether one had been predestined to eternal life or to eternal death helped to select the cultural fiction that wealth was a sign of the divine bestowal of eternal grace. Details aside, one cannot easily confute Weber's emphasis on the animistic need to know. I am reminded of Mrs Thompson's predicament in Samuel Butler's *The Way of All Flesh*. Poor Mrs Thompson is dying, and, terrified of hell, she pleads with Theobald, her minister, to save her from it. "Oh, Sir, save me, save me, don't let me go there. I couldn't stand it, Sir, I should die with fear, the very thought of it drives me into a cold sweat all over." Theobald reassures with the counsel to have faith. "But are you sure, Sir," says she, looking wistfully at him, "that He will forgive me—for I have not been a very good woman, indeed I haven't—and if God would only say 'Yes' outright with His mouth when I ask whether my sins are forgiven me. . . ."

While the task does not really fall under my purview, it will be interesting to hypothesize briefly about the selection itself of the denial of death. One possibility takes us back to the evolution of referential self-identification. From that perspective we have really taken a backward-looking stance. That is, we have related the living, representing the present, to the dead who represent a past point in time. But the persistence of relations across the line of physical death implies also that today's living will be tomorrow's dead. Thus, if yesterday's dead continue to live in a spiritual sense, tomorrow's dead shall not perish either. The denial of death would, therefore, seem to be a corollary of relational self-identification. Its evolution may have been facilitated by several selection criteria. One could have been the criterion of creature comforts; after all, self-destruction for an introspective organism that is programmed to do all it can to survive and enjoy life, must be a dreadfully uncomfortable feeling. Another may concern the predispositions of sociality. To die is not only intrinsically a nasty affair in itself; for a self-aware social being, death without a future would be awful also because one would cease thereby to be social—one would lose the support and the pleasure that derive from group living. The denial of death, thus, adds to the concept of the immortal soul the explicit notion that death is but a reunion of the generations and the peer groups, namely, the reconstitution of this world's society in the other world. Those ancestors who took the lead in expressing such ideas met with credibility in the group and with fitness-enhancing recognition.

A second, probably complementary, possibility for the selection of the denial of death returns us to the notion of mental mimicking of genetic processes that has already been relied upon in previous arguments. We first encountered the idea in Chapter 6. As consciousness evolved, the genes that programmed the organism to maximize the probability of their survival into the future generations, may also have given rise to a mental counterpart that sought personal immortality. But since there was the brute fact of physical death, as Hertz earlier put it, the wish for immortality could only have been associated with that inner, spiritual, fiction that is the soul. The present predisposition would, thus, seem to be related to the selection of the soul itself.

Ernest Becker (1973) wrote a provocative book on "the denial of death," in which he argued that "of all things that move man, one of the principal ones is his terror of death." While cautiously steering away from a rich debate (see, e.g., Zilboorg, 1943; Levin, 1951; Feifel, 1959; Choron, 1963) over the question whether the fear of death is the basic human anxiety, Becker strongly inclined to the non-Freudian position that anxiety over death is even more basic than sexual anxiety. I think the proposition can be defended. But in a sense the defense is part of the question itself. However deeply rooted in our brain, anxiety is a property of feelings and emotions over which we have a high degree of awareness. We have an intense awareness of physical death, which in principle threatens total and perennial annihilation, and anxiety is accordingly acute. To be is the atrocious problem of being, for, like the shutter in a camera, it peeps at light only to return to the steadier state of darkness. Conversely, to the extent that we have awareness of a sexual problem, the anxiety that goes with it is often associated with the feeling that what is a famine today may be a feast tomorrow. To be sure, the maximization principle inclines hard toward some notion of sexual anxiety, but mating patterns in our species suggest that it is easier to rid oneself of sexual anxiety than of anxiety over death.

The problem with the question at issue is that, from another perspective, the anxiety over sex and the anxiety over death are but two phases of the same problem. If sex is the mechanism through which our genotype succeeds or fails to achieve immortality, sex and death cannot be separated; nor can the respective anxieties. Freud and Becker were both right. Together they descried the double edge of the perennial question.

8.2.1 FURTHER CONSIDERATIONS

In a discussion of a sentiment termed "persistence of relations between the living and the dead," Pareto (1916) argues against those who believe that honoring and worshipping the dead—for example, through funerals, banquets, commemorations, sacrifices—presupposes belief in the immortality of the soul. Such phenomena, he contends, are only superficially connected with the belief. How else explain the fact that "materialists are not less punctilious than others in honouring their dead, in spite of their philosophy"? There are even such things as cemeteries for cats and dogs among people who do not credit these animals with an immortal soul. We could add, among various other things, that a certain percentage of Christians disclaim any belief in another world at the same time that they are practitioners of the faith (e.g. Lopreato and Hazelrigg, 1972). Nor is it difficult to find, among Catholics for example, individuals who never attend church, disavow any belief in the soul, and yet are religiously faithful in saying their prayers at special times of the day or the year. They often explain the apparent incongruity with such sophistries as "My dead parents would want me to do it," "It doesn't cost me anything, and I honor the memory of my dear departed ones." The ancient Semites had a proverb according to which to speak the names of the dead was to make them live again. As a minimum, to honor the dead is to continue relating to them as significant individuals.

Pareto, therefore, would seem to be on reasonable grounds when he argues that "the concept of the survival of the dead is at bottom merely an extension of another notion which is very powerful in the human being, the notion that the individuality of a person is a unit over the course of the years." Still, one senses in all this an extreme one-way reductionism, that, on the surface of it at least, amounts to a denial of a biocultural interplay. The belief in the immortality of the soul can hardly be excluded from the etiology of the cultural practices in question. The proof lies in the fact that if other animals do not possess, as it is reasonable to assume, the notion that personal individuality is a unit over the course of time (which would seem to synthesize my predispositions of *relational self-identification* and *denial of death*), it is very likely that the absence is associated precisely with the lack of a belief in the immortal soul. One cannot exist without the other. The soul is a cultural fiction, and any predispositions that operate in connection with the fiction can hardly be causally disconnected from the fiction itself and the practices and attitudes surrounding it.

8.2.2 FLUCTUATION IN THE INTENSITY OF PREDISPOSITIONS

If we now consider fluctuations in the intensity of relational self-identification and the denial of death, we may expect to find some noteworthy facts to be associated with them. When, for example, the predispositions are strongly held, we have acute consciousness of the idea of death, but we are at the same time relatively undaunted by it, for death is perceived as a mere passage from one state of life to another. "Death loses its sting when it is not conceived of as the complete end of the self " (Lessa and Vogt, 1972: 467). The final day is a natural, if dramatic, extension of the days already gone by. Hence, it too has its own dawn. Death becomes a transfer of activity from one set of relationships to another, although in many societies there are varying degrees of apprehension and fear regarding the spirits of the dead (Frazer, 1933–6; Lessa and Vogt, 1972: 467).

Likewise, the capacity to "see things" is aroused by potent states of the predispositions in question. It is as if keenness to the reality of the deceased raised a form of transcendental vision. The dead actually materialize in the mind of the living. Ghosts, then, are in one sense real: they are the tangible forms assumed by the predispositions in question. By analogy and extension, the same may be said of apparitions of deities, angels, fairies, devils, and the like. Again, it is probably safe to hypothesize that, while there is a great deal of fakery and fraud associated with seances and related phenomena, at least some of the claims made in favor of such practices may be true enough in the experiences of the participants. It certainly should be no surprise to learn that, according to a recent Gallup poll, 54 percent of the adults in the United States of America believe in angels, and about 40 percent believe in devils. One in nine believes in ghosts, and two-thirds of these (about 7 percent of the total) say that they have even seen one (*Newsweek*, 1978). Another recent survey from the Los Angeles area pinpointed a much larger percentage, a total of 44, of people who claimed encounters with others known to be dead (reported in Siegel, 1981: 65).

Relations with the dead and personal experiences with them may be expected to be particularly common and intense among preagricultural peoples (e.g. Williams, 1930; Frazer, 1933–6; Lessa and Vogt, 1972), where family relations and reciprocal behavior are "mechanical," or more purely instinctual, aspects of social solidarity (Durkheim, 1893). The best proof lies in the fact that among many such peoples death is conceived not as the occurrence of a moment but as a process that sometimes lasts numerous years. Hertz (1907–9: 29f.), for example, relates that among certain Indonesian peoples, such as the Olo Ngaju of Borneo, death in some cases does not become final in the social conscience of the living for as long as ten years after its clinical occurrence. In the meantime, the survivors pay homage to the departed through a variety of ceremonies that culminate in the final rite of the "tivah," at which time the body is permanently disposed of. Until the tivah is celebrated, and the soul moves to the celestial city of souls, the soul is believed to roam in the vicinity of the body. As a consequence, both the living and the dead have an extended period of time during which to adjust to the dramatic changes in social relations and behavior brought about by clinical death.

Everything said so far about relations between the living and the dead has in effect been filtered through the predispositions of sociality. With the social self as our starting point, it could hardly have been otherwise. Still, this perspective is bound to belittle another side of the relations in question. Across human society there is a fairly wide range of attitudes and beliefs regarding the behavior and power of the dead in relation to the living. In some places and at some times, ghosts, even those of ancestors, are evil and dangerous; in others, they are benign and beneficial. To recognize, or to believe, that the dead have their cruel and dangerous aspect is, of course, to reveal that the forces of self-enhancement are quite relevant to the religious sphere, too. As a result, the more interested the dead are believed to be in the affairs of the living, and the greater the belief that propitiation can evoke their benign side, the more prevalent is ancestor worship (Lessa and Vogt, 1972).

In Chiricahua and Mescalero Apache cultures, Morris E. Opler (1936) found ambivalent feelings toward the dead: on the one hand, mourning and grief and, on the other, dread and possibly even hatred. He cogently explained such ambivalence in terms of the nature of interpersonal relations existing among the living, the nature of their power structure, and the fear of sorcery and witchcraft. The behavior of the dead mirrored to a large extent the behavior of the living and the fear of one another's power, even within the kin group, in the context of a hostile environment. A similar finding and an essentially like explanation are offered by Melford Spiro (1952) in a study of the inhabitants of the small atoll of Ifaluk, where dread and hatred, however, are focused on only one type of ghosts: the malevolent *alusengau*. These are alleged to be (doubly) fictitious beings who conveniently absorb much of the redirected hostility of the living.

One of the most enlightening reports on the topic comes from Elizabeth Colson (1954) whose study of the Tongas shows that an individual releases several spirits at death. Some are evil; others are benign—an idea, we might add, that fits marvelously the sociological concept of the self as a compound of social beings. What is perhaps more remarkable is that the number and kinds of spirits possessed by a given deceased are determined by the social position occupied in life. In short, there is a sort of sociometric concept of the dead among the Tongas that is well documented in social science. By and large, the higher the social position, the greater one's popularity, and thus the more complex one's set of social relations.

In general, whatever the perceived nature of the ancestral spirits, they persist not only as dimensions of particular survivors but also as social forces acting on the solidarity of the living group and on the discipline of its members. That may be said of the Opler, Spiro, and Colson studies as well as of others (e.g. Bunzel, 1932; Simmons, 1942; Middleton, 1960). Thus, among the Lugbara of East Africa, where the dead are not objects of general fear, the deceased represent the ultimate sanction for proper lineage behavior. For example, they may be invoked for arbitration when the authority of the lineage elders is challenged (Middleton, 1960: ch. 2).

A phenomenon closely associated with ghosts, and much in fashion nowadays in modern society, concerns so-called after-life experiences, namely, experiences reported by individuals who in some cases have been declared dead, and yet have managed somehow to come alive. Personal reports and studies by various kinds of scholars abound (see, e.g., Kübler-Ross, 1974; Siegel and West, 1975; Noyes and Kletti, 1976; Toynbee *et al.*, 1976; Moody, 1978; Siegel, 1980, 1981). Moody (1978) has compiled an inventory of typical after-life experiences that includes: ineffability (or inability to express the experience in words), a feeling of peace and quiet, a loud ringing or buzzing noise, floating out of the body, hearing doctors pronounce the death verdict, a glowing light in a human shape, a panoramic view of one's own life, visions of great knowledge, cities of light, a realm of bewildered spirits, and meetings with guides, especially the spirits of dead relatives and friends.

R. K. Siegel (1980, 1981), an expert on hallucinations, has argued that "these deathbed visions are retrieved memory images (or fantasy images)," and found a striking similarity, indeed a virtual identity, between the near-death descriptions given by resuscitated people and the descriptions offered by persons experiencing drug-induced and other forms of hallucinations. There can be no doubt, from my perspective, that seeing and experiencing other-worldly beings and events are associated with mental states that may be termed hallucinations. When, therefore, I refer to an intensification of relational self-identification and the denial of death, I might just as easily add that such intensification may reach the point of frenzy, even hallucination. Indeed, we shall return to this context later in the chapter in connection with what will be termed the *imperative to act*.

Siegel fails to go beyond the ontogenetic level of analysis. While nothing in his reports denies the possibility of behavioral predispositions associated with hallucinations, his analysis is concerned only with more or less proximate factors, for example, visual sensations and electrical excitation of groups of cells in the visual cortex of the brain. The emphasis, therefore, is on the reaction of the optical apparatus to stimuli acting upon it. This leaves unexplained the fact, among others, that near-death experiences are not constant in time and place. The recent effervescence of American reports on the topic in question, for example, coincides with a heightening of the religious consciousness (Glock and Bellah, 1976), and that suggests an interplay between biological processes and cultural parameters, a fact which in turn suggests genetically predisposed behavior. Still, to the extent that drug-induced hallucinations are similar to the hallucinations of those who have "returned" from the dead, they may throw light on the biochemical processes that are associated with seeing ghosts under the influence of the denial of death and relational self-identification. It is probable that such predispositions reach an hallucinatory stage at the moment of dying.

We have been looking at experiences associated with a *heightening* of the predispositions of relational self-identification and denial of death. When, conversely, these forces become too weak, numbers of phenomena indicating a dread of death make their appearance. Mortality becomes an obsession, for the other world offers a void, annihilation, rather than the continuance of meaningful relationships. Modern people may suffer acutely from this affliction. The swift procession of new relationships in a rapidly changing world thrusts the individual in a continual existential crisis. The sense of continuity that in traditional society flowed from the stable manner in which we experienced our ties to others has been weakened; the reality of yesterday does not emerge as the significance of tomorrow. Indeed, yesterday has little or no bearing on tomorrow. The future becomes unknowable and capricious.

The public atmosphere in modern society is pervaded by an acute and loud hypochondria. Broadcasts, newspapers, posters, environmentalists, the good concern of entertainment stars, medical tracts, government reports, and a rapidly growing library of books warn us against all sorts of killers in the environment, and would even teach us "how to survive modern technology." Save your breath, do not waste a life or a mind, beware of VD, watch out for cancer, take care of those little cells in your ears, have your blood pressure checked—it is enough to scare us stiff. And scared we are. The skyrocketing rates of suicide attest in part to that. As Durkheim (1897) noted, the intense fear of death makes life unbearable: an endless vigil to the dreaded hour. Many cannot bear the wait. They get it over with quickly.

By the same token, as the predispositions in question weaken and the fear of death becomes debilitating, the need to believe in the Eternal and the other world is probably triggered off anew. Little wonder that as many nonbelievers approach what Malinowski termed the supreme and final crisis of life, they turn to God. The future becomes more consequential than the past and the present. The older people are, the more religious they are. Indeed, at times of crises in general, many nonbelievers come to feel like the Miriam of Nathaniel Hawthorne who whispers to Hilda, her devout friend: "I would give all I have or hope—my life, oh how freely—for one instant of your trust in God!" (Hawthorne, 1966). Men and women of the cloth are probably right: in the clutch there are few nonbelievers.

8.3 Symbolization; Reification

We have reached a point where a definition of religion, however tentative, can no longer be delayed. In the process, we shall be moving toward two new predispositions: *Symbolization* as a combiner and *Reification* as a selector. A few brief references to literature will be helpful as a beginning.

The anthropologist Clifford Geertz (1973: 90) defines religion as "(1) a system of symbols which acts to (2) establish powerful, pervasive, and long-lasting moods and motivations in men by (3) formulating conceptions of a general order of existence and (4) clothing these conceptions with such an aura of factuality that (5) the moods and motivations seem uniquely realistic." The definition is probably a bit too general; it seems it could just as easily refer to a number of other emotive-cognitive systems, including science and art. The reader will, moreover, note Geertz's consistent bias toward the position that it is symbols that establish, or determine, moods and motivations. Conversely the position of the present volume requires a systemic or mutually dependent relationship between the symbols and the "moods." Still, the emphasis on symbols does deserve attention. As Chapter 6 stressed, while there is no clearcut separation of the cultural and the biological, the cultural or symbolic dimension of social behavior plays a crucial role in the realm of religion and morality in general.

On the other side of the culture–biology theoretical axis, sociobiologist Wilson (1978: 176) emphasizes individual–group ties and group survival in his conception of religion as "above all the process by which individuals are persuaded to subordinate their immediate self-interest to the interests of the group." Wilson's approach is fairly congenial with the perspective of a number of other influential social theorists. While Pareto, for example, considered religion as too complex a cluster of facts to allow a useful definition, he (1916: 1854) nevertheless succumbed to the temptation of thinking of it as a set of phenomena "which correspond to sentiments of discipline, submission, subordination." The subordination is not only to persons, as the predispositions of hierarchy would suggest, but also, indeed especially, to symbols of group welfare. No concept of deities is necessary to grasp the essence of religion. The emphasis is on intensity of sentiment and activity in relation to group phenomena. Thus, Pareto (1916: chs 12–13) could think of revolution as a mechanism that enhances the probability of group survival by reinvigorating the leadership with persons strong in those behavioral predispositions that have been termed "the foundations of society," namely, with people who are fairly low in the social hierarchy and are energized, at least initially, by some of the least selfish of the sociality predispositions. He could add in the bargain that both a phenomenon like the French Revolution and one like the Protestant Reformation were alike revolutionary and religious in nature. "The Protestant Reformation in the sixteenth century, the Puritan Revolution in Cromwell's day in England, and the French Revolution of 1789, are examples of great religious tides originating in the lower classes and rising to engulf the sceptical [or faithless] higher classes" (Pareto, 1916: 2050). Of course, religions oscillate in the degree to which they engage individual and group discipline, and a decadence in religion ordinarily coincides with a decadence of society (Pareto, 1916: 1932). Hence the periodical religious effervescence and the revolution.

Perhaps the best-known definition of religion is the following by Durkheim (1912: 62): "A religion is a unified system of beliefs and practices relative to sacred things, that is to say, things set apart and forbidden—

beliefs and practices which unite into one single moral community called a Church, all those who adhere to them." While this definition lacks the biocultural flavor of Wilson's and Pareto's definitions, it too emphasizes individual subordination and group unity and at the same time adds explicitly the concept of the sacred that, as we shall see presently, is conceived of as a property of the society itself.

The culminating point of Durkheim's (1912: 236) theory of the *fundamentals* of religion is the argument that "The god of the clan, the totemic principle, can ... be nothing else than the clan itself, personified and represented to the imagination [symbolized] under the visible form of the animal or vegetable which serves as totem." Hence, religious symbols, the images worshipped through ritual, are transfigurations of the society itself (Crippen, 1982). From an evolutionary perspective, this conceptualization promotes a number of important considerations. Foremost is the idea that the evolution of human society has entailed the emergence of a mechanism, religion, through which the societal group makes itself vicariously the object of veneration on the part of its members. Durkheim saw in the recent history of his own French society the societal capacity to engage in self-apotheosis. Thus, he (1912: 244–5) added:

> This aptitude of society for setting itself up as god or for creating gods was never more apparent than during the first years of the French Revolution. At this time, in fact, under the influence of the general enthusiasm, things purely laical by nature were transformed by public opinion into sacred things; these were the Fatherland, Liberty, Reason. A religion tended to become established which had its dogmas, symbols, altars and feasts. It was to these spontaneous aspirations that the cult of Reason and the Supreme Being attempted to give a sort of official satisfaction.

Elsewhere he (1912: 475) asked rhetorically: "What essential difference is there between an assembly of Christians celebrating the principal dates of the life of Christ, or of Jews remembering the exodus from Egypt or the promulgation of the decalogue, and a reunion of citizens commemorating the promulgation of a new moral or legal system or some great event in the national life?"

A remarkable fact about Durkheim's theory is that it is probably the closest approximation to an evolutionary theory of religion. Equally conspicuous, however, are the facts that (1) the evolutionism is largely implicit, and (2) neither Durkheim nor other social scientists have raised the question as to the kind of animals human beings have had to become in order to be biologically predisposed toward the veneration of their society. What biocultural selections were necessary to produce beings who invented deities "in order to" revere their society?

We shall return very shortly to this problem. First, it may be helpful to endeavor a definition of religion by synthesizing, in part, elements from Geertz's, Wilson's, Pareto's, and Durkheim's own definitions. Religion may be viewed as a unified system of beliefs and practices that are ultimately motivated by certain behavioral predispositions of sociality, and proximately by the sacred symbols that manifest them. Such symbols direct the beliefs and practices toward the veneration of society through the submissive participation of the individual in a united moral community that may be called the Church.

Turning now to the evolutionary question raised just above, Durkheim himself (1912: 245f.) does make an effort to account for the manner in which the sensation of sacredness is aroused in individuals, but he shies away from an explicitly evolutionary orientation, and the result is not convincing. It is, however, exceedingly evocative. One of his strategies is to show that the clan is engaged in two forms of activity. In one, the various families constituting it fan out and wander independently in search of food. Durkheim hypothesizes that this sort of activity renders "life uniform, languishing, dull." We can add that the forces of sociality are to a large extent put to rest and their vitality imperiled as far as the clan as a whole is concerned, with the result that the dispersion of the clan nuclei represents a virtual threat to the social or organizational integrity of the clan itself.

In the other major type of activity, the clan gathers itself again, thereby dispelling the danger of disintegration, and celebrates a religious ceremony known as the *corrobbori*. "When they are once come together, a sort of electricity is formed by their collecting which quickly transports them to an extraordinary degree of exaltation." A sort of collective passion is released that for days can be restrained by nothing except exhaustion. The ecstatic, collective action of the gathering awakens within individuals "the idea of external forces which dominate them and exalt them. How could such experiences as these . . . fail [to convince] that there really exist two heterogeneous and mutually incomparable worlds? . . . The first is the profane world, the second, that of sacred things. So it is in the midst of these effervescent social environments and out of this effervescence itself that the religious idea seems to be born."

Granted that this is the origin of "the religious idea," independently of a need, a biological predisposition, to engage in intensely collective behavior, in frenzies properly speaking; we are still far from explaining the equation of God and society. The religious idea and the particular equation are not one and the same. Durkheim comes closer to the solution of the problem when he confronts the question why "the external," dominant, and exalting forces are thought of in the form of totems, namely, in the shape of animals and plants; and answers: "It is because this animal or plant has given its name to the clan and serves it as emblem." Moreover, "the clan is too complex a reality to be represented clearly in all its complex unity by such rudimentary intelligences." Thus, the totem, the flag of the clan, simplifies the conception of society and symbolically represents its sanctity.

We are still far from a reasonable solution of the problem. But in getting this far Durkheim has implicitly generated two behavioral predispositions that are crucial for an understanding of religious, indeed human, behavior. We may term them *Symbolization* and *Reification*. The problem with Durkheim's brilliant analysis lies in the indefinite meaning of "society." As Robert Bellah (1973: ix) has noted, there is "no word in Durkheim's writings more difficult, and none commoner, than 'society.' " Indeed more than difficult, it is largely incomprehensible if not altogether obstructive. Thus, it may be possible to think of society as an actual group of cooperating individuals, as a moral community, as an organization of symbols—all these meanings and some others may be found indiscriminately in Durkheim's work. But society, we should stress, is also a population bearing the marks of natural selection in the form of genetic predispositions to behave in given classes of ways. My aim in what follows is to give a plausible account of the natural selection of symbolization and reification along with their role in the evolution of the religious institution.

We must first return to relational self-identification and to George Mead's concept of "the generalized other." Mead (1934) distinguishes between the ability to relate meaningfully to (take the role of) particular persons and the capacity to take the role of the generalized other. In the former case, we take the attitudes of others toward ourselves and toward one another. In the latter case, we are additionally capable of taking the attitude of various others toward ourselves and toward one another while all are engaged *simultaneously* in a common social activity. For example, playing effectively on a football team entails taking simultaneously the role of the team as a whole—the generalized other. A team of hunters stalking a prey illustrates another case where behavior is in the form of taking the role of the generalized other. The skill is adaptive, for, the more finely tuned it is, the greater the probability of success in the hunt.

A social self in the fullest sense, and therefore a well-developed relational self-identification, requires the simultaneous absorption in one's conscience of the social psychology, the complex cooperative processes, the strengths and weaknesses, and the goals of an entire group. "It is in the form of the generalized other that the social process influences the behavior of the individuals involved in it and carrying it on, i.e., that the community exercises control over the conduct of its members."

There is reason to believe that this trait is to some extent shared not only by other primates but also by certain other animals that engage in complex and dangerous cooperative activity in the process of hunting large prey. Lions, hyenas, wolves, and African wild dogs specialize, like humans, in the art of organized hunting that, through some such process as taking the role of the generalized other, renders almost any other animal vulnerable. Certainly then, early in our history human beings should have developed something close to a full social self.

But how did that come about? Mead does not answer this phylogenetic question, but his treatment of the question at the ontogenetic level is a very good basis for reasonable speculation about the broader, evolutionary problem. Mead cogently argues that it is precisely the activity of game-playing that affords the socially maturing human being to develop the capacity to take the role of the generalized other. Hence, it is to the evolution of game-playing that we must turn to be able to conceptualize the rise of the full social self. In the process, we should be able to grasp the evolution of symbolization and reification.

It is very likely that game-playing is an ancient feature of the entire hominid line, and of primates in general (see, e.g., Altmann, 1967; Kummer, 1968, 1971; van Lawick-Goodall, 1971; Hrdy, 1977). Mounting, chasing, fleeing, grappling, harassing, derring-do (such as pulling the tail of a dominant male by a juvenile), and, most importantly, various sorts of play-fighting are very common among primates, indeed among mammals in general. For our purposes, however, it helps to think of the hunting party (or the war-party) as a team, and therefore as an organizatoin which plays a game of sorts. The evolution of the game may, therefore, be associated with the evolution of group hunting and/or war-parties.

But the game in itself did not probably suffice to give rise to the generalized other. Durkheim provided a clue to the cause of this probable fact when he noted above that even a tiny society like a clan is too complex a reality to be represented in the abstract as a generalized other in the minds of its members. One suspects that the absence of many teamwork sports which require the participation of more than a dozen persons per team is neither capricious nor accidental. Playing a game is a very complex phenomenon, and it may be excessively difficult to be a team-player in a larger group. Little wonder, too, that team spirit, which may be viewed as a dimension of the generalized-other perception, is greatly aided by all sorts of props, for example, the special song, the uniform with its special arrangement of colors, the mascot, and the ever-present flag. Come to think of it, we moderns have more totems than we know what to do with.

We are catching a glimpse, I believe, of how it became possible to take the attitude of a whole society in the early years of our evolutionary history. The aptitude came about with the conjunction of game-playing and symbolization. When the crucial game became the hunting party, and perhaps more importantly still the war-party, natural selection favored those individuals who had special aptitude at organizing the group, communicating pervasive messages, instilling an esprit de corps, intensifying the feelings of nepotism and reciprocity within the group. As leaders they were especially attractive to the members of the opposite sex.

These comments, it should be noted, apply to individuals of both genders, for hunting was not unique to the males. Nor were the females ever spared the necessity of engaging in bellicose activities; aside from at least occasional participation in the active confrontation of the enemy, many were the times when the females were surprised by the enemy in the absence of their males, and thus had to devise whatever defense they could (for a dramatic literary-paleontological treatment of this issue, see Kurtén, 1978).

In the process of organizing and energizing their group, such individuals must also have developed the skills that facilitated their work of leadership. That is, they specialized in the very same art that nowadays drives, say, the spectators of a football game into a frenzy with the presentation of the team colors, the mascot, the song. Symbolization came to the aid of communication, leadership, exhortation. These skills were, of course, generalizable from the particular to the general, for example, from the war game to the society as a whole. I, therefore, hypothesize that the more complex the social life became, the more natural selection and biocultural evolution favored such individuals as were strong in sociality predispositions and specialized in the art of reinforcing communication with symbolic representations of the complex social total, and, correlatively, of grasping effectively such symbolic representations. This is another way of saying that the predispositions of sociality came to the aid of the combinatorial and selective forces to introduce variations that may be attributed to the specific combiner that I have termed symbolization.

Unfortunately, as E. Sapir (1929), W. Percy (1958, 1961) and C. Geertz (1973), among others, have pointed out, the lack of a science of symbolic behavior is as conspicuous as is great our need for it. Such a discipline should provide specific knowledge of the evolution of the capacity to symbolize as well as the complex part played by symbols in sociocultural evolution. We do, of course, have some understanding of symbolic behavior. It may be approached by searching for a definition of symbol. The inquiry is necessarily cumbersome, for, as has been remarked by specialists of the subject, symbol, like the term culture itself, has been applied rather indiscriminately to a variety of things (e.g. Geertz, 1973: 91). Geertz prefers to view symbol as "any object, act, event, quality, or relation which serves as a vehicle for a conception—the conception is the symbol's 'meaning.' " This view of symbol is closely related to another more elaborately developed by S. Langer (1957: especially 60–6).

Geertz's approach would seem to be a bit less stringent than one proposed by Leslie White (1949: 22–33), according to whom a symbol is "a thing the value or meaning of which is bestowed upon it by those who use it." The word "thing" is intended to suggest that "all symbols must have a physical form; otherwise they could not enter our experience . . . regardless of our theory of experience." On the other hand, "physical" is broadly understood to include "the form of a material object, a color, a sound, an odor, a motion of an object, a taste," so that the difference between Geertz's position and White's may very well vanish upon close examination.

What both scholars, and many others for that matter, share without any equivocation whatsoever is their common debt to John Locke's famous phrase that symbols "have their signification from the arbitrary imposition of men." Symbols are "extrinsic sources of information in terms of which human life can be patterned—extrapersonal mechanisms for the perception, understanding, judgment, and manipulation of the world" (Geertz, 1973: 216). Again, "meaning is bestowed by human organisms upon physical things or events which thereupon become symbols" (White, 1949: 25).

Thus, a symbol can be anything: a mountain, a particular rock formation, a number, the Cross, a musical sign, a sound, a color—anything used to give meaning to experience. Accordingly, thinking consists of an organized commerce of what Mead termed "significant symbols." These refer especially to words, but also to sounds, gestures, artistic representations, mechanical devices like a thermostat, jewels, and so forth, on the basic provision that the meaning of the thing in question is not inherent in the thing; it is extrinsic, imputed to it. Thus, an inactive volcano, for example, may be seen as a mere mountain having no significance to the people who surround it beyond conveying the irregularity and possibly the hazardous nature of the terrain. But the same mountain becomes intensely symbolic if it constitutes a central element in the religious system of that people.

One other important point deserves underscoring in view of my decision to place symbolization among the combinatorial predispositions. Whatever else it may be, to symbolize certainly is *to associate or combine a thing with an idea*. It is this combination that, given our particular neurobiology and psychology, makes it possible for us to vastly expand our consciousness, our knowledge of the environment, and our manipulation of it. "There is a difference," Percy (1958—emphasis in the original) noted, "between the apprehension of a gestalt . . . and the grasping of it under its symbolic vehicle. As I gaze about the room, I am aware of a series of almost effortless acts of *matching*: seeing an object and knowing what it is. If my eye falls upon an unfamiliar something, I am immediately aware that one term of the match is missing. . . ."

There is a temptation in some quarters to attribute the capacity to symbolize to the origin and function of language. On the other hand, it can be argued that other animals have a language, too, very possibly in the absence of symbolization, so that language and symbolization are in principle independent. Indeed, Piaget (e.g. 1951, 1952, 1954, 1971), among others, has shown that the capacity to symbolize emerges in the mental processes underlying imagery of children before they have learned to speak. It is their capacity to associate objects with ideas that allegedly makes possible the learning of language. Whatever the role of language in symbolization, one thing, however, is certain: language is a great implement of symbolization. It is in itself entirely a system of symbols and at the same time a fundamental tool with which to construct other symbolic patterns of all sorts. For our purposes, therefore, we may say that symbolization has been either a by-product of the emergence of language or has been in a relationship of mutual stimulation with it.

As in so many other problems of evolutionary science, a full inquiry into the origin and function of symbolization must await further development in theoretical neurobiology. Offhand, however, in view of the above definition of the symbol, it would seem that symbolization represents a mental device whereby the brain combines abstractions or ideas with objects and thereby renders the abstractions more comprehensible and more effective as means of communication, as evokers of emotions, as stimulants of behavioral predispositions. Symbolization, thus, is a major adaptation in the fundamental sense that it brings the mind and the physical environment into an effective partnership. The immediate result is a classification of both mental images and environmental characteristics (Durkheim, 1912). Such a work of classification, further, functions as a learning tool by aiding memory and understanding with objectifications much as landmarks guide travelers to their destination. It follows that symbolization is also a device whereby we convey information and feelings to others by providing them with ready encapsulations of knowledge and affectivity (Rosch and Lloyd, 1978). Finally, through symbolization we release emotions and stimulate the activity of behavioral predispositions.

Note that objectification or *reification* is an inseparable property of symbolization. Indeed, it may properly be viewed as the other side of a double-phased process. Specifically, by providing the objective, physical, representation of the idea, reification lends durability to symbolization, and is thus its faithful selective agent.

Reification comes in degrees. The lowest lies in merely labeling an idea. Variations often persist and acquire an individuality by virtue of being given a name and by virtue of a tendency which assumes that a name always has a thing corresponding to it (Pareto, 1916: 991). Certain words may be especially effective in leading to reified selections; offhand, one would guess that the more loaded words are with behavioral predispositions, especially with sociality forces, the greater their reifying thrust. The rationale lies in the fact that predispositions of sociality inevitably entail not only feelings toward collective rules of behavior but also strong feelings toward actual objects, even people. To remain intense and vigorous, sociality requires models, prototypes of what is proper, beacons of social virtue. Thus, among all peoples, such words as Virtue, Bravery, Piety, and Fairness have been accorded such deference as is normally reserved to deities.

The highest degree of reification is found in personification, that is, in anthropomorphism, even in deification. The sequence, *in nuce*, is probably as follows. First comes an idea. Then comes the association of object or of act and idea. Third, language intervenes to classify the association. Finally, comes sex with its conceptual mechanism of gender, and reifications become anthropomorphic in nature, often veritable extra-human beings in many respects similar to men and women in the flesh. Personifications have considerable adaptive value for groups. Basically, they help to perceive and abide by group-wide norms, and thus reinforce and referee the system of reciprocal behavior.

Great writers have been aware of the fact for millennia. In the fourth book of the *Aeneid*, Dido falls in love with the foreigner Aeneas and heightens her offense by consummating an illicit "marriage" with him. Her deed sends great rumblings of indignant gossip. Virgil writes:

> Rumor goes flying
> At once, through all the Lybian cities; Rumor
> Than whom no other evil was ever swifter.
> She thrives on motion and her own momentum;
> Tiny at first in fear, she swells, colossal
> In no time, walks on earth, but her head is hidden
> Among the clouds. Her mother, Earth, was angry,
> Once, at the gods, and out of spite produced her,
> The Titans's youngest sister, swift of foot,
> Deadly of wing, a huge and terrible monster,
> With an eye below each feather in her body,
> A tongue, a mouth, for every eye, and ears
> Double that number; in the night she flies
> Above the earth, below the sky, in shadow
> Noisy and shrill; her eyes are never closed
> In slumber; and by day she perches, watching
> From tower or battlement, frightening great cities.
> She heralds truth, and clings to lies and falsehood,
> It is all the same to her. And now she was going
> Happy about her business, filling people
> With truth and lies: Aeneas, Trojan-born,
> Has come, she says, and Dido, lovely woman,
> Sees fit to mate with him, one way or another,
> And now the couple wanton out the winter,
> Heedless of ruling, prisoners of passion.

Symbolization and personification are especially active forces within a religious context. History reveals many cases of deification that fit reasonably well the above reasoning. The goddess Athena among ancient Athenians, and the goddess Roma among the Romans a little later, arose just in such a way. Another case in point is the goddess Annona among the latter people. Initially, annona referred to the food supply. In time, strong and deep-seated feelings associated with the difficulty of maintaining a food supply became a thing. Eventually, under the name of Annona, it took its place with many other things of the same kind in the Roman pantheon (Pareto, 1916: 996). The same process underlies the gradual deification of such modern concepts as Progress, Democracy, the People, Socialism.

For years now, the people of the United States of America, for example, have been bombarded with messages through the mass media that enjoin them to: brush up, America; drink up, America; shape up, America; and so forth. The process of deification of America was particularly intense during the years of preparation that culminated in the celebration of the Bicentennial in 1976. And that is certainly no surprise because this country was then, and has long been, in the throes of an intense process of historical self-identification. If we now recall the Durkheimian equation of God and society, we may conclude that such a process has been essentially religious in nature.

Newly developing personifications must normally compete with existing ones, and hence are historically precarious. Sometimes, the new remain tenuously in the experience of a minority for a brief period and then disappear. Other times, they compete more successfully and eventually replace the old personifications, and, as deifications, they may give rise to new cults. There is good reason to believe that Christianity has for some time had to compete with a variety of new cults, comprising in the whole what may be called the Democratic Religion. The process of religious transformation often presents itself in the form of "secularization." This is currently the dominant idea in the sociology of religion (e.g. Glock, 1973; Wuthnow and Glock, 1973). The prevalent view of secularization refers to the allegedly progressive narrowing of the established religions (for example, Christianity and Judaism) as societies become more complex and differentiated, especially along economic and political lines. A recent statement by Richard Fenn (1978) underscores this viewpoint by arguing that secularization is the result of concrete struggles in pluralistic societies over the secular–sacred boundaries on various fronts.

In fact, however, competition among groups is a fairly constant phenomenon in society, and the sense of the sacred that is always present comes through language that only camouflages the sacred in view of our shifting conception of it. The point really is that the "concrete struggles" are themselves sacred. That may entail a fragmentation and weakening of the prevalent sacred, but it does not substitute it with secularization. What is referred to as secularization represents a process of re-deification combined with a depression of sentiment in relation to old "deities." Moreover, like all other things, religiosity fluctuates in time (Wuthnow, 1976), but a lowering of sentiment in no way justifies conceptions of secularization.

Some years ago, Robert Bellah (1975) advanced the thesis that a religiously inspired vision or covenant has from the beginning infused American history. It has permeated both public and private behavior, cutting across the various churches and faiths, though Puritan in origin, and animating "the civil religion" of the masses, their leaders, and the intellectuals alike. This civil religion, the backbone of the American ethos and character, underwent major trials and transformations during the foundation of the Republic and the trauma of the Civil War. It is currently undergoing a third trial.

A major part of Bellah's thesis is that the Puritan conception of the covenant had both an internal and an external system. The former was based on individual propensities transcending communal purposes and obligations. The latter specified, to a degree, the cultural framework of such communal orientation. The US Constitution embodies to a large extent this communal framework. The two systems have been in competition since the beginning, though the external system prevailed initially. But already in the eighteenth century, "utilitarian thought" began advancing the cause of individual and subsocietal group interests outside the bounds of "any encompassing context of loyalties and obligations." Neither capitalism nor liberal technocracy, for example, is anchored in a society-wide accord on goals that are purely societal in scope. We have become ever more myopic, self-centered, and unimaginative. "Narrowness of vision" has been accompanied by a debunking rationalism that prevents the birth and sustenance of myths, namely, those complex symbolic constructs that engender a sense of collective drama in a people, and thus a unifying spirit. No myth—no deification, we might say—no collective purposes and solutions.

That is an eternal verity. But Bellah seems to fail to understand that such a work of societal apotheosis takes place best where the society scarcely transcends the hard-core boundaries of the kin group. Complex societies like the eighteenth-century United States of America may engage in difficult experiments to undermine self-enhancement forces and strengthen a vision of a society-wide covenant. But the vision can only be ephemeral, all the more so as the society, open to immigration, continues to differentiate along denominational, ethnic, and class lines. Bellah is discouraged by the turn of events, asserting that the "main drift is to the edge of the abyss."

Perhaps. But not entirely for the reasons adduced by him. For the covenant has *not* been broken. No such harm has come to it precisely because the covenant was in fact born riven. As long as the Puritan tradition had the political, legal, and moral power to enforce "the covenant" upon the heterogeneous peoples and subcultures that humbly came to constitute the evolving society, the fracture was easily concealed. But as "Americanization," often through coercion and humiliation, gave way to an Americanism fanned by resentment for ancient and persisting wrongs received, the fracture in the covenant became exposed under the pressure of those who had once gone along with the covenant only half-heartedly in the first place. We will return to this issue of ethnic cleavages in the next chapter.

Faith is the glory of the human spirit. But it is also the affliction of the complex social relations that through one myth after another it helps to forge. For faith is a double-edged sword. On the one side it marks the line on which people may come together to exercise their predispositions of sociality. On the other, it proceeds to shred the agreements that are initially reached, precariously, by the gathering. If religion renders sacred the moral principles of a group, it also tends to strengthen unions that are already well cemented and to fracture those that are from the start precarious.

Societies, however small—and most today are quite large—are considerably more heterogeneous entities than is normally recognized. It was the genius of an Emile Durkheim (1912) to have suggestively equated the "sacred principle" with society itself. Wherever, therefore, we find divergent moral codes, as we are likely to find them to some degree wherever we observe the game of group rivalry and alliances, there we may also expect to find *different societies* properly speaking subsisting under a single *political* umbrella. If we do not choose to so apprehend them, convenience of one sort or another cannot nevertheless hide the fact that intergroup selection is a phenomenon that takes place within what normally pass for societies as well as between societies.

The point has been proven many times in history and social thought. A convincing recent endeavor belongs to Guenter Lewy (1974). This scholar examines no less than seventeen case studies of revolutionary change, including millenarian revolts in such places and times as ancient China, ancient Palestine, and seventeenth-century England; and such modern nationalistic revolts as those in India, Ceylon, and Egypt in addition to the French Revolution and the Spanish Civil War. The record shows that religion has sometimes acted as a force for social integration. But often it has also been "an important force facilitating radical political and social change, providing the motivation, ideological justification, and social cohesion for rebellions and revolutions." The fundamental reason is that, if on the one hand religion subordinates the individual to the society, as was previously noted, on the other the individual has too myopic a vision of the boundaries of his society.

8.4 The Susceptibility to Charisma

The discussion of reification (and of personification and deification) has taken us a step beyond the mere levels of ancestral spirits, ghosts, and other such nondivine fictions. We can now go further still by introducing another predisposition of selection. It refers to what may be termed the *Susceptibility to Charisma*. Charisma is a special quality that inheres in few individuals and is associated with the performance of great deeds, such as heroism and miracles. The predisposition is intended to stress the popular tendency to respond with veneration to such individuals, create myths about them, and sometimes even transform them into deities.

The susceptibility to charisma is, thus, related to the sense of authority discussed in Chapter 5 in connection with the predisposition of deference. The two predispositions differ, however, in various respects. One is especially important. The predisposition of deference is subject to fluctuation in its intensity under the pressure of such factors as the coercive capacity of dominant individuals and the opportunities to rise in the dominance order. The greater the coercion, the more intense the deference. Likewise, the greater the opportunities to replace dominant individuals in their position of privilege, the lower the deference. The susceptibility to charisma, on the other hand, is typically unrelated to coercion and the permeability of the dominance order. Indeed, it is usually precisely the charismatic qualities of individuals that propels them into positions of dominance. In short, the predisposition of deference helps us to accommodate ourselves to a dominance structure, while the susceptibility to charisma is more like a factor of willing submission in the evolution of dominance orders.

The two predispositions together, however, help to explain an interesting phenomenon that so far has only been identified in sociology. According to Max Weber (1946), all ruling powers claim one or more of three basic types of legitimacy or authority. One is of the "legal" variety. It is characteristic of the modern bureaucracy where "a command constitutes obedience toward a norm rather than an arbitrary freedom, favor, or privilege." That is, submission or deference is based on an impersonal, rational, and functional "duty of office."

"Traditional authority," by contrast, rests on a set of attitudes at the heart of which are piety and obedience toward what actually or presumably has always been. Patriarchalism is the classical case in point. The father, the husband, the elder, the master, the prince or lord is invested with the authority to rule by an old system of sacred norms whose violation results in magical or religious evil.

A third type of authority, the "charismatic," refers to "a rule over men, whether predominantly external or predominantly internal, to which the governed submit because of their belief in the extraordinary quality of the specific person." The extraordinarily endowed person may be a sorcerer, a prophet, the leader of a victorious army or hunting expedition, a party leader, and so forth. In any case, he is the object of hero worship. This type of authority is manifestly unstable and subject to "routinization." The charisma ceases to have its normal appeal as soon as the holder ceases to produce effective magic, victories, miracles. With the death of the charismatic leader, the question of successorship arises. The question can be solved by *Kürung*, or election in terms of charismatic qualification: one charismatic leader replaces another, though typically the emergence of the more recent one is less spontaneous than was formerly the case. A second solution lies in designating a successor by consecration, as in apostolic succession. Finally, the question of successorship can be solved through the belief in hereditary charisma, as represented, for example, by subjection to the monarchy.

Note that this typology of authority is essentially acausal. It leaves unanswered the question as to why people bow to tradition and especially to charismatic qualities. To speak of belief in extraordinary qualities is to beg the question. Why the belief in the first place? Alternatively, we may say that the typology presupposes a need for charismatic leadership, but without explication of the need the typology asserts but does not explain. Particularly troublesome are the problem of succession associated with charismatic leadership and, within it, the latter two types of succession. Why should charisma persist through consecration and kingship? A possible solution to the problem lies in the recognition that, if the sentiments of deference and charisma are evolutionary selections, then, as Pareto (1916) noted, the predisposition "may to a greater or lesser extent become disengaged from the person and attached to the symbol, real or presumed, of authority." That is, the power of authority becomes an emergent property, namely, a property of the structure upon which the authority is focused. The leader is but the incarnation of such power.

That would help to explain also the importance for those in authority of keeping up appearances. They surround and invest themselves with all sorts of accoutrements whose function is to set them apart, in a manner, just as charisma does in another. The insignia, the symbols, become the outward semblance of superiority. Between the popular conception of the humble Nazarene and the observation of papal panoply within the context of a bejeweled church there is a chasm that no common logic can reconcile. The displacement of the sense of authority and the susceptibility to charisma suggest, however, that the complex of beliefs that in the Catholic (and at one time in Christians in general) define the charisma of Jesus is today satisfied by the pomp and the grandeur of the Catholic hierarchy. The occasional complaint that the church has betrayed the humility of Jesus, and the frequent explanation that the church went astray in attaching itself to a social aristocracy, introduce other interesting considerations but shoot far off the mark insofar as grasping the rise of tradition-bound authority is concerned.

These considerations help to explain also the seemingly transcendental privileges of certain individuals, for example, the Brahman in Hindu society referred to in Chapter 5. Such privileges are not unique. How could they be? After all, the sanctification and the charisma are not of the person but of the status. Where tradition is holy, the clergy has nearly always been the object of many special treatments, such as the *privilegium clericale*, whereby men of the cloth avoided punishment for a crime, at least on a first offense.

The need for charisma accounts at least in part for a variety of sociocultural variations. For example, many have been the times and places where certain individuals of exceptional accomplishments have been thought to be of divine origin. The Caesars did not invent their own divinity. The feeling that they were divine developed readily in a wildly polytheistic society that looked upon their leaders as undisputed masters of the world.

Likewise, the death of great men of history is often associated with equally great, extra-human, signs. The legend of Charlemagne's death, for example, records that the sun and the moon were darkened, and that his name vanished of its own accord from the wall of a church that he had founded. Joan of Arc has not only been the object of endless myths and even scientific curiosity; she has also become a saint. The death of Alexander the Great seemed to be mourned by the whole world. His general, Ptolemy, hijacked the funeral car so that at his own death he might be buried next to the divine leader. The august Caesar paid homage to Alexander's tomb. But more to our purpose were the hagiographies, or idolizing biographies, and the legends that followed his death. He has been saint and romantic knight. He appears in the Koran and has been honored by Jews. He is even said to have matched the Christian feat of the Assumption and ascended to heaven while still alive.

Or consider the seventh-century prophet Muhammad. One legend has it that at his birth the palace of the Persian emperor shook violently. Another claims that Muhammad's body cast no shadow and that fire could not burn his hair. Still another tells that at his birth a light appeared on his mother's breast that shone all the way to Syria. We need not linger on the usual objection that such legends are the results of rumors spread by inventive and deceptive individuals. Of course, they are. But is it not equally true that such fictions are faithfully spawned in time and place and widely accepted as facts?

We may hypothesize that when the charismatic need operates in conjunction with symbolization and reification, we take a long step upward from the mere level of the immortal soul and the world of rather ordinary spirits to the level of Supersouls and Superspirits—to the stage where folk heroes and other exceptional personages have achieved an apotheosis and constituted an entire pantheon of deities. If we, further, return to the predispositions of sociality, especially those of hierarchy, we may add that the cultural selection of divine beings has been facilitated by a transcendental transference of the human experience in a dominance order, and has often created in the world of deities an organization roughly after our own image.

The question arises as to why the susceptibility to charisma might have entered the evolutionary process. The plausible answers are many. For one, the predisposition helps to satisfy the explanatory urge. For example, in any aggregate certain individuals possess, or at least seem to possess, much greater knowledge of the human condition and its challenges, than the generality of the people. To attribute to them a superhuman status is to invest in them, as it were, as sources of knowledge and power, indeed as champions of one's own life conditions. The aleatory element is thereby fictitiously harnessed, and we come to live a more secure existence in less secure circumstances. Charismatic leaders and divine beings, therefore, give comfort, faith in one's meager abilities, and the sense of security to compete vigorously in the struggle for existence.

Related to this explanation is another which emphasizes the socially synthesizing powers of charismatic personages. Human beings have an intense awareness of the enormous power that the group has over its constituent members. We have awareness, too, that power has many uses, and can enhance our survival potential. Thus, we covet the power of the group, or in any case we seek to have access to its store. But how? On the face of it, it may seem an easy feat to conceptualize the power of the group, but in fact it is probably one of the most difficult problems confronting the human mind. For personal experience suggests that groups are mere summations of individuals; that, consequently, group power represents the sum of individual powers. Power is, thus, at once concentrated and fragmented. To submit oneself to charismatic beings may represent an effort to discard the sense of fragmented power in favor of a concentrated type. The deed is achieved by equating the group with one individual, or a small number, who have the property of exuding a sense of self-equation with the group. In this sense, the charismatic leader reflects the psychological experience of institutional emergence, and the common individual again gains a sense of security and a degree of tutelage, however fictitious, with which to more effectively wage the struggle for existence.

I conclude this section with a comment suggested by Durkheim's (1912) theory of religion. Among the functions of religion, according to this scholar, is one that helps human beings to enjoy life and avoid being oppressed by the apparent futility of a mortal existence. We achieve this feat by subconsciously identifying with the apparent immortality of our group. The hypothesis fits in the previous emphasis on the evolution of the soul, of relational self-identification, and of denial of death. It remains to be added that divine beings, often emerged through charismatic qualities and the popular susceptibility to charisma, are symbols of group immortality that reinforce one's own denial of death and one's need of immortality.

8.5 The Imperative to Act; the Need for Ritual Consumption

As noted, a full account of religion is not my goal. Durkheim's emphasis on the "elementary forms" of religion was intrinsically well chosen. The subject is extremely complex. I have deliberately pursued an elementary strategy and at the same time remained close to the highlights of Durkheim's argument. But of course I have also tried to specify in a phylogenetic sense some of the fundamental forces that are implicit in, or relevant to, Durkheim's theory. In the meantime, other scholars and other considerations have also been taken into account. The final predispositions to be called forth in connection with religious behavior refer to the *Imperative to Act* and the *Need for Ritual Consumption* (or, conveniently, just *Ritual Consumption*). The latter has already been dealt with from a different perspective in an earlier context. I continue within a Durkheimian context.

This scholar, as we have seen, maintains that religion is an institution through which society sanctifies itself. More broadly, religion is a mechanism that subordinates the individual to the group by sanctifying group values and morality through the "will" of divine and supreme forces. The discussion of symbolization, personification, and the susceptibility to charisma suggested ways whereby human beings adept at the sanctification of society could have been selected for in the course of our evolution. The present section concludes the chapter by going into behavioral predispositions whose fundamental function is to engender behavior through which the sanctity of society is maintained and periodically reinvigorated. The imperative to act, or the need to express predispositions, as it might otherwise be called, accounts for ritual, among other things, that exercises the proclivity to hold society holy. It thus adds strength to the predispositions discussed above. Ritual consumption, a related predisposition, refers to a tendency to consume certain substances for one "reason" or another; among those are substances that provide a means of communion with the society. Adhering to the notion that religious behavior enhances group survival, we may hypothesize that any variations that facilitate religious behavior are bound to be favored in evolution. The predispositions here under discussion account for many such variations.

"Powerful sentiments are for the most part accompanied by certain acts that may have no direct relation to the sentiments but do satisfy a need for action. Something similar is observable in animals. A cat moves its jaws at sight of its master; the parrot flaps its wings" (Pareto, 1916: 1089). There is no particular "reason" why a dog should wag its tail when it sees its master; similarly, there is no particular reason why at religious revivals people should scream and dance and roll on the ground (Huxley, 1935). But the impulse to do something about our innate needs is overwhelming; we express any strong emotion by action.

Behavioral predispositions and actions are joined together in a complex concatenation of actions and reactions: while the predispositions determine activity, the latter in turn reinforces the former and may call them into action even among individuals in whom they exist at a low level of intensity. Thus, not only do people often feel the need to release pent-up emotions; by engaging in certain actions they also activate emotions that had been dormant up to that point. For example, some individuals who visit revival meetings out of curiosity, or politeness to friend or kin, may find themselves swept away by the intense pitch of religious expression, and discover that they "have found Jesus." Some people go to their first football game out of civility or a desire to please a companion, and then come out of the stadium deeply at the mercy of whatever predispositions may be stimulated by the game, for example, the need for recognition and relational self-identification.

The need for activity plays an important role in certain forms of collective behavior: for example, mob behavior, student demonstrations, peace marches, indeed many kinds of social movements. If activity reinforces behavioral forces, the very fact of participating in a demonstration intensifies the predispositions behind the participation in those who had them in the first place; and it stimulates them in those who initially went along "merely for the ride" or for reasons of duty or sociability. Robert Park (1967: 48) wrote of "pervasive social excitement" as essential to the very existence of the animal herd or flock; for it facilitates "the communication of news." It also facilitates the mutual stimulation of behavioral forces and thus insures their vitality.

Konrad Lorenz (1966: ch. 6) provides an interesting perspective on the need for activity. A behavior pattern such as feeding or reproduction is always a result of a complicated system or interplay of many physiological causes or drives that have passed the test of natural selection. Such causes are sometimes related to each other in symmetrical influence; other times, one or more are overdetermining in relation to the others; still other times, some causes do their work without much influence from the rest. A given behavior pattern, therefore, represents the systemic operation of a set of behavioral forces, all of which demand expression, though to different degrees, *whether or not* they have an immediate or direct functional relationship to the behavior in question. The phenomenon is

> compatible with a widespread principle of natural economy that, for example in a dog or a wolf, the spontaneous production of the separate

impulses of sniffing, tracking, running, chasing, and shaking to death is roughly adapted to the demands of hunger. If we exclude hunger as a motive, by the simple method of keeping the dish full, it will soon be noticed that the animal sniffs, tracks, runs, and chases hardly less than when these activities are necessary to allay its hunger. Still, if the dog is very hungry, he does all this quantitatively more. Thus, though the tool [lesser] instincts possess their own spontaneity, they are driven, in this case by hunger, to perform more than they would if left alone. Indeed, *a drive can be driven* (Lorenz, 1966: 85–6—emphasis provided).

I have italicized the matter above because the statement suggests an important idea. The argument so far in this volume has been that behavioral predispositions are stimulated, and often even triggered off or released, by external factors: for example, by symbols, migration, and social mobility. But we are now obliged to consider the possibility that the repertory of behavioral predispositions contains its own internal trigger or catalyst. Such a force could be expected to be among our most ancient because we probably share it with many other animals. In that case, behavioral predispositions may be triggered off or modified in intensity by endogenous as well as exogenous forces. In the former case, the repertory achieves a degree of autonomy from behavioral and environmental factors that is especially useful to the survival of individuals and societies.

The interesting thing about symbols and other stimulators of behavioral predispositions is that they are not always present to perform their function. Or if present, they may be utterly disjoined from the predispositions. A person may, for example, be sexually starved but be completely out of contact with other, sexually stimulating, persons. The flag of a nation that practices slavery may intensify feelings of freedom in the free citizen but fail to bestir the typical slave. Yet, the sexual need and the need for freedom remain strong. Indeed, they may grow in intensity. Under these circumstances, when external agents fail to satisfy deep-rooted needs, internal forces take over. In short, organisms seem to contain self-arousal functions whose nature is not yet fully understood (e.g. Bandura, 1977: 68–72). "A drive may even become so strong that its motor responses break through in the absence of a releasing stimulus" (Tinbergen, 1951: 61).

Likewise, there are times when a society seems to flounder along in the throes of great disorder and disorganization. For example, it contains no exceptional leaders to stimulate the charismatic predisposition. Alternatively, it harbors too many, but none of them commands a large enough following to take any effective societal action. It is thus possible that, after a more or less sustained period of such chaos, the people will be impelled to break the stalemate in the competition for leadership by widening the focus of the charismatic predisposition and achieving thereby a broad consensus on a compromise leader.

Joachim Wach (1951: 32–3) suggests four universal distinguishing characteristics of the religious experience: response to ultimate reality, response to such reality as "integral persons," loyalty without peer, and "an imperative, a commitment which impels man to act." This imperative is apparently especially strong within the religious cult. "That complex of gesture, word, and symbolic vehicle which is the central religious phenomenon we call the cult," according to O'Dea (1966: 39), "is first of all an *acting out* of feelings, attitudes, and relationships." All religions either start as cults or go through an early stage known as the cult.

Within the religious context, the acting out of feelings typically takes the form of ritual. The ritual is a constant reiteration of sentiments, "a symbolic transformation of experiences that no other medium can adequately express. Because it springs from a primarily human need, it is a spontaneous activity—that is to say, it arises without intention, without adaptation to a conscious purpose" (Langer, 1957: 153, 40; for other useful sources on ritual in general, see van Gennep, 1908; Radcliffe-Brown, 1922; Mauss, 1925; Malinowski, 1948; Goffman, 1959, 1961, 1967, 1971). The ritual arises spontaneously because it is above all a "means by which the social group reaffirms itself periodically" (Durkheim, 1912: 432). Societies rely on rituals, a special organization of behaviors, to renew their social energy or solidarity and thus to maintain historical viability. There is a mutual relation between predisposition and behavior; religious sentiment incites people to perform certain rites, and the performance, in turn, intensifies the sentiment.

The imperative to act varies in intensity. It may entail a calm and thoughtful tendency, and then again it may arise to the point of exaltation, frenzy, ecstasy, even delirium. It includes chants, dances, contortions, mutilations, and a great variety of other highly energized activities. When in a state of religious ecstasy, individuals are agitated by tempestuous emotions. They are the cheer leaders, as it were, of the moral being that is the constituted society. From an evolutionary perspective, ritual behavior is a sort of blossoming of renewed energy, a resurgence of forces that, moderate or latent under normal circumstances, burst through to establish their supremacy over human conscience and, temporarily at least, over individual utilitarianism as well.

Glossolalia, that baffling and captivating trait known as speaking in tongues, is frequently an integral part of the religious ecstasy. It is a wondrous phenomenon. People who speak in tongues literally say nothing—at least nothing that is intelligible. But saying is not the function of glossolalia. The phenomenon is activity pure and simple, though it is an immensely complex kind of activity, appearances notwithstanding. It represents a veritable outpouring of deep-rooted forces, the predispositions, flooding in marvelous array the human consciousness. It is as if the hypothalamus-limbic system were utterly emptying itself out.

We must be careful to transcend the intimacy, and seeming clarity, of our own culture in order to understand the significance of the need for activity. For example, much of the current popular music is sometimes called loud. That merely describes its sound. Music is also motion. It has a rhythm and it is compelling. Indeed, "loud" music is both expression and stimulus in an orgy of bodily motions and contortions. It is almost as if the biological forces that it expresses were pouring out of those agile bodies under the irresistible exorcism of a high priest. Such music constitutes a rebellion to the circumstances of an era that, on the one hand, cruelly harnessed the human body with neckties, corsets, and the like, and on the other tethered it to the desk and the machine, depriving it of the lissomeness, health, and grace with which millions of selective years had endowed it. Little wonder that highly stirring music has come along with the jogging craze and an intense preoccupation with the health and the beauty of the body.

There is another perspective from which the need for activity can be appreciated. Many thinkers from Marx to Maslow have inquired about the mechanisms whereby the human individual attains self-fulfilment or self-actualization—how, namely, humans develop an inner sense of completeness, integrity, and satisfaction. For Marx, for example, this fulfilment of the self was attained through "voluntary" labor that was also congenial with the talents, the creative potential, and the "spiritual" needs of the individual (Marx, 1844). This view of the human being tends to postulate a fundamental, higher-order, dimension of the human psyche; and of course it isolates a particular form of activity through which that dimension is most properly expressed. Everything we know now suggests that the highest form of self-fulfilment is attained under the influence of the imperative to act. This need is a sort of spark plug for the ignition of other needs. Under its influence, the entire personality is engaged and the gamut of human experiences is sometimes synthesized in a frantic, electrifying, tumultuous convulsion of body and mind.

The imperative to act and the need for ritual consumption very often go hand in hand; indeed, it very often happens that the imperative to act may manifest itself in ritual assimilation of substances. At this juncture, if there is a reasonable basis for tying ritual consumption to a predisposition in its own right, it is more precise to say that the two behavioral forces stand to each other in a close systemic relation, and thus collaborate closely in producing variations.

Ritual consumption has already been discussed in Chapter 2. In that connection the emphasis was on demonstrating it as a cultural universal and, thus, on establishing the hypothesis that within the cultural uniformity may have lurked a universal behavioral predisposition. There is, therefore, no need here to go into detailed illustrations and other strictly factual matter. What I wish to do, rather, is to offer a few brief theoretical remarks within a religious context and from a Durkheimian vantage point.

If the Durkheimian theory is correct, the ritualistic consumption of a substance, such as a part of a totem in primitive religion or the host in the Catholic communion, is an intensely significant act through which individuals symbolically absorb or assimilate their society into their social conscience. Normally in action on a periodic basis, the behavioral predisposition behind it accomplishes two major orders of intimately related functions. In the first place, the symbolic assimilation typically takes place within a collective context and, therefore, represents an affirmation of unity by group members under the same sacred principle. To "take communion," a term widely used in ethnography, is thus to engage in a coordinated act of revitalizing the supremacy and sanctity of the organized group.

In the second place, to take communion is also a form of personal revitalization, a periodic reinforcement of one's socialization in the classical Freudian and Meadean senses of embodying the group will in one's own conscience. Closer to Durkheim's theory, communion represents a periodical rekindling of the sacred principle that resides within. As Durkheim (1912: 378–9) reasons, all members of a religious community, such as a clan, harbor a mystic substance representing, or giving rise to, the soul, which is the source of all their powers. Like all other forces, this substance is subject to a process of degradation. "Therefore the men of a totem cannot retain their position unless they periodically revivify the totemic principle which is in them; and as they represent this principle in the form of a vegetable or animal, it is to the corresponding animal or vegetable species that they go to demand the supplementary forces needed to renew this and to rejuvenate it."

In conclusion, the need to act, in view of its widespread presence in the animal kingdom, no doubt predates the need for ritualistic consumption. The latter is in all likelihood a biocultural accretion to the development of ritual behavior in general, and thus was probably selected under the ritualistic pressure of the imperative to act. Likewise, the cultural variations produced by the need to assimilate ritually are under the selective scope of the ritualistic aspect of the imperative to act. The two forces together, along with the beliefs and practices associated with them, constitute a large part of the religious experience. Such experience is above all an integrative force. That is, it cements the social group by intensifying its solidarity. In so doing, it places one group at a selective advantage in relation to another whose religious experience is less intense, and the solidarity more precarious. In the event of clashes between the two groups, the individuals of the more solidary group are themselves at a selective advantage in relation to the members of the less solidary one.

9 Evolutionary Foundations of Family and Ethnicity

This final chapter will focus on the family institution and related phenomena. Again, the aim is not an exhaustive discussion but a search for certain fundamentals that will help to elucidate in an evolutionary key the institutional complex in question. This focus will, further, foster the hypothesis that the family is the fundamental aggregate fact from which other groupings and associated facts have emerged: for example, ethnicity, ethnocentrism, and prejudice.

9.1 Homologous Affiliation and Heterologous Affiliation

My basic conceptual tool is a two-pronged or bivalent predisposition of combination that may alternatively be viewed as two distinct, though closely related, predispositions. *Homologous Affiliation* and *Heterologous Affiliation* are reasonable though not entirely satisfactory labels for them. Contrary to strategy pursued in Chapters 7 and 8, no particular selectors will be singled out to pursue the complementarity of evolutionary action. Rather, because we are essentially back to problems of sex and kin selection, I shall rely mostly on the maximization principle as a substitute for forces of selective retention. Homologous affiliation may be defined as a genetically based inclination to favor alliances and forms of reciprocal behavior with others in direct proportion to one's degree of kinship and/or phenotypical similarity to these. Heterologous affiliation refers to the opposite tendency. That is putting the matter at the level of interpersonal relations. But the predispositions tend to generalize also to behavior in general, so that they may be observed in action in all sorts of institutional contexts.

The combiners in question, or at least the cultural accretions associated with them, have often been remarked upon. Charles Sanford (1961: 20—emphasis in the original), for example, argues that people have always acted as, and forced themselves to be, "in some sense, an expression of polar opposites." In certain states of mind, they "take comfort in the *likenesses* within the texture of experience which make for routine, tradition, uniformity, authority, cosmos"; in short, they are most comfortable with variations that reinforce their sense of similarity and affiliation to given people and arrangements. In other states of mind, they are "interested in exploring *differences* in the texture of experience which make for individualism, competition, novelty, adventure, chaos"; that is, they are governed by predispositions that stress putting heterogeneous entities together.

Louis Dumont (1966) observes that a central fact of the Indian caste system is "the opposition between pure and impure" along with a vigorous tendency to associate on an intimate basis only within one or the other category. Folklore, proverbs, and maxims suggest a wealth of related phenomena: for example, "birds of a feather flock together"; *similia similibus* (similar with similar), it was said by medical experts in times not so remote. Conversely, *contraria contrariis*. Again, Sumner's already encountered distinction between in-group and out-group concerns the universal tendency to distinguish between "us" and "them" in terms of similarities in folkways or "designs for living."

In his sagacious analysis of various myths concerning the totem pairs represented by moieties, Radcliffe-Brown (1951: 116) finds a recurring theme: "The resemblances and differences of animal species are translated into terms of friendship and conflict, solidarity and opposition. In other words the world of animal life is represented in terms of social relations similar to those of human society." Appraising this argument, Lévi-Strauss (1962: 88–9) later takes one of the polarities and argues: "Consequently the division eaglehawk-crow among the Darling River tribes, with which we began, is seen at the end of the analysis to be no more than 'one particular example of a widespread type of the application of a certain structural principle' [reference to Radcliffe-Brown], a principle consisting of the union of opposites." To which he adds: "Totemism is thus reduced to a particular fashion of formulating a general problem, viz., how to make opposition, instead of being an obstacle to integration, serve rather to produce it."

In more general terms, as a sociologist has noted, "Similarity or contrast in things [in general], no matter whether real or imaginary, is a potent cause of combinations" (Parcto, 1916). It helps to explain a large class of phenomena including marital preference, the nature of relations within and between families, prejudice and discrimination, ethnic stratification, and so forth (see, e.g., Fishbein and Ajzen, 1975). We shall return to some of these topics. Again in a general key, useful evidence in favor of the hypothesized predisposition toward *homologous* affiliation comes from *Gestalt* psychology. For example, we do not see the circles and dots of a figure as a random series of events. "Rather, one's perceptions are ordered in terms of *similar* aspects of the field and circles and dots are seen as alternate rows" (Combs et al., 1976: 33).

What is the likely evolution of the predispositions under examination? The question is especially important, even though the natural selection of predispositions falls outside the purview of this book, because the answer may be particularly helpful in elucidating sociocultural facts associated with them. We must return to the maximization principle. In this connection I have argued with some reservations that insofar as natural selection implies a competition for survival in the genetic sense, there is some logical basis for postulating the maximization principle. But now, to behaviorally uphold the principle—to maximize the probability of achieving inclusive fitness—a number of activities are necessary. Access to edibles is crucial. Mating among sexual species is the *sine qua non*. But not just any kind of mating. A donkey and a horse mating together are following a self-defeating strategy to genetic fitness. Fertility among mules, the offspring of such a partnership, is so rare that among the Romans it presaged some great event and fell in the same class of such occurrences as eclipses and "divine birth."

It follows that one of the first adaptive skills essential to the individuals of various species was an innate capacity to discriminate between those individuals with whom they could productively pair up and those with whom copulation represented a genetic waste of time. We need not be concerned with the more subtle problem of mismatches between organisms belonging to the same species; they are real enough, as recessive genes can attest. The fact is that many species are so close in appearance and behavior that in principle they encourage specific transgressions. Yet, being separate species, they are genetically closed to each other: they cannot cross-mate successfully. Under the circumstances, we cannot but conclude that natural selection has worked in favor of those organisms in each species who could discriminate between a reproductive mate and a non-reproductive one. In all likelihood, that is at once the ultimate cause of the predispositions of affiliation and the root of all major forms of discrimination. On the face of it, this statement would seem to apply to homologous affiliation but not to heterologous affiliation. In a later discussion of assortative mating, we shall see, however, that the two predispositions tend to move in each other's direction in a sort of optimal behavioral equilibrium.

9.1.1 THE GENERALIZING TENDENCY

Behavioral predispositions, skills, proclivities of all sorts tend to generalize beyond their original nuclei. For example, one who favors mountain skiing is also likely to enjoy water skiing, and vice versa. One skilled in tennis is likely to have little trouble in learning to play a fair game of badminton. We might even stretch the point and speak of an associated predisposition that is reminiscent of such behavioral peculiarities as generalization and the halo effect. Physicians, for example, often benefit from a halo effect in the sense that they are widely credited with skills that go beyond their medical training, pertaining to such matters as intrafamily relations, grief, success in love, and even politics.

In view of reification and related predispositions, the halo effect need not attach to persons. To the communist, everything associated with communism is good; to the democrat, everything about democracy is pleasing. It is this tendency that often renders impossible any dialogue that might grant even a relative validity to the other's position. The history of politics, old and new, is laden with the operation of the tendency in question. It is in part also a history in which political and diplomatic heroes have achieved their eminence by virtue of either superior force behind their policies or of superior stratagem capable of producing a halo effect. Note, too, that to generalize the "good" of one often entails generalizing the "bad" of another.

Likewise, the political history of a nation is to a large extent the history of generalizing contention between political factions. For some time in Western society, "conservatives" have seen nothing but detriment in public welfare measures which for them spell out "socialism"; conversely, "liberals," Social Democrats, and the like have viewed them as categorically beneficial and as the instruments of social justice and equality. Beneath it all, there are of course powerful and divergent interests at work that concern the distribution of wealth, saving, taxation, and the like, all of which have consequences for inclusive fitness. But the main point remains: human beings are great generalizers—no doubt, I think, because our classifying tendency is strong but unimaginative. Thus, we have trouble thinking of the weather in more than four or five major categories (for example, sunny, rainy, cloudy, warm, cold). The same applies to height, beauty, taste, and numberless other phenomena.

A fertile field for the operation of combinations of both similars and opposites is magical behavior (see, e.g., Durkheim, 1912; Wallace, 1966; Lessa and Vogt, 1972). For example, it has been a widespread belief in many times and places that persons can be harmed by torturing a doll or a figurine made in their image. Voodooism is largely based on this tendency to generalize from the thing to the person. This form of witchcraft is normally attributed to certain African tribes and some of their descendants in the New World, notably in Haiti. But the practice is of all times and places. In Hector Beoce's *Croniclis*, for example, a story is related about the legendary Scottish king, Duffus, who perspired at night and was unable to rest during the day. It finally came to light, to the great relief of the king's puzzled physicians, that in a nearby town certain witches were in possession of a wax figure of the king. When they placed the royal image close to the fire, the king began to perspire; the recitation of incantations kept him from sleeping.

Homologous affiliation and the tendency to generalize combine to constitute a major source of the tendency to reason by analogy. Consequently, they perform a capital function in the production of knowledge, and thus of culture in general. Analogies play many roles. At bottom, they are tools with which the human mind constructs avenues of approach among sets of facts, systems of explanation, and the like. In so doing, the mind proposes combinations of facts that at another time or place seem to be far removed from each other on one or more significant bases. Through analogies, the mind, therefore, discovers systems of order in nature and creatively unfolds their boundaries. It is probably a truism that without analogies, science—to say nothing of such fields as philosophy and literature—would grow at a much slower pace and develop along fewer branches of endeavor. But, as was previously stated, analogies can also hinder the growth of knowledge. If offered as a means of conveying some conception of an unknown, they may be used scientifically as a way of getting from the known to the unknown. Offered as demonstration, they have not the slightest scientific value (Pareto, 1916), although, as we saw in Chapter 2, they can be exceedingly suggestive.

The production of knowledge is sometimes under the selective pressure of personification. In this case, it is by definition of a mythical, legendary nature. Common to many peoples throughout known history has been the notion that the birth of every great historical or legendary personage was the result of some divine act, or was at least attended by prodigies. Thus, Hercules was engendered by Zeus, and the feat took three nights to boot. Jesus was born of God, and had a virgin mother in the bargain. In modern times, analogous fabrications of the mind seem less outlandish because they are often presented as the result of "scientific" endeavor. Schoolchildren are given a goodly dosage of sparkling mythology about the personality and the background of such historical figures as Charlemagne, Elizabeth I, George Washington, and Karl Marx.

We now have even members of medical and related professions who specialize in psychobiographies, the dubious art of causally connecting historical events to the early biography of public figures, though some efforts seem more plausible than others (see, e.g., Padover, 1978). Needless to say, if they focus on historical scoundrels they will be sure to find knavish traits in the scoundrels' backgrounds. Richard Nixon has for some time tempted our fancy with a rich supply of vile attributions. When, conversely, these latter-day prophets of yesterday's history level their latest "scientific" tools on folk heroes, naturally psychobiography is laden with favorable, even noble, omens.

The "evil eye," attributed with greatest frequency to Catholics, is an ancient and universal trait. Hesiod, in *Opera et Dies*, states the proverb,

"You would not lose your ox if you did not have a wicked neighbour." Everywhere there have been despised sorcerers and necromancers with the alleged power to bring death to people, animals, and plants. In some societies, witches are reputed to have anomalous features, such as two pupils in each eye or a forked tongue (e.g. Simmons, 1942). The responses to them vary. Usually they are feared; less frequently, they are neutralized with other objects or acts of mysterious power; still less often, they are burnt or otherwise killed, as in medieval Europe or the early days of New England.

It is interesting to note, further, that witches are usually women. That may indicate one of the most venomous, even if subconscious, forms of male insolence. Witches may do great harm, but at a distance and surreptitiously—as contrasted, say, to (male) devils, who have such power and effrontery as to "possess" our bodies and even to figure indirectly in the Lord's prayer. Moreover, at least until recent times, witches were old and ugly, rather than young and comely, women. In that we may observe the compounding operation of a deep-rooted revulsion for the barren woman.

At all times, certain words have with amazing consistency had great fascinating power over people. Among social scientists, conflict, consensus, equality, exploitation, system, equilibrium are good candidates. The fact may help to explain the tendency to be favorably disposed toward certain theories and unfavorably inclined toward others in part on the basis of the indefinite language employed. The formal grounds have not been established, for example, for a rational choice in favor of neo-Marxian arguments (e.g. Pareto, 1916; Weber, 1922; Mills, 1956; Dahrendorf, 1959; Bottomore, 1966) or in favor of functionalist presentations (e.g. Durkheim, 1912; Davis, 1948; Parsons, 1951; Merton, 1968). Yet many of us are committed to one or the other in our lectures and writings.

What is more, concepts are attributed to the one or the other "school" categorically without much thought for the possibility that to one degree, and in some respects, they apply equally well to both. At a proximate level, that is due in part to the fact that the words used to label theories do not evoke conceptually clear and denotative meanings; they mean different things to different people—they are what Garrett Hardin (1956) termed "panchrestons." At an ultimate level, the particular attribution represents a variation under the generalized influence of homologous affiliation and heterologous affiliation. People vary in the way such terms invest their predispositions.

Nowhere are panchrestons more evident than in the use of the "system" concept in sociology and cultural anthropology. The term is typically employed synonymously with organizational concepts like group and society. Strictly speaking, however, the system refers to a mental construct, an idealized heuristic, that enjoins the theoretical determination of the manner and the degree in which aggregates feature certain properties, for example, the interdependence of their constituent parts (Henderson, 1935; Barber, 1970; Lopreato, 1971; Sztompka, 1974).

9.1.2 HOMOGAMY OR ASSORTATIVE MATING

Later I shall argue that, in conjunction with a predisposition toward "heterologous contraposition," the generalizing tendency helps to explain such phenomena as ethnic conflict and ethnocentrism. Here I continue to examine a number of implications of homologous affiliation and heterologous affiliation within the context of reproductive behavior. For example, if what we have been saying about homologous affiliation has any validity, we should be able to hypothesize that we seek to enhance our inclusive fitness not only through direct reproduction or through favoritism toward relatives, whose children bear portions of our own genotypes, but also by marrying or mating with individuals who have a greater than average coefficient of relationship ($\bar{r}$) with ourselves. That is to say that natural selection should be expected to favor the production and retention of variations, in the form of social relationships and customs, that facilitate homogamous or assortative mating, namely, reproductive partnerships which, on the basis of perceivable, or phenotypical, similarities, seem to entail a degree of genetic kinship.

On the assumption that a wife and husband are genetically total strangers, Mendel's First Law states that, through the process of cell division known as meiosis, each shares 50 percent of her/his genes with each offspring. When, conversely, two individuals mate who already have a certain percentage of genes (more precisely, alleles) in common, then each parent will share with each offspring the above 50 percent quota *plus* that portion of genes that was held in common by the two partners but was passed on to the offspring only by the mate. Of course, in due time we shall have to consider the limits beyond which this strategy becomes genetically negative. Stated otherwise, at some point homologous affiliation will have to be checked by heterologous affiliation; that will take us into the problems of inbreeding depression, incest avoidance, and incest taboo. For the present we may consider some facts and factors concerning "assortative mating."

The previously mentioned study by Thiessen and Gregg (1980) reviews much of the literature on assortative mating, both among humans and non-humans, and discusses the findings in the light of their import for the maximization principle and related evolutionary concepts. My own interest here is in *Homo sapiens* only. Assortative mating appears to be under the influence of both proximate, or cultural, factors and ultimate, or genetic, ones. It features two chief types: positive and negative. Positive assortment refers to mate selection favorably guided by phenotypic similarities, and hence is the result of homologous affiliation. Conversely, the negative variety refers to mate selection facilitated by dissimilarities (disassortative mating) and is therefore the product of heterologous affiliation. Assortative mating is a concept frequently used to comprise both types.

Apparently, the study of assortative mating is at least a century old, and from one perspective or another has interested scholars of every stripe (e.g. Pearson and Lee, 1903; Wright, 1921; Hollingshead, 1950; Clark, 1959; Beckman, 1962; Winch, 1962; Eckland, 1968, 1972; Lewontin et al., 1968; Spuhler, 1968; Novak, 1971; Shields and Gottesman, 1971; Vandenberg, 1972; Murstein, 1976; Susanne, 1977; and Thiessen and Gregg, 1980, for an excellent review and discussion). Marriage partners assort over an astonishing range of traits, including nearly all anthropometric characters, various aptitude and achievement measures, and a multitude of sociocultural and demographic indices. As a result, we can hypothesize that biocultural evolution has favored the selection of variations that both lead to and reinforce assortative mating. Some good proof may be found in neighborhood activities and socializing. Neighborhoods as units tend to be highly assorted in their own right. Club and other socializing activities going on therein, in turn, reinforce assortment by facilitating marriages within the neighborhood (e.g. Hollingshead, 1950).

By far the more common strategy is homogamy, and the positive correlations between spouses are very high for race, ethnicity, socio-economic status, religion, and political orientation. Correlations are moderate for intellectual ability, education, personality variables, and vocational interests. They are lowest on anthropometric measures and highly specialized skills.

Interestingly enough, positive assortment is common for apparently nonadaptive as well as adaptive traits, for example, for mental retardation and blindness (e.g. Penrose, 1944; Nielsen, 1964; Crow and Felsenstein, 1968; Kreitman, 1968; Reed, 1971; Ehrman, 1972; Thiessen and Gregg, 1980), indicating one or more of the following mechanisms at work, among possible others: physical propinquity as a match-maker, gradual assimilation of each other's abnormalities, a keener understanding combined with toleration of their respective abnormalities, a relative lack of choice as a result of discrimination on the part of the "normal" population, and of course plain selection for gene homology or an evolutionary mechanism operating blindly at the ultimate level of the maximization principle.

Assorting positively on nonadaptive traits is in principle a nonadaptive strategy, and thus would seem to violate the maximization principle. But everything is relative with concepts like adaptive, fit, and selected. If, for example, blind individuals are discriminated against as mates by all but other blind persons, their best, most adaptive, approach to direct or individual (as opposed to kin-enhanced) fitness is to marry amongst one another. That may be poor fitness insurance when looked at from the viewpoint of the generations to come, in the sense that the offspring, if any, may be at a selective disadvantage in comparison with the offspring of non-blind or mixed marriages. But it is still the best strategy available to them, and therefore it is adaptive within the limits of their restricted options. Indeed, it is not unlikely that this attraction between handicapped individuals reinforces their predispositions toward mutual aid or reciprocity organizations, and these in turn translate into the selection of adaptive variations at various levels of society. So, recently it has become easier for the handicapped to obtain medical services, to read without eyes, to attend institutions of higher learning, and in general to compete for self-enhancement with the rest of the population. Such developments, in turn, tend to mitigate the previously mentioned selective disadvantages of their children.

In other respects, the fitness-enhancing value of positive assortment is more evident. The behavior, for example, enhances marital stability, with the result that the offspring of the positively assorted receive better than average parental support, and are thus at a competitive advantage in their generation. In one of the more exhaustive studies available, Bentler and Newcomb (1978), for example, found that couples who had been married four or more years were more positively assorted on a host of traits than couples who cut the marriage shorter through divorce. Likewise, there is some evidence of a direct relationship between fertility and positive assortment for certain traits (see, e.g., Hill *et al.*, 1976; Bentler and Newcomb, 1978; Thiessen and Gregg, 1980).

If positive assortment tends to enhance fitness, it may be predicted that the behavior will be found among dating couples and prospective mates as well as among married people. In a study of college student pairs, Bersheid and Walster (1974) found that there was a very high similarity in attractiveness within each pair. More importantly, the more equally attractive the couples were, the more likely they were to be engaged in intimate contacts. It is no great mystery, therefore, if college fraternities, sororities, cooperatives, and the like sometimes remind one of the old ethnocentric saying, "All Chinese look alike to me." Such associations are sociocultural selections that facilitate assortative mating, *inter alia*, and are brought about at least in part by homologous affiliation under the pressure of the maximization principle.

The evidence is overwhelming in favor of the hypothesis that the maximization principle works to enhance inclusive fitness by increasing homozygosity, up to a point. Available data and evolutionary processes require that we stress the qualifier *up to a point*: when homozygosity, or the union of gametes that are identical for one or more pairs of alleles, is carried beyond a certain point, there are grave genetic penalties to be incurred. And here is where heterologous affiliation comes into the picture as a complement to homologous affiliation.

The genetic penalties in question concern *inbreeding depression*. This phenomenon refers to a tendency to lowered fitness due to a high inbreeding coefficient (F), namely, the condition whereby identical alleles, or gene forms, are found on the same locus on a pair of chromosomes, due to common descent. Thus, F is related to r, the coefficient of relationship. The greater the r between two sexual partners, the higher the F, and the greater the probability that the identical genes will be among the many lethal variants that have been isolated for the human species (e.g. McKusick, 1975). In a sizable study of Japanese marriages, showing a comparison of offspring from first-cousin marriages with offspring from a control group, Schull and Neel (1965) found significantly more types of depression or abnormality in the experimental group than in the control group. Among the forms of depression was prenatal and postnatal mortality. Similar results have been reported from Ceylon (Reid, 1976), India (Rao and Inbaraj, 1979), Sweden (Alstrom, 1977), and Czechoslovakia (reported in Wilson, 1978), among other places. Inbreeding depression may be expected to be severest in matings between first-degree relatives (for example, brother–sister, parent–child) (Wright, 1921; Spuhler, 1967). Alstrom (1977) reports that the incidence of recessive disorders among the offspring of such matings is twice as great as that for half-siblings, and four times greater than the rate for first cousins. These results fit Mendel's First Law to an astonishing degree, inasmuch as first-degree relatives have 50 percent of their genes in common; half-siblings, one-half as many; and cousins, one-fourth that amount.

Thiessen and Gregg (1980) reach the conclusion that to optimize fitness through assortative mating, the best strategy is to select a mate who is phenotypically similar to family members but is not a member of the family group. "In the vernacular," they (1980: 123) add colorfully, "an individual male may avoid mating with his mother but still 'want a girl just like the girl that married dear old dad.' " Available studies show various cultural rules that support this statement of evolutionary strategy. Positive assortative mating on various traits is encouraged among non-relatives and restrained among related couples (e.g. Goode, 1970; Crognier, 1977; Jakobi and Marquer, 1977; Neel, 1978). Homologous affiliation and heterologous affiliation would, thus, seem to combine to produce sociocultural selections that facilitate the maximization of inclusive fitness by increasing the rate of genetic homology among the mates while at the same time reducing the danger of inbreeding depression.

Assortative mating cuts across a variety of other phenomena. One is incest, and we shall presently return to that. Another concerns kinship classifications or categories. Such classifications are universal, and may be one cultural product of a broad biological tendency to classify observations of all sorts along various dimensions. A more nearly complete theory of human nature might, therefore, have to include some such thing as the need to classify in the overall repertory. Certainly, if the maximization principle has implications for assortative mating, as argued above; and if assortment is adaptive when it stops short of the penalties of inbreeding depression; it follows that natural selection favored individuals who could symbolize in such a way as to construct and obey classifications and classification rules that, within a mating context, would maximize gene homology, and thus inclusive fitness, without incurring a heavy risk of inbreeding depression. Kinship classifications may be cultural by-products of the maximization principle.

In an important paper that reveals an extraordinary keenness about the interplay of biology and culture, Robin Fox (1979) has cogently argued that kinship systems, in one of their aspects, represent systems of assortative mating; and kin classifications are "flexible means of adjusting the categories of marriageable and unmarriageable kin." Such classifications are, therefore, cultural mechanisms whereby breeding populations achieve the degree of specificity and flexibility that is required to enhance inclusive fitness through a degree of homozygous reinforcement, and at the same time avoid the dangers of excessive inbreeding. Indeed, an analysis of various systems of classifying kin (by degree, by descent group, and by category) leads Fox (1979: 139) to the conclusion that kin classifications may be viewed as "various ways of combining an outbreeding tendency with a closely endogamous tendency: to avoid sibling, and often first-cousin marriages, but to encourage, or render inevitable, second- and other cousin marriage."

9.2 Incest Avoidance

9.2.1 INCEST TABOO AND EXOGAMY

To mate with a close relative (parent, child, sibling) is to commit incest. The incest taboo, a rule that prohibits such matings, is a cultural universal. Very rarely do human beings approve of mating between two individuals whose coefficient of relationship is larger than $1/8$. When the rule is broken, the probability of defective offspring is greatly increased. This form of reduced fitness is known as "inbreeding depression." It is not unreasonable to hypothesize, therefore, that somewhere along our phylogenetic history natural selection favored the rise of a behavioral predisposition, which may be termed *Incest Avoidance*, "in order to" avoid the penalties of inbreeding depression. More precisely, those individuals and groups that evolved incest avoidance must have been at a selective advantage over those that failed to do likewise. As a result, incest avoidance is a biological feature of all human populations. Indeed, the predisposition is very probably an ancient selection, for it is common to many species, as we shall presently see.

As a cultural fact, the incest taboo is a much more recent selection. Still, with the advent of culture, it could hardly fail to emerge, for the taboo is a reinforcer of the biological predisposition and, thus, a form of insurance against inbreeding depression. Gradually, the taboo surrounded itself with any number of variations that facilitated general acquiescence in it. Many societies have legal statutes that prohibit incestuous marriage. Others have evolved norms that threaten great harm against those who would violate the rule.*

The incest taboo is one of the most discussed problems in sociocultural science. The biological guarantee of incest avoidance and the evolutionary connection of the latter to the incest taboo entered the sociological imagination at least as early as Henry Morgan's *Ancient City* (1877). The insight was further developed by Edward Westermarck (1891) in *The History of Human Marriage* with prescient consideration of what is now widely recognized as the proximate mechanism whereby incest avoidance is practiced. We shall return to this topic presently with a discussion of incest-like avoidance even when the incest taboo is missing and inbreeding depression is irrelevant. For the time being, it bears noting that, despite a solid theoretical basis and mounting empirical evidence in favor of Morgan's hypothesis (e.g. Wolf, 1966; Harris, 1968; Shepher, 1971, 1979; Parker, 1976; Wolf and Huang, 1980; Thiessen and Gregg, 1980), social scientists have with few exceptions (e.g. Fox 1967, 1972, 1979; Lindzey, 1967; Fortes, 1969; Ember, 1975; Parker, 1976) dismissed or altogether made sport of the hypothesis (e.g. Starcke, 1889; Tylor, 1889; Fortune, 1932; White, 1948; Lévi-Strauss, 1949; Schneider, 1968; Cohen, 1978).

There are two major sociocultural theories of incest avoidance and taboo competing with the one based on natural selection and inbreeding depression. One, roughly known as the family-solidarity hypothesis, was initiated by C. N. Starcke (1889). The other, known widely as alliance theory, seems to have been first proposed in that same year by Edward Tylor (1889). The first was further buttressed by Malinowski (1927) and Parsons and Bales (1955), among others. It argues, basically, that incest avoidance and the incest taboo developed as mechanisms of conflict regulation within the nuclear family. The rationale is that conflict would be greatly exacerbated, and solidarity or kin reciprocity accordingly reduced, if mother and daughter had to compete for the sexual attentions of father

---

* Occasional incest has been practiced, and is observable, perhaps in all societies. Moreover, as a culturally accepted practice, marriage between individuals related by an $r \geqslant {}^1/_4$ has been featured by the Incas of Peru, the royalty of Ancient Egypt, certain tribes of Uganda and the Sudan, Thailand, and the indigenous Hawaiians, among a few others. Even so, however, culturally accepted incestuous marriage has been the exception rather than the rule, being reserved for those at the top of the dominance order where the males have also practiced polygyny with partners from outside the kin group (see, e.g., Roscoe, 1923a, 1923b; Herskovits, 1938; Rowe, 1946; Harris, 1968).

and son(s), and if, correlatively, father and son(s) had to compete for the like attentions of mother and daughter(s).

Such an argument is rightly considered superficial and specious by some scholars (e.g. van den Berghe, 1980). While sexual competition may indeed add to intra-family conflict, it does not follow that the family as a social group could cope with this sort of strife less efficiently than it does with other types of stress germane to it. Indeed, as the discussion of sociality showed, conflict and consensus are really two sides of the same coin, and at close examination conflict may even be a force of consensus (see also, e.g., Marx, 1847; Simmel, 1955; Coser, 1956). Family-solidarity theory, thus, is founded on a misconception of the role of dissensus in the sociocultural system as well as on faulty knowledge of evolutionary biology, as will become evident.

Alliance theory was further developed especially by Lévi-Strauss (1949) in his famous work on the elementary structures of kinship, although in the introduction to the English translation of this work, Lévi-Strauss (1969) modified his position to view the incest taboo as a fact related to both biology and culture. Further, the theory has widened its focus so that it has become a historical theory of marriage patterns, especially exogamy, between groups rather than a theory of incest avoidance and taboo properly speaking. It states, in a nutshell, that exogamy, or marriage outside one's group, however varied the coefficients of relationship may have been therein, developed in the human past as a way of establishing important intergroup ties of alliance, commerce, and reciprocity. As Lévi-Strauss put it, the "prohibition of incest is a rule of [intergroup] reciprocity." Accordingly, exogamy may also be viewed as one catalyst of civilization, for it broadened social horizons, facilitated reciprocal favoritism beyond nepotistic favoritism, stimulated the division of labor, and facilitated the diffusion of knowledge, thereby stimulating sociocultural variation and selection.

This theory is not completely incompatible with a biocultural theory of exogamy, and indeed it may be an important complement to it. We shall return to it presently. Still, to the extent that alliance theory leaves out the workings of natural selection, it is necessarily specious. Furthermore, while exogamy may forge alliances, it is equally true that it is frequently the cause of intergroup feuds. As Yueh-hwa Lin (1947: 101) put it for the Lolo people of China, "it frequently happens in Lolo society that today two clans may become united as affinal relatives on friendly terms with each other, and then tomorrow, because of some misunderstanding [typically involving interclan marriages], they quarrel, fight, kill each other, and eventually become clan enemies."

In any case, both sociocultural theories of incest avoidance and taboo (and exogamy) fail to cope with the fundamental problem of inbreeding depression mentioned above. Let us reiterate the basic argument: The greater the inbreeding, the lower the inclusive fitness because excessive inbreeding increases inbreeding depression or the probability of harmful recessive alleles appearing in homozygous form. Thus, societies that did not somehow develop or adopt incest avoidance and the accompanying taboo could be expected to have been unfavorably selected in the course of evolution, with the result that today's societies are the descendants of those that did elude incestuous marriages.

Fortunately, this is an area of evolutionary science where the facts supporting the theory are quite compelling. Several studies have shown that there are very deleterious genetic effects of incestuous unions (e.g. Schull and Neel, 1965; Adams and Neel, 1967; Seemanová, 1971; Stern, 1973). Adams and Neel (1967), for example, found that among offspring of incestuous relations the rate of death and/or major birth defects was 1 in 3. Conversely, in the normal control group it was 1 in 18. Interestingly enough, these researchers were able to find that the offspring of first-cousin unions have only a 4–5 percent greater chance than the normal population to suffer from birth defects, supporting thereby the widespread tendency among preagrarian peoples to either permit or encourage marriages between blood first cousins.

A fascinating question remains: at the behavioral level—at the level of the actual choice of a sexual mate—how did individuals "learn" to correctly choose certain individuals and reject certain others as mates? This is the question alluded to above in connection with Edward Westermarck's theory of incest avoidance. It is also a question that underscores a crucial issue of biocultural science: statements of ultimate causation, such as the relationship between incest avoidance and inbreeding depression, sometimes make little sense apart from statements of proximate causation. There may seem to be an irony here in view of my celebration of ultimate causation. But the irony vanishes in the concomitant exuberance over the hypothesis of biocultural interplay.

Studies of marriage in Israeli kibbutzim have revealed behavior that mimics the incest taboo (Talmon, 1964; Shepher, 1971, 1979; Beit-Hallahmi and Rabin, 1977; Kaffman, 1977). Shepher (1971), for example, found that among 2,769 marriages recorded, none had been contracted by couples who had been raised together by the same nursemother surrogate in the same communal children's house during the first six years of life. The finding is remarkable because most of the individuals involved were not kin; furthermore, there was no cultural rule against marriages between them. Interviews, however, revealed that such potential marriage partners loved each other like brother and sister. It thus appears that a sexual aversion automatically develops toward those with whom we spend the first years of our life in intimate contact, whether related or not. Such aversion is termed "negative imprinting" by Shepher. Whatever we call it, we are dealing with an epigenetic, or quasi-ultimate proximate, type of causation that performs at the service of the maximization principle.

Another set of related findings concerns the traditional custom among poor Chinese families of selling or giving away infant daughters to richer families. Here the girls were often raised as prospective brides for the natural sons. Wolf (1966) and Wolf and Huang (1980) found that such intended marriages either failed to take place or resulted in reduced fertility. Thus, Wolf ascertained in a Taiwanese village that in 17 out of 19 such cases the intended spouses failed to carry out family intention, and the two cases involved brides who had entered the grooms' family at an age past infancy. In a broader study of Taiwan, spanning a century ending in 1945, Wolf and Huang (1980) further discovered that the fertility of "adopted" wives was significantly lower than average, whereas, when adopted girls grew to marry outside the adoptive families, their fertility was normal.

Note, in keeping with the hypothesis of biocultural interplay that, as van den Berghe (1980: 154) puts it, such cultural arrangements for child-rearing have the effect of fooling Mother Nature by producing incest-like avoidance. That gives us a good clue to the question raised just above as to how people know to practice incest avoidance. The lethal effects of incestuous relations are so common and evident that some have been tempted to conclude that incest avoidance and the incest taboo are the direct results of such awareness (e.g. Ember, 1975). Such an hypothesis cannot be dismissed out of hand. On the other hand, the findings from the Israeli kibbutzim counsel in favor of a less rationalistic interpretation. What seems to be involved is the development, during the first years of life, of a strong sexual inhibition toward those with whom we grow up that both reflects and triggers off the predisposition of incest avoidance. In short, intimate and sustained contact at an early age leads us to love others like brothers and sisters. A sexual union calls into action a different sort of attraction.

This interpretation is supported by evidence showing that the basic predisposition to avoid incestuous relations is prehuman in origin. It is becoming increasingly clear that incest avoidance, especially between parent and offspring, is common to mammals and many other vertebrates (see, e.g., Hill, 1974; Bischof, 1975; Parker, 1976). Several studies of nonhuman primates reveal that mating between mother and son is simply not in the nature of things. The same applies, possibly to a lesser extent, to father–daughter and brother–sister relations (e.g. Tokuda, 1961–2; Sade, 1968; van Lawick-Goodall, 1971; Jolly, 1972; Parker, 1976).

Similar evidence is available for many species of vertebrates, especially rodents and birds, where sexual inhibition has been observed between littermates and between clutchmates as well as between parents and offspring (e.g. Scott, 1964; Eisenberg, 1967; Lorenz, 1970; Hill, 1974; Eibl-Eibesfeldt, 1975). What is especially interesting is that these comparative findings support the negative imprinting hypothesis, namely the biocultural hypothesis that a life-cycle mechanism both reflects and triggers off the predisposition of incest avoidance. Thus, Lorenz (1970) reports that birds will not mate with each other if they have been closely associated during the early period of development. Again, rats and mice reared together form bonds that "interfere with sexual behavior later" (Scott, 1964; also Hill, 1974). Returning to primates, B. K. Alexander (1970) reports that a male macaque from an Oregon colony reared a young female and in the process displayed many of the behaviors typical of a mother. When the adopted child reached sexual maturity, "father" and "daughter" had sexual relations with others in the troop but maintained a strictly "platonic" relationship with each other.

This is a good place to return to the problem of exogamy, which in an ethnographic context may be defined as a behavioral practice whereby the members of one clan marry members of a different clan. The tendency in social science has been to treat the phenomenon as an extension of incest avoidance. But recently van den Berghe (1980: 155f.) has argued in favor of a "fundamental" difference between the two phenomena and, defining exogamy as a cultural rule prohibiting marriage between any two members of a lineage or clan, has claimed that exogamy "accompanied the invention of unilineal descent and plant domestication." The reasons given for drawing a basic difference between exogamy and incest avoidance are not, however, entirely cogent. He (1980: 156) states, for example, that incest avoidance is "natural (though it often takes a cultural expression in incest *taboos*), while exogamy is a cultural phenomenon (though it often uses the *idiom* of biological descent)."

I am not certain that the distinction between "natural" and "cultural" is very helpful in the present context. Humans have typically lived in societies of 25–30 individuals for most of their evolutionary history (e.g. Pfeiffer, 1977). Such microsocieties either practiced outbreeding or they ran into the problem of inbreeding depression, for after all they were cases of extended families. Outbreeding should not, moreover, have been very difficult to practice. If societies remained more or less constant in size, any population increase would have favored scission, migration, and very probably the multiplication of interbreeding clans or lineages. Accordingly, the practice of exogamy is just as natural as incest avoidance, provided that exogamy maximizes, as it does, the probability of avoiding mates with whom the coefficient of relationship exceeds $1/8$.

The association of exogamy, unilineal descent, and the domestication of animals and plants is especially problematic. Domestication places the emergence of exogamy toward the late neolithic period, namely, some 10,000 years ago, give or take a few. As noted, however, group life of a form that required coping with the dangers of inbreeding is much older; and whether or not we apply the label "clan" to characterize such group organization cannot alter the substance of the issue in question. Indeed, I would argue in favor of the opposite hypothesis that it was precisely the presence of interclan contacts, and thus the movement away from excessive inbreeding, that provided our distant ancestors with a sort of laboratory wherein to discover the basis of the incest taboo as a cultural rule. In short, natural selection *combined* with intergroup contacts to establish incest avoidance on a systematic basis through the practice of exogamy. There are probably few clearer cases of ultimate–proximate causality, or of biocultural interplay, than the present one. The fact that exogamy became more common relatively recently, among horticultural and pastoralist societies, as van den Berghe (1980: 157) suggests, is important, but for another reason.

The domestication of plants and animals constituted a great technological revolution that, through the attendant increase in the food supply, resulted in a population explosion, *inter alia* (Pfeiffer, 1977; Lenski and Lenski, 1978). Demographic pressure, in turn, expanded the number of societies and, therefore, the opportunities for exogamous arrangements. Finally, population increase and the emergence of new societies were facilitated by a lowered rate of inbreeding and its lethal consequences. There have thus been complex biological and cultural factors associated in the related phenomena of incest avoidance, incest taboo, and exogamy.

To place the origin of exogamy at a moment so near in evolutionary history is to miss the deep and extensive roots of human history. What was said above is probably proof enough for this statement. The hypothesis of a much more ancient origin of exogamy is favorable, however, on other grounds. Pusey (1980) reports, for example, that free-ranging female chimpanzees at the Gombe National Park in Tanzania in effect practiced both incest avoidance and what may properly be termed a form of chimpanzee exogamy. Thus, upon reaching sexual maturity, not only did they sever relations with their brothers and other previously intimate males; with few exceptions, they also transferred more or less permanently to other chimpanzee troops where, during prior visits, they had met sexually attractive males. In view of the very close genetic similarity obtaining between human beings and chimpanzees, there is reasonable ground to hypothesize that exogamy is prehuman in origin and that the two forms of it among humans and chimpanzees are homologous in nature.

It follows, moreover, that there is some validity to alliance theory, and, correspondingly, that the usual sociobiologist's emphasis on natural selection for purposes of the problem at hand may be plainly excessive. Without some such things as intergroup alliances and commerce, natural selection probably would have had no chance to produce enough incest avoidance to add materially to the factors behind the steady population growth in the history of *Homo sapiens*. Wilson (1978: 38) appears to be guilty of exaggeration, therefore, when he argues, for example, that alliances and other exogamy-related phenomena were "felicitous results of outbreeding, but they are more likely to be devices of convenience, secondary cultural adaptations that made use of the inevitability of outbreeding for direct biological reasons." In a sense, the argument is irrefutable. Let's face it: our ancestors either practiced incest avoidance through some arrangement or other, or their genes simply failed to replicate at a self-sustaining rate. Still, if interest in the biocultural hypothesis is to be productive, a livelier systemic view of the biological and the cultural is essential for the problem at hand. Such a view suggests that the road to incest avoidance was strewn with countless errors and population extinctions. Errors, however, became fewer, and incest avoidance was promoted to incest taboo as the conditions for exogamous practices came to flourish.

The cultural factor is not nearly as important to a theory of human behavior as environmentalists claim. Still, it is very probably much more significant than sociobiologists are still able to imagine. Consider another fact about incest. Wilson (1978: 36) points out that in general "mother–son intercourse is the most offensive, brother–sister intercourse somewhat less and father–daughter intercourse the least offensive." This pattern does not make sense to me from the vantage point of inbreeding depression and the cultural reflexes of that phenomenon. From this perspective, mother–son intercourse is likely to take place at a time when the mother is either barren or at least reaching that biological state. Conversely, this natural form of contraception is less likely to be present in father–daughter and brother–sister intercourse. I refrain from speculating about the psychocultural factors possibly associated with the facts presented by Wilson, but the fact remains that from a purely biological perspective the pattern is a bit surprising.

9.3 Women, Men; Marriage and Children

There is old evidence, abundantly and recently confirmed (e.g. Rosaldo and Lamphere, 1974), that women have been rather universally perceived as being in a general way inferior to men. And they have been treated accordingly. Contrary to some romantic misconceptions, women have had with few exceptions the worst of it in the most primitive societies. At the turn of this century, the Yakut women of Siberia, for example, occupied a slave-like position in the system of property relations; a husband had almost absolute rights over her, including the right to beat her and to put her out to hire, though he could not sell or kill her with impunity (Kharuzin, 1898). Even after the advent of the Soviet regime, the Yakuts considered women unclean and unworthy of touching certain sacred objects (Popov, 1933), much as many totemic peoples had considered women as profane (Durkheim, 1912).

Such facts, and worse ones, have been fairly typical of human society in general, and they are probably still far from extinct. From the viewpoint of evolutionary science, I find them entirely absurd, though perhaps not altogether immune to explanation. What is totally mystifying is the position of some critics of human sociobiology that this discipline supports the anti-feminist prejudices of the male gender (e.g. SSGSP, 1976).

It would be preposterous to argue in a formal sense that in the calculus of evolution either gender is in a general sense superior to the other, although there are specific dimensions along which superiority can be attributed now to one gender, now to the other—for example, speed in motion and care of the newborn. In a partly ideological key, however, if it is possible to argue that human sociobiology is anti-feminist, as the Sociobiology Study Group of Science for the People (1976) has done, it is equally easy to argue in favor of the contrary proposition. One could even stretch the point and argue that, from the viewpoint of evolutionary biology in general, men are to women as dirt is to gold. The fact that men, like most other males in the animal kingdom, have with some exceptions lorded it over their female partners does not necessarily gainsay that proposition. It merely shows the great weight of physical dominance and the relatively crippling effects of such female conditions as pregnancy and lactation. Female subordination results not from biological inferiority but from the cultural repercussions of differential biological functions. As a consequence, since to an extent the word and the pen are becoming "mightier than the sword," the hypothesis of a developing symmetrical relationship between men and women, if not an outright reversal of dominance, can no longer be easily discounted. Certainly, sex roles will never again be the same; the evidence is mounting that the changes are considerable (e.g. Kanowitz, 1969; Huber, 1973; Leach, 1980; Reiss, 1980; Tax, 1980; Dawson Scanzoni and Scanzoni, 1981; Hutter, 1981; Lamanna and Riedmann, 1981).

The pro-feminist hypothesis could, moreover, lean on some interesting facts. Women are intrinsically healthier and physio-anatomically better constructed than men. The reason is simple: they have been selected to be the bearers and socializers of the generations that give continuity to the species. For the fundamental task of living, women are the stronger, not the weaker, sex. Their greater longevity is not an accident of the division of labor and other environmental factors, although these most assuredly have a bearing on it. It is much more a reflection of a superior phylogenetic product—by which I mean especially a more complex and more finely tuned reproductive organism.

There is some basic evidence to support this position. Trivers and Willard (1973) have predicted that among vertebrates in general, and especially among mammals and birds, the more precarious the physical condition of females, the greater the production of female as contrasted to male offspring. The prediction holds for humans as well as for a number of other species. There are various reasons for this finding. To the point in the present context is the fact that male fetuses have a greater mortality rate under conditions of adversity than do female fetuses (Wilson, 1978: 39).

For dramatic proof of the possible evolutionary superiority of women one might look at the special nature of their reproductive system. On the average a woman produces about 400 eggs in the course of a lifetime, each of which is approximately 85,000 times larger in size than a sperm. By contrast, it is not an exaggeration to say that a young and healthy man is capable of spewing out hundreds of millions of sperms in the course of a single day. Consider, then, that if tomorrow, by an unhappy and unlikely event, the world were to be rid of all men but one, within a relatively short period of time the size and composition of today's world population would be fairly faithfully reproduced. Given present scientific developments and *in vitro* conception, the lucky fellow would not even have to strain himself in endless mating exercises. Reflect, conversely, on the evolutionary future of a human society consisting of billions of men and only one reproductive woman. Chances are that it would soon become extinct. At best, it would take thousands of years to reach again its present size and composition.

But all this is polemic. Let us turn to more serious issues. Given that women have been constructed according to the criterion of gamete scarcity, and given the maximization principle, what can be predicted about male–female role differences within the family context? The most basic fact is that in every human society the major responsibility for child-rearing, at least in the first half dozen years, devolves on women. This greater maternal investment in children is typical of mammals in general. There is a double but indivisible reason for this circumstance, at least for our species. In the first place, at least until recently, the mother has been responsible for the nutrition needs of the infant with her very body. In the second place, this role fits fairly well her evolutionary interests, for, as we have noted, her investment in a child is very great, and there is no one better suited to safeguard that investment than herself. Little wonder then that modern attempts to free mothers of their traditional role, like the communal child care instituted in Israeli kibbutzim, sometimes run into female resistance (Tiger and Shepher, 1976). Little wonder, too, that in modern societies, where women have been treated more like mothers than like concubines or baby-makers, divorce decrees have typically favored women in the assignment of children. Legal variations have thus been selected in keeping with evolutionary logic, at least in this crucial respect.

Taking a wider zoological view, we note that the difference between males and females in the degree to which they defend or enhance their fitness via care of their offspring is so great that, in many species where parental care is typical, the father plays little or no role at all in the upbringing of his children. From the male viewpoint there is a variety of reasons for this circumstance. One, as we have seen, is related to the fact that paternity is often associated with extreme male–male competition and, thus, with an unusually large number of progeny among the successful competitors. The matter relates to the phenomenon of polygyny, or a mating pattern between one male and multiple females, and we shall turn to that presently. A second, more fundamental, reason suggests in my view that the very fact which makes the male compete for multiple mating partners, namely, his great mass of sex cells, is also responsible for a psychology that *tends to* conceive of offspring as cheap issue and therefore relatively negligible. Perhaps more fundamental still is the fact that there is a major difference between males and females: the latter are much more sure of their maternity than the former are of their paternity. *Pater semper incertus*, the ancient Romans used to say about our species—paternity is always doubtful. Earlier still, in Homer's *Odyssey*, Telemachus converses as follows with Athena about his father:

> My mother saith that he is my father;
> For myself I know it not,
> For no man knoweth who hath begotten him.

These considerations, and especialy the fundamental difference in the production of sex cells existing between males and females across the animal kingdom, suggest also a radical difference between the two sexes in their respective mating strategies. Briefly stated, males have tended to pursue a multiple-mate strategy, or polygyny. Females, by contrast, have tended to insure their inclusive fitness through hypergamy, or mating with males who dispose of superior influence, power, and material resources, and thus give greater promise of enhancing the offspring's probability of survival, their reproduction, and their transmission of parental genotype to the future. Quantity is the male strategy. Females have evolved to emphasize quality. Should we wish to return to the ideological controversy, this would be another way of comparing dirt with gold.

There are various forms of marriage or mating strategies in nature. Monogamy, or one-partner marriage or mating, is probably a more recent evolutionary arrangement than polygamy, that is, marriage or mating to multiple males or females, and it tends to occur when selection and ecological pressures force mating pairs to establish more or less permanent bonds by equalizing parental investment. The father in effect is forced to pursue a quality strategy. Among the most crucial pressures are: (1) extreme scarcity of valuable resources in the given territory along with the attendant necessity that parents cooperate in defending them against possible intruders; (2) such severity of the physical environment that both parents are needed by the offspring to cope with it; and (3) early breeding (Wilson, 1975b: 327, 330–1).

Monogamy is widespread among birds. About 90 percent of all bird species are estimated to be monogamous, at least during a given breeding season; a few even form lifelong pair bonds (e.g. Crook, 1965). Such a strategy seems to be instigated by scarcity of nest sites or food sources (Lack, 1968) along with the high metabolic rate of birds and the necessity, therefore, to satisfy the offspring's voracious appetite. In general, the greater the need of parental cooperation for the survival of the clutch, the greater the monogamous tendency. It follows that the males of some species are torn between a polygynous strategy and a monogamous one. Long-billed marsh wrens, for example, have a strong polygynous tendency, and, having engaged in one mating, the males get busy soliciting additional mates. When, however, the search proves futile, or the breeding season approaches the end, they mend their prodigal ways and become diligent partners in the task of provisioning their offspring (Verner, 1964).

In general, however, in view of the great difference in gamete production between males and females, along with the phenomenon of anisogamy, referring to the difference in size or biomass investment between the two sex cells, polygamy is the most widely diffused arrangement in the animal kingdom. A number of general conditions favor polygamy, of which polygyny is the more common form. They include: (1) superabundance of food; (2) precocial young, or early maturation; and (3) sexual bimaturism and long life (Wilson, 1975b: 327–30).

For example, when the newborn are capable of a high degree of independence, and mothers can guide them to well-protected and rich feeding areas, the need for the participation of fathers is reduced, and predictably these devote much time to other activities, including competition for additional mates. This condition obtains among many mammals, and is especially marked among herbivores. Accordingly polygyny is widespread here (e.g. Eisenberg, 1966). Female mammals further reduce the necessity of paternal assistance by their unique adaptation for nourishing the young with milk. At the same time, grazing by the young can be done independently of either parent. It is interesting to note, however, that mammalian monogamy is most common among carnivores, such as foxes and coyotes, where the father can help the young with food that is scarce and hard to secure (Barash, 1977).

A preference for limited polygyny has been typical of human society, too. Thus, Murdock's (1967) *Ethnographic Atlas* shows that only about 13 percent of nearly 1,000 preagrarian societies are classifiable as strictly monogamous (see also Lenski and Lenski, 1978: 122, 358), while about 76 percent allow polygyny. The same data indicate that in the majority of societies marriage patterns have evolved toward various cultural selections that culminate in an economic transaction. The practice is especially common in advanced horticultural societies where, almost without exception, marriageable females are considered a valuable economic asset, and men must compete for them with a bride price, perhaps the most notable such cultural selection. While family networks often come to the aid of indigent would-be husbands, often too the best and the most brides go to the wealthiest, with the result that some men remain bachelors. Indeed, the practice of polygyny has been so tenacious in horticultural societies that apparently it has been one major cause of the Christian missionaries' failure to christianize sub-Saharan Africa, or, conversely, one chief factor in the African resistance to the onslaught of European religion (Lenski and Lenski, 1978: 412).

Polygyny is not limited to preagrarian peoples. Muslim countries, representing a large portion of the human population, have with few exceptions allowed it. Only industrial societies presumably do not approve of it (e.g. Lenski and Lenski, 1978: 358). Such a development would seem to represent a radical departure from the mammal trend and, to a lesser extent, even from our own earlier cultural eras. This is strong evidence that changing economic conditions and cultural selection in general can intervene to modify the operation of biological forces. Furthermore, it is undeniable that human beings are strongly inclined to form pair bonds that are intense in the quality of the relationship, and endure years rather than seasons.

Still, raw biological forces seem to have a way of retaining at least a degree of influence by hook or by crook. Thus, if on the one hand we have evolved such variations as legal strictures against any form of polygamy, on the other we have legal loopholes with which to bypass the restrictions. Divorce and remarriage constitute polygamy in all but the name. Indeed, it corresponds to what has been termed "serial" polygamy, that is, a mating arrangement whereby a person secures multiple spouses, one at a time. And we should not disregard the question of marital infidelity, which is by all accounts quite considerable. Under the circumstances, given the high rate of divorce and remarriage in industrial societies, the really remarkable change that has taken place is not the alleged disappearance of polygyny but rather the emergence of *de facto* polyandry, or marriage to multiple husbands, on a rather equal footing with polygyny. Until recent times, *de jure* polyandry was found in less than 1 percent of human societies, and *de facto* polyandry was subject to severe punitive action (Murdock, 1967). Perhaps it is true that industrial society is a great equalizer in many respects. What seems even more certain is that traditional sex roles have been changing, and women who seek greater equality of rights, responsibility, and obligation with men would seem to have not only nature as a witness but also the support of a likely new trend in biocultural evolution.

Indeed, it is quite possible that women are getting to be in that proverbial position wherein they can have the cake and eat it, too. If *de facto* polyandry places them on an equal footing with men, hypergamy puts them in a privileged position. We have already noted that in many species most males are non-reproductive. Polygyny among human beings amounts to the same result on a reduced scale. It also implies a tendency on the part of women, in cases of heterogamy, or cross-status marriage, to marry men of superior social, political, economic status: to practice hypergamy. Thus, the emphasis on quality, as contrasted to quantity, in the production of gametes reflects on the cultural emphasis on quality through the institution of hypergamy. To the extent that we may think of status differences in terms of the predisposition of heterologous affiliation, it is this combiner that has worked together with the maximization principle to produce the selection of hypergamous mating. Returning to previously treated kinds of behavioral predispositions, we may add that hypergamy has been selected also under the influence of the climbing maneuver. It is, therefore, all the more lamentable that, with few exceptions, studies of social mobility have focused almost exclusively on male mobility.

Heterogamy has been documented in various places and times, for example, Ciceronian Rome (Cowell, 1956); English society of various periods (Habbakuk, 1950; Holmes, 1957; Stone, 1960–1; Thrupp, 1962; Foster, 1968); fifteenth-century Spain (Mariéjol, 1961); eighteenth-century France and earlier (Barber, 1955; Bloch, 1970); nineteenth-century India (Weber, 1946) and earlier (Dickemann, 1979); mandarin and pre-revolutionary China (Hsu, 1948; Dickemann, 1979); and contemporary Irish (Arensberg and Kimball, 1940), Greek (Friedl, 1962), and Italian societies (Hazelrigg and Lopreato, 1972; see also, for various societies, Burgess and Locke, 1953; Berent, 1954; Miller, 1954; Lipset and Bendix, 1959; Goode, 1966; Blake, 1974; and Murstein, 1976).

Social scientists tend to view patterns of interclass marriage as the fundamental barometer of social class closure or openness. Thus, for Schumpeter (1951) the "most important symptom [of interclass fluidity] is the ease or difficulty with which members of different classes contract legally and socially recognized marriages" (see also Barber, 1955; Tilly, 1964; Lopreato, 1967). The observation is especially useful for purposes of understanding the distribution of social honor or prestige in society. It is probably less appropriate for a grasp of the relationship existing between marriage and politico-economic phenomena (see Weber, 1946: ch. VII, for an important distinction between social, economic, and political or legal dimensions of social stratification). When it is in people's economic interest to marry outside of their class, however far down they have to reach in a social sense, they are likely to penetrate social boundaries rather readily without thereby leaving the walls of social barriers in ruins.

The important point for our purposes is that the several scores of studies available on the relationship of social class and marriage show that most marriages are homogamous, or more specifically isogamous: they take place between members of the same class. However, when heterogamy is observable, "men tend to marry down and women tend to marry up" (Burchinal, 1964; but cf. Laumann, 1966; Rubin, 1968). That is, hypergamy prevails over hypogamy.

Social science provides several explanations for the relative predominance of hypergamy. A common one states that hypergamy is a result of male social dominance. Male rank in the social stratification allegedly determines the rank of the family as a whole; therefore, a man has little or nothing to lose by choosing a mate of lower position (e.g. Merton, 1941; Goode, 1966; van den Berghe, 1978b), while his wife and children stand to gain from that strategy. A second explanation argues in effect that, to the extent that heterogamy is unavoidable, it is more likely to be of the hypergamous than of the hypogamous variety because "husbands feel their male dominance threatened by a wife of higher class status" (van den Berghe, 1978b: 129).

The mixture of explanations is a healthy sign of recognition that complex phenomena require recourse to complex types and levels of theoretical achievements. Certainly the above explanations are entirely compatible with, if not altogether complementary to, the biocultural explanation alluded to in the foregoing pages. To begin with, scholars have noted that hypergamy is even more widespread in other social animals, especially among birds and mammals, than in *Homo sapiens*. Consider that among females there is little variation in the frequency of mating. Among males, conversely, large, healthy, dominant individuals mate more frequently than smaller, less healthy, and submissive ones (e.g. Trivers and Willard, 1973; Barash, 1977; Wilson, 1978).

Herein lies the ultimate, though partial, explanation of animal hypergamy. For it turns out that according to the logic of natural selection, and some basic differences between males and females, the latter are "more fussy than males about whom they mate with" (Dawkins, 1976: 175). Hence, part of the explanation lies in the mating preferences of females. Women should be selected for rating men in large part by the reproductively relevant resources these control.

By now we are familiar with the underlying logic. The gist of it is roughly as follows. Natural selection, it will be recalled, entails the proposition that organisms are, with some possible exceptions, programmed to maximize their inclusive fitness, namely, the transmission of their genes to future generations through offspring and/or other relatives. Now, if men could in principle fertilize millions of eggs a day, whereas women can only accommodate roughly one sperm per year, and for a limited number of years, we should expect the two sexes to have evolved partly different perspectives on reproduction. Specifically, women have more to lose from reproductive mistakes: from matings that decrease the probability of their giving birth to healthy offspring and nurturing them to sexual maturity. On the average, therefore, they may be expected to be more careful, more choosy, in the search for a mate. It follows that, since socioeconomic status or a high position in a dominance order is in the typical case directly related to the probability of offspring survival, to the extent that heterogamy exists, women should feature a greater tendency toward upward mobility (hypergamy) than downward mobility (hypogamy).

We have been speaking, however, at an extremely high level of abstraction—at the level of the ultimate biological causation. But we have also embraced the thesis of coevolution, or the interplay of biology and culture. Thus, not only should we be receptive, cautiously, to various explanations of hypergamy but also we should not be surprised to encounter cases of hypogamy in time and place. Moreover, for sociocultural reasons alone we should expect to find that, as Rubin (1968) has noted in an analysis of Blau and Duncan's (1967) data on the American occupational structure, heterogamy in general is likelier between some occupational groups than between others. In Italian society, for example, there is a formidable social barrier between farming and nonfarming occupations, with the former having a much lower status than the latter (Lopreato, 1967). As a result, very few men of nonfarm social origin marry farmers' daughters (Lopreato and Hazelrigg, 1972). We may be biologically programmed to follow self-seeking strategies, but culture, like a harness on a horse, rarely gives biological forces free rein. The leash pulls both ways. On the other hand, it is also true that traditionally Italian farmers have not been an economically powerful class; consequently, there has been little biologically relevant reason for nonfarming men to marry socially downward.

Still, in favor of the coevolution hypothesis, let us consider the following. In view of the scarce and precious nature of female gametes, we would have to predict from a purely biological perspective that the bride price is a very widely diffused variation in human society. To be sure, it is not uncommon. But neither is it universal. In many societies, the pattern is entirely missing. In others, we find the very opposite, namely, what amounts to a groom price. Returning to Italy, for example, at least in the south of that society the groom has traditionally had few burdensome economic obligations toward the marriage contract, whereas the bride has typically been expected to "bring a dowry" that represents a considerable investment in time and money on the part of her family. In cases of marriage to a *socially* superior man, the size of the dowry may be expected to increase significantly in value, for many a man of superior status has found a *sistemazione* (security for the future) precisely by exchanging social status for material resources.*

The factors behind the phenomenon of the bride dowry, as contrasted to the bride price, are numerous. Probably they also vary in time and place. For the case of southern Italy, and in recent times, one would have to consider at least the following. War and the emigration of single men (of which the latter was, in turn, partly a result of poor economic conditions) have resulted in a scarcity of grooms. Therefore, the absence of polygyny, which would have to be explained in part by recourse to complex religious and historical factors, has tended to intensify competition among females for marriage partners, thereby reversing to an extent the more common pattern found in nature. Moreover, a long history of feudal abuses against the genetic fitness of the common man exacerbated male jealousy, resulting in an extreme emphasis on family honor, which has been assessed largely in terms of the absolute certainty of chastity among unmarried women. Consequently, marriageable daughters have constituted a tremendous psychological burden on the family, and the response has been to get rid of it at any cost.

But note, in conclusion of this section, that only for strictly analytical purposes can we separate the cultural from the biological. Thus, the low supply of marriageable men has been a result of war and territorial behavior, in the form of emigration, which are in part rooted in biological factors. Again, the emphasis on honor is only apparently a purely cultural fact. At close examination, it appears to be associated with the drive toward maximizing one's fitness, which, as noted, was hindered by the feudal aristocracy.

9.4 Heterologous Contraposition

9.4.1 THE GENERALIZATION OF KIN SELECTION: ETHNICITY AND ETHNOCENTRISM

Social scientists almost to a person have understood family and ethnicity to be based on fundamentally different principles of affiliation. A few, conversely, have gone so far as to view nations as well as ethnic groups as descent groups, and thus as biological extensions of the family aggregate (e.g. Francis, 1966; Keyes, 1976). I reject the descent hypothesis on grounds that, while early in the formation of ethnic groups and even nations biological kindredness may be a crucial factor of affiliation, in time

---

* The distinction in social science between hypergamy and hypogamy is not always appropriate. It tends to confuse two important but different concepts, namely, social status, which is a weaker source of fitness, and social class (or material resources), which is a better predictor of genetic fitness.

there typically occurs such a mixture of lineages that the concept of kinship loses most or all of its biological relevance.

Still, since the drive toward the enhancement of inclusive fitness is a very powerful human motivation, there is little ground to doubt that the favoritism that is observable within the kin network tends to generalize to larger aggregates, with the result that these feature much behavior that makes them appear as extensions of the family group. Indeed, we have little or no choice but to think in these terms, and the option was removed by the facts of assortative mating, *inter alia*. This phenomenon proves that kinship is really a matter of very wide-ranging degree, and people tend to favor, if not entirely fabricate, even the slightest coefficients of relationship. The predisposition toward homologous affiliation, or the parochial inclination, seems then to radiate far afield from the narrow family context.

To a degree, we may therefore call on kin selection to aid in the task of grasping certain fundamental forms of behavior that concern ethnicity and ethnic groups. I must stress the term *fundamental*. Kin selection theory cannot explain everything. My interest continues to focus on elementary forms, on aspects of behavior that may be associated, however tentatively, with the principles of kin selection. The wealth of behaviors left out concern both additional loci of potential scientific propositions and the great array of exceptions that usually surround scientific laws and underscore the need for additional contingencies or qualifiers.

Pettigrew (1978: 25), among others, has recently remarked on the universality or near-universality of ethnicity and the special difficulty of defining the phenomena that are associated with it. Universality translates into a great variety of forms in which ethnicity is experienced, and that certainly complicates matters. There is the further question as to whether ethnicity should be defined objectively or subjectively (e.g. Barth, 1969). An objective definition poorly recommends itself because typically, at least in contemporary society, individuals represent ethnic compounds rather than pure ethnic strains. That is a problem of boundaries (Pettigrew, 1978). Fortunately, a subjective definition is likely to encompass, at least in part, the objective facts, and that would seem to favor either a subjective definition or one constituting a mixture of the objective and the subjective.

I conceive of ethnicity as a property of groupings which, because of their large size, render the formal coefficient of relationship of little or no practical use, but which nevertheless constitute roughly what geneticists call breeding populations wherein assortative mating is practiced. The individuals constituting such ethnic groups belong or believe that they belong to a line of people who have at least one major feature in common, whether biological or cultural, for example, tribal membership, nationality, religion. Such a feature stimulates homologous affiliation, and is typically experienced as comforting to personal psychology, as enhancing the probability of marital success and continuity, and as supportive of longstanding traditions (the ethnic "subculture"). Examples of ethnic groups in a society like the United States of America are the Irish, among whom we might further distinguish between Catholics and Protestants, the Poles, the Germans, the Italians, the Norwegians, and so forth. There are few if any nation states nowadays that do not constitute ethnic mosaics (for some general statements on ethnicity, see Glazer and Moynihan, 1963; Gordon, 1964, 1978; Newman, 1973; Greeley, 1974; Dinnerstein and Reimers, 1975; Geschwender, 1978; Bahr et al., 1979; Luhman and Gilman, 1980; McLemore, 1980).

Homologous affiliation is based on the positive emotion of attraction (see Fishbein and Ajzen, 1975: ch. 6, for a discussion and review of this topic). However, it does not operate in a vacuum, for the salience of attraction is in part a reflection of its opposite emotion, avoidance. Indeed, it is probably true that the greater the level of attraction within a given ethnic group, the greater the propensity toward avoidance of other ethnic groups. In short, there is another side to homologous affiliation, and we may conveniently term it the predisposition of *Heterologous Contraposition*, or just *Contraposition*. By contraposition is meant a biologically based predisposition to have negative attitudes toward others on the basis of real or perceived differences that are typically, though not exclusively, of an ethnic nature.

Ethnicity is a filter for the dialectical tension existing between homologous affiliation and heterologous contraposition. The acid test of this proposition may be found precisely in a complex society like the United States of America. Yinger (1976) estimates that in 1957 close to two-fifths of all religious intermarriages that could have occurred did in fact take place. The figure may be considered large or small according to one's own predilection, but the fact remains, if the estimate is correct, that the majority of marriages took place between Protestants and Protestants, Catholics and Catholics, Jews and Jews. Indeed, within these religious categories, there has been a considerable degree of avoidance in the selection of marriage mates based on differences in nationality origin. Among Catholics, for example, the Irish, the Italians, the Poles continue to show a marked preference for marriage partners of similar national origin (e.g. Abramson, 1973).

In a general sense, the relationship between homologous affiliation and heterologous contraposition may summarily be expressed as follows: the more phenotypically (and, by implication, genotypically) similar two persons or groups are to each other, the more they tend to attract each other and to practice mutual favoritism; conversely, the more phenotypically dissimilar they are, the more they tend to avoid each other and, as the occasion arises, oppose each other's aims, moral standards, and the like.

This conceptualization introduces the well-known concepts of prejudice, discrimination (the active phase of prejudice), and ethnocentrism (Sumner, 1906; Allport, 1954; Levine and Campbell, 1972). Prejudice, discrimination, and ethnocentrism (along with xenophobia and parochialism) are, therefore, fundamental properties of heterologous contraposition. Briefly, prejudice represents an emotional aversion to a person or group on the basis of such morally and/or phenotypically relevant criteria as religious, ethnic, racial, and class differences. For example, at the turn of the century, and to a lesser extent even more recently, immigrants to the United States of America from southeastern Europe were considered undesirable candidates for American citizenship because of differences in religion, tradition, and physical type. Discrimination refers to behavior, typically based on prejudice, that is consciously or unconsciously motivated to exclude another from access to desirable and fitness-enhancing resources. For example, the US immigration laws reject a group in favor of another strictly on the basis of ethnic differences. Ethnocentrism concerns a complex of attitudes whereby members of one group assess the cultural, social, biological attributes of another group in terms of their own "superior" attributes and consequently find the latter wanting, inferior, even despicable. The socialist system, for example, is "superior" to the capitalist system—and conversely, depending on the point of view (for an excellent propositional analysis of Sumner's celebrated concept of ethnocentrism from the perspective of various types of social theory, see Levine and Campbell, 1972).

Returning to the question of marital preference within religious categories and, within these between members of similar national origin, we note that the three concepts just defined are very closely related. Moreover, they tend to generalize to multiple phenotypical traits. The findings, for example, suggest prejudice and ethnocentrism along both religious and nationality-group lines. Some scholars have tended to deny this connection between prejudice* and religion; and indeed the available facts are a bit difficult to decipher. The basic problem may lie in the seemingly curvilinear nature of the relationship between the two variables. Prejudice appears to be lowest both among the least religious and among the most religious. It allegedly reaches its maximum among people in the middle (e.g. Allport, 1963; also, Greeley, 1972). The latter category, however, represents a majority of the people in most times and places.

From our perspective, these data on religiosity would suggest that in a minority of the population, religion may engender a degree of beneficence toward others that, as noted in Chapter 6, may even reach the level of ascetic altruism. Alternatively, or additionally, the allegedly lower rate of prejudice among the least religiously inclined is associated with highly educated individuals who are strong in reciprocal favoritism and the

---

* The term prejudice may be used conveniently as a synthesis of all three related concepts.

"soft" type of "altruism." They are thus experts in the art of a global cultural perspective, or a high degree of cultural relativism mixed with a flair for the sophistries that most effectively hide the bane of prejudice. Moreover, such people are the most mobile, both socially and geographically, and thus have fewer opportunities to exercise the prejudice that along various dimensions is associated with narrow favoritism.

In a classic study of ethnicity in the United States of America, Milton Gordon (1964, also 1978) found that there is much more acculturation in this country than assimilation; that is, there is a movement toward cultural homogeneity along with a tendency in each ethnic group to marry endogamously and at the same social class level, the "ethclass" (for an analogous and updated statement, see Newman, 1973). These findings are to an extent questioned by Weinfeld's study of Toronto (referred to in Pettigrew, 1978: 30). Although a more recently constructed ethnic mosaic than the typical city in the United States of America, Toronto allegedly features an assimilationist tendency: "structural" separateness, or ethnic exogamy, declines directly with education and with each generation. The study, however, uncovers a strong "affective ethnicity," that is, a tendency toward ethnic self-identification combined with support for cultural pluralism (for a broader view of the ethnic situation in Canada, see the collection by Driedger, 1978). Analogous findings are available for the United States of America (e.g. Greeley, 1974; Alba, 1976). As Pettigrew (1978: 31) puts it, the core of ethnicity may be its "affective basis." Ethnocentrism is a good label for this idea, and the support for cultural pluralism may at least in part constitute insurance against the prejudice and discrimination of others.

When viewed in affective terms, ethnicity facilitates understanding it as a fluctuating phenomenon. The intensity of the emotional attachment to the people who once were a tighter breeding population rises and falls in view of any number of environmental factors. "At just the moment when most social scientists were anticipating the rapid extinction of the remaining significant differences among white ethnic groups," McLemore (1980: 325) has recently written, "these distinctions seemed suddenly to revive with startling intensity." Such a recrudescence has led some observers to overstate the case. A heading in one of M. Novak's (1971) books puts it bluntly as follows for the United States of America: "not a melting pot—a jungle." The reasons adduced for this "ethnic affirmation" or "emergent ethnicity" are numerous. They are all useful. With few exceptions, however, they also steer away from considerations of ultimate causation. Accordingly, they leave untouched the plain fact that, for one "reason" or another, ethnic behavior, and ethnic conflicts, have a relentlessly devilish way of effervescing time and time again throughout this Mother Earth of ours (e.g. Gelfand and Lee, 1973).

A recent study of ethnic intensification in the United States of America reviews many of the sociological explanations of the phenomenon and finds that the traditional explanatory emphasis has been on "the importance of the portable heritage which a group brings from one generation and place to another" (Yancey et al., 1976). By contrast, these authors argue that ethnic culture crystallizes as ethnic heritage and persists in time to varying degrees in association with macrosocial structures, in particular with "the structural parameters characterizing working-class life generally"—for example, residential stability or segregation and related occupational positions.

At a proximate level, both point and counterpoint make good sense. The same may be said of arguments emphasizing the fact that, as Novak (1971) again puts it, for many recent citizens of the United State of America, growing up in America was "an assault" upon their "sense of worthiness." Thus, one is required to agree at least in part with Lewontin's (1979) earlier objection that such behavioral categories as dominance and tribalism are "historically and ideologically conditioned constructs." Still, Novak comes close to a much needed ultimate type of explanation. For if, for example, growing up in America wreaked havoc with fundamental social needs—for example, the need for recognition—sooner or later such a need had to be fulfilled. On this basis alone, as those who yesterday were grossly humiliated and now attained the objective accomplishments on which rested the hauteur of their proud degraders, they could easily be expected to get good and angry and yell, "I, a Pole, a black, a Jew, am as good as you ever could hope to be."

"But much more than pride was involved here. The new groups being revived (or formed) were very concerned about the rapid changes taking place in American life and the effects of these changes on their lives. This concern was particularly visible among white ethnics of lower-middle-class status, those who worked mainly in blue-collar jobs." This statement by the sociologist McLemore (1980: 325–6) is closely related to the argument posed by Yancey and his associates (1976), among others. It, however, adds to their explanation, and to his own, more proximate, ones, a dimension that helps to cast better light on ethnic revivals in all times and places. Consider that natural selection implies the corollary that individuals and groups are constantly on guard against any change in the distribution of the existing pie that is not to their own advantage. After all, a redistribution that favors others is indication enough that natural selection operates in their favor. Not to resist a change, real or perceived, that is detrimental to one's own share is to contravene the direction indicated by the maximization principle.

In particular, the "new ethnics" were worried that the gains of black Americans, which *appeared* to be enormous in recent years, would be solely at their expense (Novak, 1971: 14). As McLemore (1980: 326) puts it in a broader key:

Most of the large-scale changes of the 1960s were perceived by working-class ethnic Americans as threatening to undermine their hard-won gains. Many white ethnics feared that the rules of "making it" in America had been changed. The War on Poverty, the apparent success of confrontation politics, and the growing opposition to the war in Vietnam all created the impression that hard work, obedience to the law, and patriotism were no longer highly valued in America. They felt betrayed and cheated.

But let us beware not to make too much of the behavior of those urban, working-class, white ethnics. Among the virtues of McLemore's volume is a tendency to treat facts with the scientific respect that they deserve. There is, however, another tendency in academia—liberalism, compassion, and "objectivity" notwithstanding—to take easy shots at working-class people, just as there is an inclination to view prejudice (that of others, of course) as nothing less than a mental disease. And yet, have you noticed those bumper stickers that in the United States of America proffer the invitation to "secede" from the Union? They are fairly common in the states that are rich in mineral resources, but they reflect no working-class, white-ethnic, phenomenon. Many Americans claim that there is "an OPEC within." Some even speak of a veritable civil war being waged with taxes rather than muskets, and with the North at the losing end this time around. The source of contention is the "severance tax," or duties levied on exports of oil, natural gas, and coal from one state to another. In truth, severance taxes are nearly a century and a half old, but only in recent years has a total of at least thirty-four states joined "the internal OPEC club." Those who suffer from such taxes naturally speak of a tax warfare and even of "the largest transfer of wealth in the history of this country." There is here evidence of parochialism on a large scale that is truly exacerbated by predispositions of self-interest.

Sometimes, parochialism is greatest there where it is most vociferously condemned. Thus, New York, the "Big Apple," is the *non plus ultra* among cities, except that on occasion it is strenuously challenged by San Francisco and Chicago. There are few communities that in one way or another do not consider themselves distinct and superior to others. City magazines, professional publicists, and the ever-present sports teams are at once stimulants and manifestations of the tribal inclination. Self-pride, of course, goes along with other-denigration. So, New Jersey is "the sticks" for New Yorkers. Texas is still where cowboys shoot you down without provocation; but Fort Worth is out in the woods as far as contiguous Dallas is concerned. St Paul is the poor cousin of Minneapolis. If that is not enough, consider the pecking order existing among the academic departments of colleges and universities. And so it goes.

9.5 Conclusion

For too long have the social and biological sciences steered almost entirely separate courses. For much longer still they will continue to look at behavior with specialized skills, interests, and perspectives that both reflect and justify their unique and respective contributions to behavioral science. But in many respects and along many paths they can no longer avoid theoretical convergences. The gradual development of behavioral biology, underscored of late by the emergence of sociobiology, is but one major sign of the convergences now being forged on many fronts, even if in some cases without awareness of the fact by the forgers themselves.

One may say of sociobiology many things. But what the history of science will record most unequivocally is that it sought to expand the theoretical competences of evolutionary behavioral biology by venturing into the sphere of what may be the most challenging animal species on earth. Emboldened by the powerful principles of Darwinian and Mendelian science, sociobiology has approached social science in a somewhat condescending fashion—or so it has seemed to some. But beneath condescension there is frequently an eager attempt to understand facts that echo from afar. Ultimately, if I may use this term in its conventional sense, sociobiology covets the help to grasp the special mysteries of *Homo sapiens* as well as the privilege of showing the way to the profoundly historical linkage existing between the symbolical animal and the rest of the kingdom. In the process, sociobiology deserves credit for promoting the study of coevolution, and that is the beginning of a major breakthrough in the history of science.

On their part, the social sciences have most assuredly earned the credentials with which to oblige their sister disciplines in evolutionary biology. Only social science, for example, can provide the input necessary to add the cultural dimension to the study of behavior evolution. At the same time, the social sciences stand to profit in a great many ways from a theoretical cross-fertilization. Evolutionary biology dramatizes the importance of time and change in the human condition. It demonstrates the enormous value of the comparative method. It introduces us to the productive logic of ultimate causation. Above all, it sensitizes us to the necessity of a theory of human nature, thereby in part guiding us to the rediscovery of some of the keenest insights of classical social science and to the principle that human behavior is directed in a deeply historical way. Whether or not we wish to take a strictly genetic orientation to the biogram, its search will be invaluable for parsimoniously classifying the great wealth of human behaviors and, therefore, for discovering laws of human behavior and sociocultural formations.

Harking back to subjects treated just above, it seems unlikely that without the insights of evolutionary biology social science can understand the almost paradoxical fact that the global society, of late so energetically stimulated by the communication revolution, among other things, can subsist side by side with the tendency toward fragmentation and "pseudo-speciation" (Erikson, 1966; Lorenz, 1973; Campbell, 1979)—with prejudice, parochialism, and ethnocentrism. Self-enhancement forces are ever at work to sustain the primeval tendency to pull tight the circle, to restrict the range, to think in terms of in-groups and out-groups. It is remarkable, but apart from evolutionary biology again incomprehensible, to learn from "attribution theory" that in our global society attributions of characteristics to the out-group are not only negative in nature. It is also true that, as Pettigrew (1978: 39–40) has succinctly put it, "negative acts of the outgroup are more often seen as caused by the distinctive biological origins of the outgroup, while positive acts are more likely to be explained away either situationally, and motivationally, or exceptionally."

Is there cause for despair from a biocultural viewpoint? I think not. As we come to understand the ultimate causes of human behavior and their fluctuating manifestations, we should also be better able to guide ethical principles toward the modification of environmental, proximate factors that help to account for the ancient assaults on fundamental ethical aspirations. We should also be able to waste less time and resources casting blame in this and that direction that ends by inflaming divergences, animosities, suspicions, conflicts. Explanation by reference to extremely narrow proximate causes is more than just poor science. Frequently it is also so much name-calling and gossip that unwittingly pit one group against another. For example, it is hardly fair to working-class ethnics, or helpful to black Americans for that matter, to surreptitiously imply that the former represent the major and a deliberate impediment in the proper and fair advancement of the latter. Prejudice and discrimination are nobody's monopoly. Furthermore, they can also be institutional in nature.

Feagin and Feagin (1978: 20–1) define "discrimination American style" as "actions or practices carried out by members of dominant groups, or their representatives, which have a differential and negative impact on members of the subordinate groups." They convincingly show that of late direct or overt discrimination has abated in the United States of America; but the more subtle, covert *institutional* type has increased. Institutional discrimination is embedded in organization or community structures and rules. Unlike the direct or individual variety, it is largely unintentional. When it is no longer chic or forgivable to be a blatant racist, we somehow manage to hide our parochialism under mountains of paper and cleverly drafted regulations. Out of the door and in through the window!

Biocultural science has sometimes been accused of racism, conservatism, prejudice, and all sorts of related evils. Social scientists have been among the most sanctimonious. It is a sad irony, therefore, that they have, in turn, charged one another of confirming with their insidious ideologies

Marx's famous aphorism that the prevailing ideas of a given epoch are the ideas of the dominant social class. In short, social scientists have not seldom accused each other of "bourgeois" orientations. I do not know which of the two charges is the more justified. I do, however, have ample reason to hold that biocultural science and Marxist social thought converge on one crucial point: The struggle for scarce resources is intense, and those who in most periods of history wage war with the greatest relish are not those who have less but those who have more. It is a fact of life that we work harder at defending and enlarging rich endowments of resources than at gaining control over them. To lose what we value and already own is more grievous than to fail to attain what we value but do not possess. To skid is more painful than not to rise.

To return to discrimination, it is, therefore, reasonable to conclude that, contrary to general findings, it is greatest not at the bottom or even in the middle of the stratification system but at the top—where possession and defense are at their highest level. As a matter of fact, it makes little sense to assess discrimination by asking such conventional "scale" questions as "Do you strongly believe that your church represents the one true religion?," "Do you feel that you would like or dislike person X?," "Do you feel that you would enjoy working with person Y?," and so forth. Such queries have a perverse tendency to zero in on the underbelly of the relatively dispossessed social classes. They reflect victimization, our tendency to pick on the weak and in the process to avoid inquiry into the more remote and concealed data. That is also another way of revealing our previously noted addiction to sophistries rather than to institutionally fastened behaviors.

The sophistic inclination is double-edged. On the one hand, we make a fetish of the verbal behavior of the lowly masses in areas that are hardly critical for the fundamental problem of the existence, which is competition for scarce and fitness-enhancing resources. On the other hand, we fail to appreciate that *noblesse oblige*, that those who lead in the race can easily afford to show tolerance precisely in the relatively insignificant areas of behavior. Better education and the special circumstances of their life conditions make them masters of the sophistry. Besides, why should they care about such things as churches and who sits next to whom at a conveyor belt? But try asking them about momentous matters of war and peace, the international arms bazaar, transactions involving billions of dollars, affirmative action, the Equal Rights Amendment, and the like. Look under the pseudological veneer of their bland pronouncements and see how thick self-interest and prejudice turn out to be.

The best measure of institutional and other types of discrimination is the tenacity with which privileges are defended by hook or by crook. The most efficacious antidote to discrimination is organization and agitation. This is not intended as a call to arms, particularly since what is antidote today will be venom tomorrow. When the humble inherit the earth they have a strange way of inheriting haughtiness as well. My intention is merely to give a name to the evolutionary game. It is inherent in the concept of competition. *Free* competition—that is at once the game and the fundamental principle of the ethics implicit in biocultural science.

Glossary

Adaptation The functional interdependence of biology (genetics) and environment in the process of evolution. Also the process whereby any genetic or sociocultural item tends to increase fitness to different degrees among the members of a population.

Adaptive Pertaining to any genetic or sociocultural item that tends to increase fitness to different degrees among the members of a population.

Allele One of a pair of genes that compete at a given chromosomal locus.

Altruism See Ascetic altruism.

Analogy A similarity between two or more units, for example, behaviors, that is due to evolutionary convergence or chance rather than to a common genetic origin.

Anthropocentrism A tendency to consider the human species so unique that it cannot be compared in any way to other species. Also the tendency to view and evaluate the behavior of other animals according to criteria typical of human beings.

Ascetic altruism Behavior, consciously or unconsciously motivated, that potentially reduces the inclusive fitness of the actor and potentially increases the fitness of other(s).

Assortative mating Marriage or mating with a phenotypically similar individual (positive assorting) or with a phenotypically dissimilar individual (negative assorting).

Behavioral biology In the narrower sense: ethology. In the broader sense: that branch of evolutionary biology that studies animal (including human) social behavior, especially from the viewpoint of the maximization principle.

Behavioral predisposition A genetically based tendency to be socialized to, and behave according to, certain major sociocultural patterns called universals.

Biocultural evolution A process of sociocultural transformation ultimately under the control of behavioral predispositions of variation and selective retention and proximately under the influence of various sociocultural and other environmental factors.

Biogram See Human nature.

Charisma A special quality of certain individuals based on their capacity for such things as great deeds, miracles, heroism, and rhetoric. The charisma may be displaced to a role, for example, that of king.

Cheating The more or less deliberate failure to reciprocate toward benefactors within a system of reciprocal exchanges.

Chromosome A rodlike structure in the nucleus of a cell, bearing part of the genes of the cell.

Climbing maneuver A behavioral predisposition that impels toward taking selfish advantage of opportunities to rise in a social hierarchy. The opportunities are frequently provided at least in part by collective action, and accordingly the predisposition typically cloaks itself with egalitarian sophistries.

Coevolution A term synonymous with biocultural evolution; hence, a process of evolution in which genetic and cultural factors interact.

Combinations Products of the mind, that is, cultural traits, that have emerged under the influence of combinatorial (or innovative) predispositions (the combiners) and are more or less long-lasting.

Combiners Behavioral predispositions ultimately responsible for the production of new sociocultural items, or combinations (which are termed variations or selections if they endure, and are incorporated in the preexisting sociocultural repertory).

Creature comforts Amenities of life, such as comfortable and speedy means of transportation, that tend to make for bodily and/or mental well-being.

Deference A behavioral predisposition that impels toward submissive behavior in relation to dominant others and, thus, typically reduces one's genetic fitness as well as one's creature comforts. Also an ultimate cause of dominance orders.

Denial of death A behavioral predisposition that impels individuals to believe in the immortality of the soul and to seek a happy and social existence for it.

Discrimination Behavior, typically based on prejudice, that is consciously or unconsciously motivated by the desire to exclude others from access to desirable and fitness-enhancing resources.

Distribution A property of a unit, for example, genotype or variation, whereby a given society or population has more or less of that unit than another population. For example, the distribution of technological variations is richer in modern than in agrarian society.

Dominance order A system of superordination and subordination whereby some individuals or groups have more power, and thus easier access to fitness-enhancing resources, than others.

Domination A behavioral predisposition that impels to achieve power over others and, thus, typically to enhance one's genetic fitness as well as one's creature comforts. Also an ultimate cause of dominance orders.

Ethnicity A property of groups that tends to equate them with breeding populations, that is, endogamous groupings. The property may be based on such factors as a degree of biological affinity, common religion, and national origin.

Ethnocentrism A complex of attitudes whereby members of one group consider themselves superior to another group on the basis of their own conception of what is socially, culturally, biologically good or right.

Ethology   A branch of evolutionary biology; more specifically, the study of whole patterns of animal behavior with a focus on the analysis of adaptation and the evolution of the patterns. See also Behavioral biology.

Evolution   See Biocultural evolution.

Exogamy   The practice of marrying outside one's own group, which may be a clan, an ethnic group, a social class, and so forth.

Explanatory urge   A behavioral predisposition that impels the individual to explain the facts and fictions of the existence; hence a tendency to find causes, often by hook or by crook.

Fitness   The measure of relative genetic success in the process of natural selection. The more favorably selected one is, the fitter one is.

Gene   A segment of chromosome in the DNA molecule, and the basic unit of heredity.

Genotype   The genetic constitution of any given individual.

Group selection   The differential contribution of genotypes and/or sociocultural variations, by diverse populations, to the world population and/or sociocultural repertory of future generations.

Haplodiploidy   The method of reproduction and sex determination in which males are derived from haploid (that is, unfertilized) eggs, while females derive from diploid (that is, fertilized) eggs.

Heterogamy   Marriage or mating between two or more individuals of unequal status with respect to fitness-enhancing resources.

Heterologous affiliation   A behavioral predisposition that impels toward variations and alliances, including marriage contracts, and other forms of behavior in inverse proportion to the degree of similarity—for example, in kinship or phenotype—between the potential associates or partners.

Heterologous contraposition   A behavioral predisposition whereby individuals tend to have negative attitudes toward others on the basis of real or perceived differences, often of an ethnic nature. It may be considered the ultimate cause of prejudice, discrimination, ethnocentrism.

Homogamy   Marriage or mating with a phenotypically similar individual. Less broadly: marriage or mating with an individual of the same social class, ethnic group, religion, and so on. See also Assortative mating.

Homologous affiliation   A behavioral predisposition that impels toward variations and alliances, including marriage contracts, and other forms of behavior in direct proportion to the degree of similarity—for example, in kinship or phenotype—between the potential associates or partners.

Homology   A similarity between two or more units, for example, behaviors, that is due to a common genetic origin.

Homozygous Pertaining to homozygosity or the union of sex cells that are identical for one or more pairs of alleles.

Human nature A set of genetically based behavioral predispositions evolved by natural selection at least in part under the pressure of sociocultural evolution.

Hypergamy The marriage or mating of a female to a male who possesses superior fitness-enhancing resources.

Hypertrophy The extreme growth of an originally simpler phenomenon. For example, the complex Catholic liturgy of today is an hypertrophy of early Christian rituals.

Hypogamy The marriage or mating of a female to a male possessing inferior fitness-enhancing resources.

The Imperative to act A behavioral predisposition that impels individuals to express their other behavioral predispositions. It thus produces behavior that stimulates predispositions. It also stimulates predispositions even in the absence of external or environmental stimulation.

Inbreeding depression The lowering of genetic fitness due to homozygosity, as in incestuous mating.

Incest avoidance A behavioral predisposition whereby individuals tend to avoid mating with relatives closer than first cousins.

Incest taboo A cultural universal prohibiting marriage between relatives closer than first cousins.

Inclusive fitness The sum of fitness deriving from one's own reproductive behavior plus the reproductive behavior of one's blood kin; hence the total effect of kin selection with reference to any given individual.

Individual selection In a biocultural key: a process of variation and selective retention wherein individuals contribute differentially to the future sociocultural and genetic repertoires of a given society. See Natural selection.

Intensity A property of a unit, for example, behavioral predispositions, universals, or variants, whereby in a given individual or population that unit is more salient than in another. For example, reciprocal behavior is more marked in agrarian than in industrial society.

Intersocietal selection See Group selection.

Intra-societal selection See Individual selection.

Kin selection Natural selection viewed from the vantage point of the kin group; hence, the differential contribution of genes to future generations by individuals who tend to enhance their own reproductive success and/or the reproductive success of their blood kin. See Inclusive fitness.

Maximization principle Otherwise known as the optimality principle, this central proposition of sociobiology specifies the principle of natural selection in a behavioral key approximately as follows: To the extent that organisms are under the influence of natural selection, they tend to behave so as to maximize their inclusive fitness.

Monogamy Marriage or mating between one male and one female.

Multiplier effect Pertaining to a process whereby every active force produces more than one change. For example, the invention of the automobile has had multiple consequences.

Mutation An error of replication in the DNA molecule; more broadly: a discontinuous change in the genotype of an organism.

Natural selection According to Darwin: the differential contribution of offspring to the next generation by genetically different individuals belonging to the same population. The broader concept: the differential contribution of genes (through offspring or other relatives) to the next generation by genetically different individuals belonging to the same population.

The Need for conformity A behavioral predisposition that impels, in varying degrees, to engage in behavior that sustains cultural norms, including rules of dominance and reciprocity orders, and to demand that others do likewise.

The Need for recognition A behavioral predisposition that impels toward the vain or proud advertisement of one's qualities in the pursuit of various sorts of adaptive gain. Sometimes, it becomes directed exclusively toward nonmaterial gain—for example, immortality of fame—and hence loses its adaptive value.

The Need for ritual consumption A behavioral predisposition that impels individuals to consume certain substances, for example, human flesh, in a ritual ceremony in the belief that special qualities may thereby be derived from such substances.

The Need for self-purification A behavioral predisposition that impels to acts of expiation "in order to" restore a sense of personal well-being that was lost, or is imagined to have been lost, through the violation of a moral rule, such as a taboo.

The Need for vengeance A behavioral predisposition that impels an individual or group to harm another as a reaction to a wrong, real or imaginary, previously received.

Nepotistic favoritism Behavior that benefits a relative, and thus indirectly oneself, as predicted from the concepts of kin selection and inclusive fitness.

Ontogeny The development of an organism through its entire life cycle.

Phenotype The observable features of an organism—for example, hair color, general appearance, behavior—resulting from the interplay of genotype and environment.

Phenotypical favoritism Behavior in favor of another based on phenotypical similarities, as represented, for example, by assortative mating, or homogamy.

Phylogeny The evolutionary history of a particular group of organisms; more broadly: the family tree showing the history of speciation.

Polyandry The marriage or mating of one female to multiple males.

Polygamy Marriage or mating of one male or one female to multiple partners.

Polygyny The marriage or mating of one male to multiple females.

Population A number of individuals of the same species, normally occupying a clearly delimited space at a given time, who practice a high degree of endogamy, or within-population mating. The population may be an ethnic group, a national society, indeed even a whole species.

Power The capacity of an individual or group within a dominance order to impede the access to fitness-enhancing resources by others and to facilitate it for oneself.

Prejudice An emotional aversion to a person or group on the basis of such morally and/or phenotypically relevant criteria as religious, ethnic, racial, and class differences.

Reciprocal favoritism Behavior that is mutually beneficial to two or more individuals within a context of reciprocal exchanges.

Reciprocation A behavioral predisposition that impels to engage in reciprocal behavior, that is, in a system of mutual exchanges which are more or less equally adaptive for the participants in the exchange.

Reification A behavioral predisposition that impels the individual to transform an idea or abstraction into a thing, such as a person-like deity.

Relational self-identification A behavioral predisposition to relate oneself to others and to understand oneself in terms of the relationships experienced with others.

Selections See Variations.

Selectors Behavioral predispositions that are ultimately responsible for the retention and institutionalization of combinations.

Social hierarchy See Dominance order.

Socialization A process of learning whereby the mind translates the directives and constraints of behavioral predispositions into behavior that develops along the broad channels of cultural transmission known as cultural universals.

Sociocultural evolution See Biocultural evolution.

Soul The allegedly indestructible phase of a person, probably evolved as a moral substitute of genetic action under the influence of the natural selection of self-deception.

Speciation The process of genetic diversification in a population whereby one or more new species emerge from it.

Species A population or set of populations that is genetically closed; that is, breeding across species does not take place; or, if it does happen, it tends to reduce fitness through parent or offspring sterility.

The Susceptibility to charisma A behavioral predisposition that impels individuals to respond with submission and veneration to certain individuals possessed of charisma as a personal quality or as a quality attaching to their social position.

Symbol Anything that evokes a response on the basis of meaning that is culturally imputed to it, as contrasted to a genetically programmed response to it.

Symbolization A behavioral predisposition that impels the individual to facilitate communication through the invention of symbols, that is, through the association of a thing and an idea.

Systemic immanence A system property whereby change is due to dynamics internal to systems, and tends to be directed by existing system conditions.

Temporecentrism A tendency to neglect the historical nature of phenomena, and thus to give excessive weight to present factors in attempts to explain those phenomena.

Territoriality A behavioral predisposition that impels toward the acquisition and defense of space containing fitness-enhancing resources; more broadly: the acquisitive impulse, or the sense of property. The predisposition is typically associated with aggressive behavior.

Territory An area occupied rather exclusively and defended by an individual or group.

Universal A major sociocultural pattern that is rooted in one or more behavioral predispositions, and may be found in all human societies.

Variant An aspect of culture, smaller than the universal, that tends to vary greatly in time and from one society to another.

Variations Products of the mind, that is, cultural traits, that have emerged under the influence of combinatorial (or innovative) predispositions (the combiners) and have been retained under the influence of selective predispositions (the selectors); inclusive of variants.

Victimization A behavioral predisposition that impels individuals and groups to sacrifice each other's fitness through various devices, for example, exploitation of resources, legal abuses, slavery, and human sacrifice.

References

Abramson, J. H. (1973), *Ethnic Diversity in Catholic America* (New York: Wiley).
Adam, H. (1971), *Modernizing Racial Domination* (Berkeley, Calif.: University of California Press).
Adams, M., and J. V. Neel (1967), "Children of incest," *Pediatrics*, 40: 55–61.
Adams, R. N. (1975), *Energy Structure: A Theory of Social Power* (Austin, Texas: University of Texas Press).
Adler, A. (1932), *What Life Should Mean to You* (London: Allen & Unwin).
Alba, R. D. (1976), "Social assimilation among American Catholic national-origin groups," *American Sociological Review*, 41: 1030–46.
Alcock, J. (1975), *Animal Behavior: An Evolutionary Approach* (Sunderland, Mass.: Sinnauer Associates).
Alexander, B. K. (1970), "Parental behavior of adult male Japanese monkeys," *Behavior*, 36: 270–85.
Alexander, K. L., M. Cook, and E. L. McDill (1978), "Curriculum tracking and educational stratification: some further evidence," *American Sociological Review*, 43: 47–66.
Alexander, R. D. (1961), "Aggressiveness, territoriality, and sexual behaviour in field crickets," *Behavior*, 17: 130–223.
Alexander, R. D. (1971), "The search for an evolutionary philosophy of man" *Proceedings of the Royal Society of Victoria*, 84: 99–120.
Alexander, R. D. (1974), "The evolution of social behavior," *Annual Review of Ecology and Systematics*, 5: 325–83.
Alexander, R. D. (1975), "The search for a general theory of behavior," *Behavioral Science*, 20: 77–100.
Alexander, R. D. (1979), "Evolution and culture," pp. 59–78 in *Evolutionary Biology and Human Social Behavior: An Anthropological Perspective*, ed. N. A. Chagnon and W. Irons (North Scituate, Mass.: Duxbury Press).
Alland, A., Jr (1967), *Evolution and Human Behavior* (Garden City, NY: Natural History Press).
Alland, A., Jr (ed.) (1968), *War* (Garden City, NY: Natural History Press).
Allee, W. C. (1938), *Cooperation among Animals* (New York: Schuman).
Allen, E., *et al.* (1975), "Against 'sociobiology,' " *The New York Review of Books*, 13 November: 182, 184–6.
Allport, G. W. (1954), *The Nature of Prejudice* (Reading, Mass.: Addison-Wesley).
Allport, G. W. (1963), "Behavioral science, religion, and mental health," *Journal of Religion and Health*, 2: 187–97.
Alstrom, C. H. (1977), "A study of incest with special regard to the Swedish penal code," *Acta Psychiatrica Scandinavica*, 56: 357–72.
Altmann, S. A. (1962), "A field study of the sociobiology of rhesus monkeys, *Macaca mulatta*," *Annals of the New York Academy of Sciences*, 102: 338–435.
Altmann, S. A. (ed.) (1967), *Social Communication among Primates* (Chicago: University of Chicago Press).
Ammerman, A. J., and L. L. Cavalli-Sforza (1973), "A population model for the diffusion of early farming in Europe," pp. 343–57 in *The Explanation of Culture*

Change: Models in Prehistory, ed. C. Renfrew (Pittsburgh, Pa: University of Pittsburgh Press).
Amory, C. (1947), *The Proper Bostonians* (New York: Dutton).
Anderson, E. (ed.) (1961), *The Letters of Beethoven* (London: Macmillan).
Anderson, J. R. (1980), *Cognitive Psychology and its Implications* (San Francisco: Freeman).
Anderson, R. C. (1967), "On genetics and sociology (II)," *American Sociological Review*, 32: 997–9.
Annuario Pontifico (1980) (Vatican City: Statistical Bureau).
Arens, W. (1979), *The Man-Eating Myth: Anthropology and Anthropophagy* (New York: Oxford University Press).
Arensberg, C. M., and S. T. Kimball (1940), *Family and Community in Ireland* (Cambridge, Mass.: Harvard University Press).
Argyris, C. (1973), "Personality and organizational theory revisited," *Administrative Science Quarterly*, 18: 141–67.
Arieti, S. (1976), *Creativity: The Magic Synthesis* (New York: Basic Books).
Aron, R. (1953), "Social structure and the ruling class," pp. 567–77 in *Class, Status and Power: A Reader in Social Stratification*, ed. R. Bendix and S. M. Lipset (Glencoe, Ill.: The Free Press).
Asimov, I. (1980), "What's fueling the popular science explosion?", *Saturday Review*, August: 22–6.
Back, K. W. (1950), "The exertion of influence through social communication," pp. 21–36 in *Theory and Experiment in Social Communication*, ed. L. Festinger, K. W. Back, S. Schacter, H. H. Kelley, and J. Thibaut (Ann Arbor, Mich.: University of Michigan Research Center for Dynamics).
Badillo, G., and C. D. Curry (1976), "The social incidence of Vietnam casualties," *Armed Forces and Society*, 2: 397–406.
Bahr, H. M., B. A. Chadwick, and J. H. Stauss (1979), *American Ethnicity* (Lexington, Mass.: D. C. Heath).
Baltzell, E. D. (1958), *Philadelphia Gentlemen* (Glencoe, Ill.: The Free Press).
Bandura, A. (1969), *Principles of Behavior Modification* (New York: Holt, Rinehart & Winston).
Bandura, A. (1977), *Social Learning Theory* (New York: General Learning Press).
Bandura, A., and P. G. Barab (1971), "Conditions governing nonreinforced imitation," *Developmental Psychology*, 5: 244–55.
Banfield, E. C. (1958), *The Moral Basis of a Backward Society* (Glencoe, Ill.: The Free Press).
Barash, D. P. (1977), *Sociobiology and Behavior* (New York: Elsevier).
Barber, B. (ed.) (1970), *L. J. Henderson on the Social System* (Chicago: University of Chicago Press).
Barber, E. G. (1955), *The Bourgeoisie in 18th Century France* (Princeton, NJ.: Princeton University Press).
Barnett, H. G. (1953), *Innovation: The Basis of Cultural Change* (New York: McGraw-Hill).
Barth, F. (ed.) (1969), *Ethnic Groups and Boundaries: The Social Organization of Cultural Differences* (Boston, Mass.: Little, Brown).
Barzun, J. (1937), *Race: A Study in Modern Superstition* (New York: Harcourt Brace).
Bateson, G. (1963), "The role of somatic change in evolution," *Evolution*, 17: 529–39.
Becker, E. (1973), *The Denial of Death* (New York: The Free Press).
Beckman, L. (1962), "Assortative mating in man," *Eugenics Review*, 54: 63–7.
Beecher, H. K., A. S. Keats, F. Mosteller, and L. Lasagna (1953), "The effectiveness of oral analgesics (morphine, codeine, acetylsalicylic acid) and the problem of

placebo 'reactors' and 'nonreactors,' " *Journal of Pharmacology and Experimental Therapeutics*, 109: 393–400.

Beit-Hallahmi, B., and A. I. Rabin (1977), "The kibbutz as a social experiment and as a child-rearing laboratory," *American Psychologist*, 32: 532–41.

Bell, C. (1916), *Art* (London: Chatto & Windus).

Bell, D. (1973), *The Coming of Post-Industrial Society* (New York: Basic Books).

Bellah, R. N. (1964), "Religious evolution," *American Sociological Review*, 29: 358–74.

Bellah, R. N. (ed.) (1973), *Emile Durkheim on Morality and Society* (Chicago: University of Chicago Press).

Bellah, R. N. (1975), *The Broken Covenant: American Civil Religion in Time of Trial* (New York: Seabury).

Bellamy, E. (1889), *Looking Backward* (Boston, Mass.: Houghton Mifflin).

Bendix, R. (1960), *Max Weber: An Intellectual Portrait* (Berkeley, Calif.: University of California Press).

Benedict, R. (1934), *Patterns of Culture* (New York: Houghton Mifflin).

Benedict, R. (1952, orig. 1943), *Thai Culture and Behavior: An Unpublished War-Time Study Dated September, 1943* (Ithaca, NY: Cornell University, Southeast Asia Program).

Bentler, P. M., and M. D. Newcomb (1978), "Longitudinal study of marital success and failure," *Journal of Consulting and Clinical Psychology*, 40: 1053–70.

Berent, J. (1954), "Social mobility and marriage: a study of trends in England and Wales," pp. 321–38 in *Social Mobility in Britain*, ed. D. Glass (London: Routledge & Kegan Paul).

Berlyne, D. E. (1971), *Aesthetics and Psychobiology* (New York: Appleton-Century-Crofts).

Bernstein, I. S., and L. G. Sharpe (1966), "Social roles in a rhesus monkey group," *Behavior*, 26: 91–104.

Bersheid, E., and E. Walster (1974), "Physical attractiveness," pp. 157–215 in *Advances in Experimental Social Psychology*, ed. L. Berkowitz (New York: Academic Press).

Bierstedt, R. (1950), "An analysis of social power," *American Sociological Review*, 15: 730–8.

Bischof, N. (1975), "The comparative ethology of incest avoidance," pp. 37–67 in *Biosocial Anthropology*, ed. R. Fox (New York: Wiley).

Bishop, C. A. (1970), "The emergence of hunting territories among Northern Ojibwa," *Ethnology*, 9: 1–15.

Bishop, C. A. (1974), *The Northern Ojibwa and the Fur Trade: An Historical and Ecological Study* (New York: Holt, Rinehart & Winston).

Bitterman, M. (1965), "Phyletic differences in learning," *American Psychologist*, 20: 396–410.

Blake, J. (1974), "The changing status of women in developed countries," *Scientific American*, 231: 136–47.

Blassingame, J. W. (1972), *The Slave Community* (New York: Oxford University Press).

Blau, P. M. (1964), *Exchange and Power in Social Life* (New York: Wiley).

Blau, P. M., and O. D. Duncan (1967), *The American Occupational Structure* (New York: Wiley).

Bloch, M. (1961, orig. 1940), *Feudal Society* (London: Routledge & Kegan Paul).

Bloch, M. (1970), *French Rural History* (Berkeley, Calif.: University of California Press).

Block, N. (ed.) (1980), *Readings in Philosophy of Psychology* (Cambridge, Mass.: Harvard University Press).

Blumberg, B. S., and J. E. Hesser (1976), "Anthropology and infectious disease,"

pp. 260–94 in *Physiological Anthropology*, ed. A. Damon (Oxford: Oxford University Press).
Blute, M. (1976), "Review of E. O. Wilson, *Sociobiology: The New Synthesis*," *Contemporary Sociology*, 5: 727–31.
Blute, M. (1979), "Sociocultural evolutionism: an untried theory," *Behavioral Science*, 24: 46–59.
Boehm, C. (1978), "Rational preselection from hamadryas to *Homo sapiens*: the place of decisions in the adaptive process," *American Anthropologist*, 80: 265–96.
Bonham, G. S. (1977), "Who adopts: the relationship of adoption and social-demographic characteristics of women," *Journal of Marriage and the Family*, 39: 295–306.
Boorman, S. A., and P. R. Levitt (1972), "Group selection on the boundary of a stable population," *Proceedings of the National Academy of Sciences*, 69: 2711–13.
Boorman, S. A., and P. R. Levitt (1973), "Group selection on the boundary of a stable population," *Theoretical Population Biology*, 4: 85–128.
Boorman, S. A., and P. R. Levitt (1980), *The Genetics of Altruism* (New York: Academic Press).
Bottomore, T. B. (1966), *Classes in Modern Societies* (New York: Random House).
Bowers, W. J. (1974), *Executions in America* (Lexington, Mass.: D. C. Heath).
Bowles, S., and H. Gintis (1976), *Schooling in Capitalist America: Educational Reform and the Contradictions of Economic Life* (New York: Basic Books).
Brandon, S. G. F. (1962), *Man and His Destiny in the Great Religions* (Manchester: Manchester University Press).
Braudel, F. (1982, orig. 1979), *The Wheels of Commerce* (New York: Harper & Row).
Brinton, C. (1938), *The Anatomy of Revolution* (New York: Norton).
Brinton, C. (1959), *A History of Western Morals* (New York: Harcourt Brace).
Brodbeck, M. (1959), "Models, meaning and theories," pp. 373–403 in *Symposium on Sociological Theory*, ed. L. Gross (Evanston, Ill.: Row, Peterson).
Brown, J. L. (1964), "The evolution of diversity in avian territorial systems," *Wilson Bulletin*, 76: 160–9.
Brown, J. L. (1969), "Territorial behavior and population regulation in birds: a review and re-evaluation," *Wilson Bulletin*, 81: 293–329.
Brown, J. L. (1975), *The Evolution of Behavior* (New York: Norton).
Bruce, H. M. (1960), "A block to pregnancy in the mouse caused by proximity to strange males," *Journal of Reproductive Fertility*, 1: 96–102.
Bullock, T. H., R. Orkand, and A. Grinnell (1977), *Introduction to Nervous Systems* (San Francisco: Freeman).
Bumpass, L. L., and C. F. Westoff (1970), *The Later Years of Childbearing* (Princeton, NJ: Princeton University Press).
Bunzel, R. (1932), "Zuni Katchinas," pp. 837–1086 in *The 47th Annual Report of the Bureau of American Ethnology* (Washington, DC: Bureau of American Ethnology).
Burchinal, L. G. (1964), "The premarital dyad and love involvement," pp. 632–74 in *Handbook of Marriage and the Family*, ed. H. T. Christensen (Chicago: Rand McNally).
Burgess, E. W., and H. J. Locke (1953), *The Family* (New York: American).
Burton, F. D. (1972), "The integration of biology and behavior in the socialization of *Macaca sylvana* of Gibraltar," pp. 29–62 in *Primate Socialization*, ed. F. E. Poirer (New York: Random House).
Bury, J. B. (1955, orig. 1932), *The Idea of Progress* (New York: Dover).
Busino, G. (1967), *La Sociologie de Vilfredo Pareto* (Geneva: Droz).
Butler, J. S. (1980), *Inequality in the Military: The Black Experience* (Saratoga, Calif.: Century Twenty-One).

Campbell, B. (1966), *Human Evolution* (Chicago: Aldine).
Campbell, D. T. (1960), "Blind variation and selective retention in creative thought as in other knowledge processes," *Psychological Review*, 67: 380–400.
Campbell, D. T. (1965), "Variation and selective retention in sociocultural evolution," pp. 19–49 in *Social Change in Developing Areas*, ed. H. R. Barringer, G. I. Blanksten, and R. W. Mack (Cambridge, Mass.: Schenkman).
Campbell, D. T. (1972), "On the genetics of altruism and the counter-hedonic components in human culture," *Journal of Social Issues*, 28: 21–37.
Campbell, D. T. (1974), "Unjustified variation and selective retention in scientific discovery," pp. 139–61 in *Studies in the Philosophy of Biology*, ed. F. S. Ayala and T. Dobzhansky (New York: Macmillan).
Campbell, D. T. (1975), "On the conflicts between biological and social evolution and between psychology and moral tradition," *American Psychologist*, 30: 1103–26.
Campbell, D. T. (1979), "Comments on the sociobiology of ethics and moralizing," *Behavioral Science*, 24: 37–45.
Caplan, A. L. (ed.) (1978), *The Sociobiology Debate* (New York: Harper & Row).
Carlin, J. E., J. Howard, and S. L. Messinger (1966), "Civil justice and the poor," *Law and Society Review*, 1: 85–9.
Carpenter, C. R. (1958), "Territoriality: a review of concepts and problems," pp. 224–50 in *Behavior and Evolution*, ed. A. Roe and G. G. Simpson (New Haven, Conn.: Yale University Press).
Carroll, J. W., D. W. Johnson, M. E. Marty, and G. Gallup, Jr (1978), *Religion in America, 1950 to the Present* (New York: Harper & Row).
Caso, A. (1953), *El Pueblo del Sol* (México: Fondo de Cultura Económica).
Cavalli-Sforza, L. L., and M. Feldman (1973), "Models for cultural inheritance I. Group mean and within group variation," *Theoretical Population Biology*, 4: 42–55.
Chagnon, N. A. (1968), *Yanomamö, the Fierce People* (New York: Holt, Rinehart & Winston).
Chagnon, N. A., and W. Irons (eds) (1979), *Evolutionary Biology and Human Social Behavior* (North Scituate, Mass.: Duxbury Press).
Chance, M. R. A. (1962), "Social behaviour and primate evolution," pp. 84–130 in *Culture and the Evolution of Man*, ed. M. F. Ashley Montague (New York: Oxford University Press).
Chance, M. R. A. (1970), "The nature and special features of the instinctive social bond of primates," pp. 17–33 in *Social Life of Early Man*, ed. S. L. Washburn (Chicago: Aldine).
Charnov, E. L., and J. R. Krebs (1975), "The evolution of alarm calls: altruism or manipulation?", *American Naturalist*, 109: 107–12.
Cherns, A. (ed.) (1980), *Quality of Working Life and the Kibbutz Experience: Proceedings of an International Conference in Israel, June 1978* (Norwood, Pa: Norwood Editions).
Childe, V. G. (1951), *Social Evolution* (London: Watts).
Chomsky, N. (1965), *Aspects of the Theory of Syntax* (Cambridge, Mass.: MIT Press).
Chomsky, N. (1972), *Language and Mind* (New York: Harcourt Brace Jovanovich).
Chomsky, N. (1973), *For Reasons of State* (New York: Pantheon).
Chomsky, N. (1975), *Reflections on Language* (New York: Random House).
Choron, J. (1963), *Death and Western Thought* (New York: Collier).
Clark, P. J. (1959), "The heritability of certain anthropometric characters as ascertained from measurements of twins," *Human Biology*, 31: 121–37.
Clarke, D. L. (1978), *Analytical Archaeology*, 2nd edn, rev. B. Chapman (New York: Columbia University Press).

Cloak, F. T., Jr (1975), "Is a cultural ethology possible?", *Human Ecology*, 3: 161–82.
Cochrane, S. H. (1979), *Fertility and Education: What Do We Really Know?* (Baltimore, Md: Johns Hopkins University Press).
Cohen, E. (1976), "Environmental orientations: a multidimensional approach to social ecology," *Current Anthropology*, 17: 49–70.
Cohen, J. (1959), "The psychology of luck," *Advancement of Science*, 16: 203.
Cohen, J. (1980), "Rational capitalism in Renaissance Italy," *American Journal of Sociology*, 85: 1340–55.
Cohen, R. (1962), "The strategy of social evolution," *Anthropologica*, 4: 321–48.
Cohen, Y. (1978), "The disappearance of the incest taboo," *Human Nature*, 1: 27–8.
Coles, R. (1966), "Psychiatrists and the poor," pp. 181–90 in *Poverty in the Affluent Society*, ed. H. H. Meissner (New York: Harper & Row).
Collins, R. (1975), *Conflict Sociology* (New York: Academic Press).
Colson, C. W. (1977, orig. 1976), *Born Again* (New York: Bantham).
Colson, E. (1954), "Ancestral spirits and social structure among the Plateau Tonga," *International Archives of Ethnography*, XLVII, pt I: 21–68.
Combs, A. W., A. Richards, and F. Richards (1976), *Perceptual Psychology: A Humanistic Approach to the Study of Persons* (New York: Harper & Row).
Comte, A. (1875–7), *System of Positive Polity* (London: Longmans, Green).
Comte, A. (1896), *The Positive Philosophy of Auguste Comte* (free translation and abridgement by H. Martineau), Vol. III (London: Bell).
Connolly, W. (1970), "On 'interests' in politics," pp. 259–77 in *The Politics and Society Reader*, ed. I. Katznelson, G. Adams, P. Brenner, and A. Wolfe (New York: McKay).
Cooley, C. H. (1922), *Human Nature and the Social Order* (New York: Charles Scribner's Sons).
Coser, L. A. (1956), *The Functions of Social Conflict* (New York: The Free Press).
Count, E. W. (1958), "The biological basis of human sociality," *American Anthropologist*, 60: 1049–85.
Count, E. W. (1973), *Being and Becoming Human: Essays on the Biogram* (New York: Van Nostrand Reinhold).
Coursey, D. G., and C. K. Coursey (1971), "The new yam festivals of West Africa," *Anthropos*, 66: 444–84.
Cowan, W. M. (1979), "The development of the brain," *Scientific American*, 241: 112–33.
Cowell, F. R. (1956), *Cicero and the Roman Republic* (Harmondsworth: Penguin).
Cox, H. (1977), quoted in *Time*, 28 March, p. 81.
Crippen, T. (1982), "Born-again politics: persistence of revival traditions," unpublished PhD Dissertation (Austin: Department of Sociology, University of Texas at Austin).
Crippen, T., and J. Lopreato (1981), "Dimensions of social mobility and political behavior," *Journal of Political and Military Sociology*, 9: 149–61.
Crognier, E. (1977), "Assortative mating for physical features in an African population from Chad," *Journal of Human Evolution*, 6: 105–14.
Crook, J. H. (1965), "The adaptive significance of avian social organizations," *Symposia of the Zoological Society of London*, 18: 237–58.
Crook, J. H. (1973), "The nature and function of territorial aggression," pp. 183–220 in *Man and Aggression*, ed. M. F. A. Montagu (London: Oxford University Press).
Crook, J. H. (1981), *The Evolution of Human Consciousness* (Oxford: Oxford University Press).
Crow, J. F., and J. Felsenstein (1968), "The effect of assortative mating on the genetic composition of a population," *Eugenics Quarterly*, 15: 85–97.

Dahrendorf, R. (1959), *Class and Class Conflict in Industrial Society* (Stanford, Calif.: Stanford University Press).
Damas, D. (ed.) (1969), Conference on Band Societies, National Museum of Canada, *Bulletin* no. 228.
Darlington, C. D. (1969), *The Evolution of Man and Society* (New York: Simon & Schuster).
Darwin, C. (1958, orig. 1859), *On the Origin of Species* (New York: New American Library, Mentor Books).
Darwin, C. (1871), *The Descent of Man and Selection in Relation to Sex* (New York: Appleton), 2 vols.
Darwin, C. (1965, orig. 1872), *The Expression of Emotions in Man and Animals* (Chicago: University of Chicago Press).
Davie, M. R. (1929), *The Evolution of War* (New Haven, Conn.: Yale University Press).
Davie, M. R. (1936), *World Immigration* (New York: Macmillan).
Davies, J. C. (1962), "Toward a theory of revolution," *American Sociological Review*, 27: 5–19.
Davis, K. (1948), *Human Society* (New York: Macmillan).
Davis, K. (1959), "The myth of functional analysis as a special method in sociology and anthropology," *American Sociological Review*, 24: 757–72.
Davis, K., and W. E. Moore (1945), "Some principles of stratification," *American Sociological Review*, 10: 242–9.
Dawidowicz, L. S. (1975), *The War against the Jews, 1933–1945* (New York: Holt, Rinehart & Winston).
Dawkins, R. (1976), *The Selfish Gene* (London: Oxford University Press).
Dawkins, R. (1979), "Twelve misunderstandings of kin selection," *Zeitschrift für Tierpsychologie*, 51: 184–200.
Dawson Scanzoni, L., and J. Scanzoni (1981), *Men, Women, and Change: A Sociology of Marriage and Family* (New York: McGraw-Hill).
Deacon, A. B. (1934), *Malekula: A Vanishing People in the New Hebrides* (London: Routledge).
DeFries, J. C., and G. E. McClearn (1970), "Social dominance and Darwinian fitness in the laboratory mouse," *American Naturalist*, 104: 408–11.
Deutsch, H. (1965), *Neuroses and Character Types* (New York: International University Press).
DeVore, B. I. (ed.) (1965), *Primate Behavior* (New York: Holt, Rinehart & Winston).
DeVore, B. I. (1971), "The evolution of human society," pp. 297–311 in *Man and Beast: Comparative Social Behavior*, ed. J. F. Eisenberg and W. S. Dillon (Washington, DC: Smithsonian Institute Press).
DeVore, B. I., and S. L. Washburn (1963), "Baboon ecology and human evolution," pp. 93–108 in *Man in Adaptation: The Biosocial Background*, ed. Y. A. Cohen (Chicago: Aldine).
Dewey, J. (1929), *The Quest for Certainty: A Study of the Relation of Knowledge and Action* (New York: Minton, Balch).
Dickemann, M. (1979), "Female infanticide, reproductive strategies, and social stratification: a preliminary model," pp. 321–67 in *Evolutionary Biology and Human Social Behavior: An Anthropological Perspective*, ed. N. A. Chagnon and W. Irons (North Scituate, Mass.: Duxbury Press).
Dinnerstein, L., and D. M. Reimers (1975), *Ethnic Americans: A History of Immigration and Assimilation* (New York: Dodd, Mead).
Djilas, M. (1957), *The New Class: An Analysis of the Communist System* (New York: Praeger).

Dobzhansky, T. (1937), *Genetics and the Origin of Species* (New York: Columbia University Press).
Dobzhansky, T. (1962), *Mankind Evolving* (New Haven, Conn.: Yale University Press).
Dobzhansky, T. (1967), "On types, genotypes, and the genetic diversity in populations," pp. 1–18 in *Genetic Diversity and Human Behavior*, ed. J. N. Spuhler (Chicago: Aldine).
Dobzhansky, T. (1974), "Chance and creativity in evolution," pp. 307–38 in *Studies in the Philosophy of Biology*, ed. F. J. Ayala and T. Dobzhansky (New York: Macmillan).
Dobzhansky, T., F. J. Ayala, G. L. Stebbins, and J. W. Valentine (1977), *Evolution* (San Francisco: Freeman).
Dohrenwend, B. P., B. S. Dohrenwend, M. S. Gould, B. Link, R. Neugebauer, and R. Wunsch-Hitzig (1980), *Mental Illness in the United States: Epidemiological Estimates* (New York: Praeger).
Douglas, F. (1968, orig. 1855), *My Bondage and My Freedom* (New York: Arno Press).
Douglas, M. (1968), "Pollution," pp. 336–41 in *International Encyclopedia of the Social Sciences*, ed. D. L. Sills, Vol. XX (New York: Crowell Collier).
Douglas, M. (1975), *Implicit Meanings: Essays in Anthropology* (London: Routledge & Kegan Paul).
Driedger, L. (ed.) (1978), *The Canadian Mosaic: A Quest for Identity* (Toronto: McClelland & Stewart).
Dumond, D. (1975), "The limitation of human population: a natural history," *Science*, 187: 713–21.
Dumont, L. (1970, orig. 1966), *Homo Hierarchicus* (Chicago: University of Chicago Press).
Durham, W. H. (1976), "Resource competition and human aggression, Part I: A review of primitive war," *Quarterly Review of Biology*, 51: 385–415.
Durham, W. H. (1978), "Toward a coevolutionary theory of human biology and culture," pp. 428–48 in *The Sociobiology Debate*, ed. A. L. Caplan (New York: Harper & Row).
Durham, W. H. (forthcoming), "Coevolution and law: the new yam festivals of West Africa," in *Law, Biology and Culture*, ed. M. Gruter and P. Bohannon.
Durkheim, E. (1933, orig. 1893), *The Division of Labor in Society* (New York: Macmillan).
Durkheim, E. (1951, orig. 1897), *Suicide* (New York: The Free Press).
Durkheim, E. (1965, orig. 1912), *The Elementary Forms of the Religious Life* (New York: The Free Press).
Durkheim, E. (1953), *Sociology and Philosophy* (Glencoe, Ill.: The Free Press).
Dyson-Hudson, R., and E. A. Smith (1980, orig. 1978), "Human territoriality: an ecological reassessment," pp. 367–93 in *Selected Readings in Sociobiology*, ed. J. H. Hunt (New York: McGraw-Hill).
Eaton, G. G. (1976), "The social order of Japanese Macaques," *Scientific American*, 235: 96–106.
Eckland, B. K. (1967), "Genetics and sociology: a reconsideration," *American Sociological Review*, 32: 173–94.
Eckland, B. K. (1968), "Theories of mate selection," *Eugenics Quarterly*, 15: 71–84.
Eckland, B. K. (1972), "Evolutionary consequences of differential fertility and assortative mating in man," pp. 293–305 in *Evolutionary Biology*, ed. T. Dobzhansky (New York: Appleton-Century-Crofts).
Eckland, B. K. (1976), "Darwin rides again," *American Journal of Sociology*, 82: 692–7.

Edelman, G. M., and V. B. Mountcastle (1978), *The Mindful Brain: Cortical Organization and the Group-Selective Theory of Higher Brain Function* (Cambridge, Mass.: MIT Press).

Ehrman, L. (1972), "Genetics and sexual selection," pp. 105–35 in *Sexual Selection and the Descent of Man, 1871–1971*, ed. B. Campbell (Chicago: Aldine).

Eibl-Eibesfeldt, I. (1975), *Ethology: The Biology of Behavior* (New York: Holt, Rinehart & Winston).

Eisenberg, J. F. (1966), "The social organization of mammals," *Handbuch der Zoologie*, 10: 1–92.

Eisenberg, J. F. (1967), "A comparative study in rodent ethology with emphasis on evolution of social behavior," *Proceedings of the U.S. National Museum* (Smithsonian Institution), 122: 1–49.

Eisenberg, J. F., and R. E. Kuehn (1966), "The behavior of *Ateles geoffroyi* and related species," *Smithsonian Miscellaneous Collections*, 151: 1–63.

Eisenstadt, S. N. (ed.) (1970), *Readings in Social Evolution and Development* (London: Pergamon).

Eisenstadt, S. N. (1978), *Revolution and the Transformation of Societies* (New York: The Free Press).

Elkin, A. P. (1954), *The Australian Aborigines* (Sidney: Angus Robertson).

Ember, M. (1975), "On the origin and extension of the incest taboo," *Behavior Science Research* (Human Relations Area Files, New Haven, Conn.), 10: 249–81.

Emlen, S. T. (1976), "An alternative case for sociobiology," *Science*, 192: 736–38.

Emlen, S. T. (1980), "Ecological determinism and sociobiology," pp. 125–50 in *Sociobiology: Beyond Nature/Nurture?*, ed. G. W. Barlow and J. Silverberg (Boulder, Colo: Westview Press).

Engels, F. (1978, orig. 1883), "Speech at the graveside of Karl Marx," pp. 681–2 in *The Marx–Engels Reader*, ed. R. C. Tucker, 2nd edn (New York: Norton).

Erikson, E. (1966), "Ontogeny of ritualization in man," *Philosophical Transactions of the Royal Society in London*, Series B, 772, 251: 337–49.

Errington, P. L. (1963), *Muskrat Populations* (Ames, Iowa: Iowa State University Press).

Esser, A. H. (ed.) (1971), *Behavior and Environment: The Use of Space by Animals and Men* (New York: Plenum).

Essock-Vitale, S. M., and M. T. McGuire (1980), "Predictions derived from the theories of kin selection and reciprocation assessed by anthropological data," *Ethology and Sociobiology*, 1: 233–43.

Evans-Pritchard, E. E. (1940), *The Nuer: A Description of the Modes of Livelihood and Political Institutions of a Nilotic People* (London: Oxford University Press).

Evans-Pritchard, E. E. (1956), *Nuer Religion* (Oxford: Clarendon).

Farb, P. (1973), *Word Play* (New York: Knopf).

Feagin, J. R. (1975), *Subordinating the Poor* (Englewood Cliffs, NJ: Prentice-Hall).

Feagin, J. R., and C. B. Feagin (1978), *Discrimination American Style: Institutional Racism and Sexism* (Englewood Cliffs, NJ: Prentice-Hall).

Feder, H. M. (1966), "Cleaning symbioses in the marine environment," pp. 327–80 in *Symbiosis*, Vol. 1, ed. S. M. Henry (New York: Academic Press).

Feifel, H. (ed.) (1959), *The Meaning of Death* (New York: McGraw-Hill).

Feldman, M. W., and L. L. Cavalli-Sforza (1976), "Cultural and biological evolutionary processes, selection for a trait under complex transmission," *Theoretical Population Biology*, 9: 238–59.

Feldman, M. W., and L. L. Cavalli-Sforza (1979), "Aspects of variance and covariance analysis with cultural inheritance," *Theoretical Population Biology*, 15: 276–307.

Fenn, R. K. (1978), *Toward a Theory of Secularization* (Storrs, Conn.: Society for the Scientific Study of Religion).

Festinger, L. (1950), "Informal social communication," *Psychological Review*, 57: 781–2.
Festinger, L. (1957), *Theory of Cognitive Dissonance* (Stanford, Calif.: Stanford University Press).
Festinger, L., K. W. Back, S. Schacter, H. H. Kelley, and J. Thibaut (eds) (1950), *Theory and Experiment in Social Communication* (Ann Arbor, Mich.: University of Michigan Research Center for Dynamics).
Feuer, L. S. (1969), *The Conflict of Generations* (New York: Basic Books).
Firth, R. W. (1939), *Primitive Polynesian Economy* (London: Routledge & Kegan Paul).
Firth, R. W. (1959), *Economics of New Zealand Maori* (Wellington: R. W. Owen, Government Printer).
Firth, R. W. (1963), "Offering and sacrifice: problems of organization," *Journal of the Royal Anthropological Institute*, 93: 12–24.
Fischer, M. M. J. (1980), *Iran: From Religious Dispute to Revolution* (Cambridge, Mass.: Harvard University Press).
Fischer-Galati, S. (ed.) (1979), *The Communist Parties of Eastern Europe* (New York: Columbia University Press).
Fishbein, M., and I. Ajzen (1975), *Belief, Attitude, Intention and Behavior: An Introduction to Theory and Research* (Reading, Mass.: Addison-Wesley).
Fisher, R. A. (1930), *The Genetical Theory of Natural Selection* (Oxford: Clarendon).
Flannery, R. (1953), *The Gros Ventre of Montana: Part I, Social Life* (Washington, DC: Catholic University of America).
Flew, A. (1967a), *Evolutionary Ethics* (New York: Macmillan).
Flew, A. (1967b), "Evolutionary ethics," pp. 31–51 in *New Studies in Ethics*, ed. W. Hudson (New York: St Martin's Press).
Flew, A. (1972), "Immortality," pp. 139–50 in *The Encyclopedia of Philosophy*, ed. P. Edwards (New York: Macmillan).
Flink, J. J. (1970), *America Adopts the Automobile 1895–1910* (Cambridge, Mass.: MIT Press).
Fodor, J. A. (1975), *The Language of Thought* (New York: Crowell).
Fodor, J. A. (1981), "The mind–body problem," *Scientific American*, 244: 114–23.
Forbes, E. (ed.) (1967), *Thayer's Life of Beethoven* (Princeton, NJ: Princeton University Press).
Fortes, M. (1969), *Kinship and the Social Order* (Chicago: Aldine).
Fortune, R. F. (1932), "Incest," *Encyclopedia of the Social Sciences*, 7: 620–2.
Fortune, R. F. (1935), *Manus Religion* (Philadelphia, Pa: Memoirs of the American Philosophical Society, Vol. III).
Foster, G. M. (1965), "Peasant society and the image of limited good," *American Anthropologist*, 67: 293–315.
Foster, G. M. (1973), *Traditional Societies and Technological Change* (New York: Harper & Row).
Foster, J. (1968), "Nineteenth-century towns: a class dimension," pp. 281–99 in *The Study of Urban History*, ed. H. J. Dyos (New York: St Martin's Press).
Fox, R. (1967), *Kinship and Marriage* (Harmondsworth: Penguin).
Fox, R. (1972), "Alliance and constraint: sexual selection in the evolution of human kinship systems," pp. 282–331 in *Sexual Selection and the Descent of Man 1871–1971*, ed. B. Campbell (Chicago: Aldine).
Fox, R. (1975), "Primate kin and human kinship," pp. 9–35 in *Biosocial Anthropology*, ed. R. Fox (New York: Wiley).
Fox, R. (1979), "Kinship categories as natural categories," pp. 132–44 in *Evolutionary Biology and Human Social Behavior: An Anthropological Perspective*, ed. N. A. Chagnon and W. Irons (North Scituate, Mass.: Duxbury Press).

Francis, E. K. (1966), *Interethnic Relations: An Essay in Sociological Theory* (New York: Elsevier).
Frazer, Sir J. G. (1890), *The Golden Bough: A Study in Comparative Religion* (New York: Macmillan).
Frazer, Sir J. G. (1933–6), *The Fear of the Dead in Primitive Religion* (London: Macmillan), 3 vols.
Freedman, D. G. (1979), *Human Sociobiology* (New York: The Free Press).
Freeman, D. (1966) "Social anthropology and the scientific study of behavior," *Man*, 1 (n.s.): 330–42.
Freud, S. (1931, orig. 1913), *Totem and Taboo* (New York: New Republic).
Freud, S. (1946, orig. 1930), *Civilization and its Discontents* (London: Hogarth).
Freud, S. (1938), *The Basic Writings of Sigmund Freud* (New York: Random House).
Friedl, E. (1962), *Vasilika: A Village in Modern Greece* (New York: Holt, Rinehart & Winston).
Friedrichs, R. W. (1960), "Alter versus Ego: an exploratory study of altruism," *American Sociological Review*, 25: 496–508.
Fromm, E. (1947), *Man for Himself* (New York: Rinehart).
Fromm, E. (1961), "Introduction" to E. Fromm (ed.), *Marx's Concept of Man* (New York: Ungar).
Gadgil, M. (1975), "Evolution of social behavior through interpopulation selection," *Proceedings of the National Academy of Sciences*, 72: 1199–201.
Gallup, G., Jr (1977), "U.S. in early stage of religious revival?", *Journal of Current Social Issues*, 14: 50–5.
Gallup, G. G. (1977), "Self-recognition in primates: a comparative approach to the bidirectional properties of consciousness," *American Psychologist*, 32: 229–38.
Gamson, W. A. (1968), *Power and Discontent* (Homewood, Ill.: Dorsey Press).
Gans, H. J. (1973), *More Equality* (New York: Pantheon).
Garcia, J., F. R. Ervin, and R. Koelling (1966), "Learning with prolonged delay of reinforcement," *Psychonomic Science*, 5: 121–2.
Garcia, J., and R. Koelling (1966), "Relation of cue to consequence in avoidance learning," *Psychonomic Science*, 4: 123–4.
Gardner, R. A., and B. T. Gardner (1971), "Two-way communication with an infant chimpanzee," pp. 117–84 in *Behavior of Non-Human Primates*, ed. A. Scheier and F. Stollinitz (New York: Academic Press).
Gazzaniga, M. S., and J. E. Ledoux (1978), *The Integrated Mind* (New York: Plenum).
Gebhard, P. H. (1979), Communication to *Time* magazine, 28 May.
Geertz, C. (1966), "Religion as a cultural system," pp. 1–46 in *Anthropological Approaches to the Study of Religion*, ed. M. P. Banton (London: Tavistock).
Geertz, C. (1973), *The Interpretation of Cultures* (New York: Basic Books).
Geiger, T. (1949), *Die Klassengesellschaft in Schmelztiegel* (Cologne: Kiepenheur).
Geis, G., and R. F. Meier (1977), *White Collar Crime* (New York: The Free Press).
Geist, V. (1971), *Mountain Sheep: A Study in Behavior and Evolution* (Chicago: University of Chicago Press).
Gelfand, D. E., and R. D. Lee (eds) (1973), *Ethnic Conflicts and Power: A Cross-National Perspective* (New York: Wiley).
George, H. (1879), *Progress and Poverty* (New York: Appleton).
Geschwender, J. A. (1978), *Racial Stratification in America* (Dubuque, Iowa: William C. Brown).
Geschwind, N. (1979), "Specialization of the human brain," *Scientific American*, 241: 180–99.
Ghiselin, M. T. (1974), *The Economy of Nature and the Evolution of Sex* (Berkeley, Calif.: University of California Press).

Giddens, A. (1973), *The Class Structure of the Advanced Societies* (London: Hutchinson).
Giddings, F. H. (1896), *The Principles of Sociology* (New York: Macmillan).
Gillespie, C. C. (1960), *The Edge of Objectivity: An Essay in the History of Scientific Ideas* (Princeton, NJ: Princeton University Press).
Girard, R. (1965), *Deceit, Desire and the Novel: Self and Other in Literary Structure* (Baltimore, Md: Johns Hopkins University Press).
Glazer, N., and D. P. Moynihan (1963), *Beyond the Melting Pot* (Cambridge, Mass.: MIT Press).
Glock, C. Y. (ed.) (1973), *Religion in Sociological Perspective: Essays in the Empirical Study of Religion* (Belmont, Calif.: Wadsworth).
Glock, C. Y., and R. N. Bellah (eds) (1976), *The New Religious Consciousness* (Berkeley, Calif.: University of California Press).
Gluckman, M. (1965), *Politics, Law and Ritual in Tribal Society* (New York: Mentor Books).
Goffman, E. (1959), *The Presentation of Self in Everyday Life* (Garden City, NY: Doubleday).
Goffman, E. (1961), *Encounters: Two Studies in the Sociology of Interaction* (Indianapolis, Ind.: Bobbs-Merrill).
Goffman, E. (1967), *Interaction Ritual* (Garden City, NY: Doubleday).
Goffman, E. (1969), *Strategic Interaction* (Philadelphia, Pa: University of Pennsylvania Press).
Goffman, E. (1971), *Relations in Public* (New York: Basic Books).
Goldman, N. (1970), "Differential selection of juvenile offenders: analysis of the individual communities," pp. 96–104 in *The Politics of Local Justice*, ed. J. R. Klonoski and R. I. Mendelsohn (Boston, Mass.: Little, Brown).
Goode, W. J. (1966), "Family and mobility," pp. 582–601 in *Class, Status, and Power*, ed. R. Bendix and S. M. Lipset (New York: The Free Press).
Goode, W. J. (1970), *World Revolution and Family Patterns* (New York: The Free Press).
Gordon, M. M. (1964), *Assimilation in American Life* (New York: Oxford University Press).
Gordon, M. M. (1978), *Human Nature, Class, and Ethnicity* (New York: Oxford University Press).
Gould, S. J. (1978, orig. 1976), "Biological potential vs. biological determinism," pp. 343–51 in *The Sociobiology Debate*, ed. A. L. Caplan (New York: Harper & Row).
Gould, S. J. (1977), *Ever Since Darwin: Reflections in Natural History* (New York: Norton).
Gouldner, A. W. (1970), *The Coming Crisis of Western Sociology* (New York: Basic Books).
Gouldner, A. W. (1973), *For Sociology: Renewal and Critique in Sociology Today* (New York: Basic Books).
Grace, J. (1975), *Domestic Slavery in West Africa: With Particular Reference to the Sierra Leone Protectorate, 1896–1927* (New York: Barnes & Noble).
Granovetter, M. (1979), "The idea of 'advancement' in theories of social evolution and development," *American Journal of Sociology*, 85: 489–515.
Greeley, A. M. (1971), *Why Can't They Be Like Us?* (New York: Dutton).
Greeley, A. M. (1972), *The Denominational Society: A Sociological Approach to Religion in America* (Chicago: Scott, Foresman).
Greeley, A. M. (1974), *Ethnicity in the United States* (New York: Wiley).
Greenberg, J. H. (1957), *Essays in Linguistics* (New York: Wenner-Gren Foundation).
Greene, G. (1951), *The End of the Affair* (New York: Viking Press).

Gregory, M. S., A. Silvers, and D. Sutch (eds) (1978), *Sociobiology and Human Nature* (San Francisco: Jossey-Bass).

Grobstein, C. (1973), "Hierarchical order and neogenesis," pp. 31–47 in *Hierarchy Theory: The Challenge of Complex Systems*, ed. H. H. Pattee (New York: Braziller).

Grossberg, S. (1978), "A theory of human memory: self-organization and performance of sensory-motor codes, maps, and plans," *Progress in Theoretical Biology*, 5: 233–74.

Guhl, A. M. (1968), "Social inertia and social stability in chickens," *Animal Behavior*, 16: 219–32.

Guhl, A. M., N. E. Collias, and W. C. Allee (1945), "Mating behavior and the social hierarchy in small flocks of white leghorns," *Physiological Zoology*, 18: 365–90.

Guhl, A. M., and G. J. Fisher (1969), "The behavior of chickens," pp. 515–53 in *The Behavior of Domestic Animals*, ed. E. S. E. Hefez (Baltimore, Md: Williams & Wilkens).

Gusinde, M. (1937), *The Yahgan: The Life and Thought of the Water Nomads of Cape Horn* (Mödling bei Wien: Anthropos-Bibliothek).

Habbakuk, H. J. (1950), "Marriage settlement in the eighteenth century," *Transactions of the Royal Historical Society*, 32: 15–30.

Haggett, P. (1972), *Geography: A Modern Synthesis* (New York: Harper & Row).

Haldane, J. B. S. (1932), *The Causes of Evolution* (New York: Harper).

Halley, P. (1979), *Lest Innocent Blood Be Shed* (New York: Harper & Row).

Hamblin, R. L., J. L. L. Miller, and D. E. Saxton (1979), "Modeling use diffusion," *Social Forces*, 57: 799–811.

Hamilton, W. D. (1964), "The genetical theory of social behaviour: I and II," *Journal of Theoretical Biology*, 7: 1–52.

Hamilton, W. D. (1972), "Altruism and related phenomena, mainly in social insects," *Annual Review of Ecology and Systematics*, 3: 193–232.

Handlin, O. (1951), *The Uprooted* (New York: Grosset & Dunlap).

Hardin, G. (1956), "Meaninglessness of the word protoplasm," *Scientific American*, 82: 112–20.

Hardin, G. (1977), *The Limits of Altruism: An Ecologist's View of Survival* (Bloomington, Ind.: Indiana University Press).

Hardy, A. (1964), *The Living Stream* (New York: Harper & Row).

Harris, M. (1968), *The Rise of Anthropological Theory* (New York: Crowell).

Harris, M. (1974), *Cows, Pigs, Wars, and Witches; the Riddles of Culture* (New York: Random House).

Harris, M. (1977), *Cannibals and Kings: The Origin of Culture* (New York: Random House).

Harris, M. (1979), *Cultural Materialism: The Struggle for a Science of Culture* (New York: Random House).

Hawthorne, N. (1966), *The Marble Faun* (New York: Airmont).

Hayes, K. J., and C. Hayes (1952), "Imitation in a home-raised chimpanzee," *Journal of Comparative and Physiological Psychology*, 45: 450–9.

Hazelrigg, L. E., and J. Lopreato (1972), "Heterogamy, inter-class mobility, and socio-political attitudes in Italy," *American Sociological Review*, 37: 264–77.

Hediger, H. P. (1970), "The evolution of territorial behavior," pp. 34–59 in *Social Life of Early Man*, ed. S. L. Washburn (Chicago: Aldine).

Hegel, G. W. F. (1975, orig. 1835), *The Philosophy of Fine Art* (New York: Hacker Art Books).

Heider, F. (1958), *The Psychology of Interpersonal Relations* (New York: Wiley).

Henderson, L. J. (1935), *Pareto's General Sociology* (Cambridge, Mass.: Harvard University Press).

Henderson, N. D. (1982), "Human behavior genetics," *Annual Review of Psychology*, 33: 403–40.
Herskovits, M. J. (1938), *Dahomey* (New York: Augustin).
Hertz, R. (1960, orig. 1907–9), *Death and the Right Hand* (Aberdeen: Aberdeen University Press).
Hill, C. T., Z. Rubin, and L. A. Peplau (1976), "Breakups before marriage: the end of 103 affairs," *Journal of Social Issues*, 32: 147–68.
Hill, J. L. (1974), "Peromyscus: effect of early pairing on reproduction," *Science*, 186: 1042–4.
Hinde, R. A. (1970), *Animal Behavior: A Synthesis of Ethology and Comparative Psychology* (New York: McGraw-Hill).
Hinde, R. A. (1974), *Biological Bases of Human Social Behavior* (New York: McGraw-Hill).
Hodge, R. W., and D. J. Treiman (1968), "Class identification in the United States," *American Journal of Sociology*, 73: 535–47.
Hofstadter, R. (1955), *Social Darwinism in American Thought* (New York: Braziller).
Hogbin, H. I. (1934), *Law and Order in Polynesia* (London: Christophers).
Hogg, G. (1966), *Cannibalism and Human Sacrifice* (New York: Citadel Press).
Hollingshead, A. B. (1949), *Elmtown's Youth* (New York: Wiley).
Hollingshead, A. B. (1950), "Cultural factors in the selection of marriage mates," *American Sociological Review*, 15: 619–27.
Hollingshead, A. B., and F. C. Redlich (1958), *Social Class and Mental Illness* (New York: Wiley).
Holmes, G. A. (1957), *The Estates of the Higher Nobility in Fourteenth Century England* (Cambridge: Cambridge University Press).
Homans, G. C. (1961), *Social Behavior: Its Elementary Forms*, 1st edn (New York: Harcourt, Brace & World).
Homans, G. C. (1964), "Bringing men back in," *American Sociological Review*, 29: 809–18.
Homans, G. C. (1967), *The Nature of Social Science* (New York: Harcourt, Brace & World).
Homans, G. C. (1974), *Social Behavior: Its Elementary Forms*, 2nd edn (New York: Harcourt Brace Jovanovich).
Homans, G. C., and C. P. Curtis, Jr (1934), *An Introduction to Pareto: His Sociology* (New York: Knopf).
Hooglund, E. J. (1982), *Land and Revolution in Iran, 1960–1980* (Austin, Texas: University of Texas Press).
Horn, J. M., J. C. Loehlin, and L. Willerman (1979), "Intellectual resemblance among adoptive and biological relatives: the Texas Adoption Project," *Behavior Genetics*, 9: 177–201.
Houghton, J., and J. Lopreato (1981), "Protestantism and capitalism: a biosocial emendation of Weber's thesis," *Revue Européene des Sciences Sociales*, 19: 55–78.
Howitt, A. W. (1904), *The Native Tribes of Southeast Australia* (New York: Macmillan).
Hrdy, S. B. (1974), "Male-male competition and infanticide among the langurs (*Presbytis entellus*) of Abu, Rajasthan," *Folia Primatologica*, 22: 19–58.
Hrdy, S. B. (1977), *The Langurs of Abu* (Cambridge, Mass.: Harvard University Press).
Hsu, F. L. K. (1948), *Under the Ancestor's Shadow* (New York: Columbia University Press).
Hubel, D. H. (1979), "The brain," *Scientific American*, 241: 44–53.
Huber, J. (ed.) (1973), *Changing Women in a Changing Society* (Chicago: University of Chicago Press).

Hubert, H., and M. Mauss (1964, orig. 1898), *Sacrifice: Its Nature and Function* (Chicago: University of Chicago Press).
Hull, D. L. (1978), "Scientific bandwagon or traveling medicine show?," pp. 136–63 in *Sociobiology and Human Nature*, ed. M. Gregory, A. Silvers, and D. Sutch (San Francisco: Jossey-Bass).
Hullum, J. R. (1980), "Human nature and society in evolutionary context: convergence between sociobiology and early sociological thought," unpublished Ph.D. dissertation (Austin: Department of Sociology, University of Texas at Austin).
Hume, D. (1896, orig. 1739), *A Treatise of Human Nature* (London: Oxford University Press).
Hunt, J. H. (ed.) (1980), *Selected Readings in Sociobiology* (New York: McGraw-Hill).
Hutter, M. (1981), *The Changing Family: Comparative Perspectives* (New York: Wiley).
Huxley, A. (1935), "Review of Pareto's 'The Mind and Society,' " *New York Herald Tribune Books*, 9 June, pp. 1–8.
Huxley, J. (1942), *Evolution: The Modern Synthesis* (New York: Harper).
Huxley, J. (1958), "Introduction to the Mentor edition," pp. ix–xv in C. Darwin, *On the Origin of Species* (New York: Mentor Books).
Huxley, T. H. (1971, orig. 1863), *Man's Place in Nature* (Ann Arbor, Mich.: University of Michigan Press).
Huxley, T. H. (1892), *Collected Essays* (London: Macmillan).
Inkeles, A. (1950), "Social stratification and mobility in the Soviet Union: 1940–1950," *American Sociological Review*, 15: 465–79.
Irons, W. (1975), "The Yomut Turkmen: a study of social organization among a central Asian Turkic speaking population," *Anthropological Paper No. 58* (Ann Arbor, Mich.: University of Michigan, Museum of Anthropology).
Irons, W. (1979a), "Cultural and biological success," pp. 257–72 in *Evolutionary Biology and Human Social Behavior*, ed. N. A. Chagnon and W. Irons (North Scituate, Mass.: Duxbury Press).
Irons, W. (1979b), "Natural selection, adaptation, and human social behavior," pp. 4–39 in *Evolutionary Biology and Human Social Behavior: An Anthropological Perspective*, ed. N. A. Chagnon and W. Irons (North Scituate, Mass.: Duxbury Press).
Jackson, E.F., and H.J. Crockett, Jr (1964), "Occupational mobility in the United States: a point estimate and trend comparison," *American Sociological Review*, 29: 5–15.
Jacobson, M. (1978), *Development Neurobiology* (New York: Plenum).
Jakobi, L., and P. Marquer (1977), "A study of the relationships between assortative mating and exogamy in two series of French married couples," *Journal of Human Evolution*, 6: 115–22.
James, J. J. (1872), *A History of the Hawaiian Islands* (Honolulu: Hitchcock).
James, W. (1890), *Principles of Psychology* (New York: Holt), 2 vols.
Jellinck, E. M. (1946), "Clinical tests on comparative effectiveness of analgesic drugs," *Biometrics Bulletin*, 2: 87–91.
Jolly, A. (1972), *The Evolution of Primate Behavior* (New York: Macmillan).
Jones, R. A. (1977), *Self-Fulfilling Prophesies: Social, Psychological, and Physiological Effects of Expectancies* (Hillsdale, NJ: Lawrence Erlbaum Associates).
Jones, T. A. (1978), "Modernization and education in the U.S.S.R.," *Social Forces*, 57: 523–46.
Jones, W. (1939), *Ethnography of the Fox Indians* (Washington, DC: Government Printing Office).
Kaffman, M. (1977), "Sexual standards and behavior of the kibbutz adolescent," *American Journal of Orthopsychiatry*, 47: 207–17.
Kahl, J. (1957), *The American Class Structure* (New York: Rinehart).
Kamazani, R. K. (1980), "Iran's revolution: patterns, problems and prospects," *International Affairs*, 56: 443–57.

Kandel, E. R. (1976), *Cellular Basis of Behavior: An Introduction to Behavioral Neurobiology* (San Francisco: Freeman).
Kanowitz, L. (1969), *Women and the Law: Unfinished Revolution* (Albuquerque, New Mexico: University of New Mexico Press).
Kant, I. (1960, orig. 1763), *Observations on the Feeling of the Beautiful and Sublime* (Berkeley, Calif.: University of California Press).
Karlin, S. (1978), "Comparisons of positive assortative mating and sexual selection models," *Theoretical Population Biology*, 14: 281–312.
Kaufmann, J. H. (1967), "Social relations of adult males in a free-ranging band of rhesus monkeys," pp. 73–98 in *Social Communication among Primates*, ed. S. A. Altmann (Chicago: University of Chicago Press).
Kawai, M. (1958), "On the system of social ranks in a natural troop of Japanese monkeys," *Primates*, 1–2: 11–30.
Kawai, M. (1965), "Newly acquired pre-cultural behavior of the natural troop of Japanese monkeys on Koshima Islet," *Primates*, 6: 1–30.
Keller, A. G. (1915), *Societal Evolution: A Study of the Evolutionary Basis of the Science of Society* (New York: Macmillan) (rev. 1931).
Keller, S., and M. Zavalloni (1964), "Ambition and social class: a respecification," *Social Forces*, 43: 58–70.
Keyes, C. F. (1976), "Towards a new formulation of the concept of ethnic group," *Ethnicity*, 3: 202–13.
Kharuzin, N. (1898), "The juridical customs of the Yakut," *Etnograficheskoe Obozrenie* 10, kn. 37: 37–64.
King, G. E. (1976), "Society and territory in human evolution," *Journal of Human Evolution*, 5: 323–32.
Knox, I. (1936), *The Aesthetic Theories of Kant, Hegel, and Schopenhauer* (New York: Columbia University Press).
Kohlberg, L. (1969), "Stage and sequence: the cognitive developmental approach to socialization," pp. 347–80 in *Handbook of Socialization Theory and Research*, ed. D. A. Goslin (Chicago: Rand McNally).
Kohler, W. (1925), *The Mentality of Apes* (New York: Harcourt, Brace).
Konishi, M. (1965), "The role of auditory feedback in the control of vocalization in the white-crowned sparrow," *Zeitschrift für Tierpsychologie*, 22: 770–83.
Kreitman, N. (1968), "Married couples admitted to mental hospital," *British Journal of Psychiatry*, 114: 699–718.
Kroeber, A. (1908), *Ethnology of the Gros Ventre* (New York: American Museum of History).
Kropotkin, P. (1955, orig. 1902), *Mutual Aid: A Factor of Evolution* (Boston, Mass.: Extending Horizons Books).
Kübler-Ross, E. (1974), *Questions and Answers on Death and Dying* (New York: Macmillan).
Kuffer, S. W., and J. G. Nicholls (1976), *From Neuron to Brain: A Cellular Approach to the Function of the Nervous System* (New York: Sinauer Associates).
Kuhn, T. S. (1957), *The Copernican Revolution: Planetary Astronomy in the Development of Western Thought* (Cambridge, Mass.: Harvard University Press).
Kuhn, T. S. (1962), *The Structure of Scientific Revolutions* (Chicago: University of Chicago Press).
Kummer, H. (1968), *Social Organization of Hamadryas Baboons* (Chicago: University of Chicago Press).
Kummer, H. (1971), *Primate Societies* (Chicago: Aldine).
Kunkel, J. H. (1970), *Society and Economic Growth* (London: Oxford University Press).
Kurtén, B. (1980, orig. 1978), *Dance of the Tiger: A Novel of the Ice Age* (New York: Berkley Books).

Kushner, G., M. Gibson, J. Gulick, J. J. Honigmann, and R. Nonas (1962), *What Accounts for Socio-Cultural Change? A Propositional Inventory* (Chapel Hill, NC: University of North Carolina Press).

Lack, D. (1954), *The Natural Regulation of Animal Numbers* (Oxford: Clarendon Press).

Lack, D. (1968), *Ecological Adaptations for Breeding in Birds* (London: Methuen).

Lamanna, M. A., and A. Riedman (1981), *Marriages and Families: Making Choices throughout the Life Cycle* (Belmont, Calif.: Wadsworth).

Lamarck, J.-B. de (1963, orig. 1809), *Zoological Philosophy* (New York: Hafner).

Lang, O. (1946), *Chinese Family and Society* (New Haven, Conn.: Yale University Press).

Langer, S. (1957), *Philosophy in a New Key* (Cambridge, Mass.: Harvard University Press).

Langton, J. (1979), "Darwinism and the behavioral theory of sociocultural evolution: an analysis," *American Journal of Sociology*, 85: 288–309.

Laqueur, W. (1981), *The Terrible Secret* (Boston, Mass.: Little, Brown).

Lasagna, L., F. Mosteller, J. M. von Felsinger, and H. K. Beecher (1954), "A study of the placebo response," *American Journal of Medicine*, 16: 770–9.

Laumann, E. O. (1966), *Prestige and Association in an Urban Community* (Indianapolis, Ind.: Bobbs-Merrill).

Lauter, P., and F. Howe (1970), *The Conspiracy of the Young* (New York: World Publishing).

Lavandera, B. (1977), "Context, meaning and the Chomskyan notion of creativity," *American Anthropologist*, 79: 638–41.

Leach, E. (1976), *Culture and Communication. The Logic by Which Symbols Are Connected: An Introduction to the Use of Structuralist Analysis in Social Anthropology* (Cambridge: Cambridge University Press).

Leach, W. (1980), *True Love and Perfect Union: The Feminist Reform of Sex and Society* (New York: Basic Books).

Leavitt, G. (1977), "The frequency of warfare: an evolutionary perspective," *Sociological Inquiry*, 47: 49–58.

LeBoeuf, B. J. (1974), "Male-male competition and reproductive success in elephant seals," *American Zoologist*, 14: 163–76.

Lee, R. B. (1972), "The !Kung Bushmen of Botswana," pp. 327–68 in *Hunters and Gatherers of Today*, ed. M. G. Bicchieri (New York: Holt, Rinehart & Winston).

Lee, R. B., and B. I. DeVore (eds) (1968), *Man the Hunter* (Chicago: Aldine).

Lehrman, D. S. (1970), "Semantics and conceptual issues in the nature-nurture problem," pp. 17–52 in *Development and Evolution of Behavior*, ed. L. R. Aronson, D. Lehrman, and J. Rosenblatt (San Francisco: Freeman).

Lenneberg, E. H. (1964), "A biological perspective on language," pp. 65–88 in *New Directions in the Study of Language*, ed. E. H. Lenneberg (Cambridge, Mass.: MIT Press).

Lenski, G. E. (1966), *Power and Privilege* (New York: McGraw-Hill).

Lenski, G. E. (1976), "History and social change," *American Journal of Sociology*, 82: 548–64.

Lenski, G. E. (1978), "Marxist experiments in destratification: an appraisal," *Social Forces*, 57: 364–83.

Lenski, G. E., and J. Lenski (1978), *Human Societies: An Introduction to Macrosociology*, 3rd edn (New York: McGraw-Hill).

Lessa, W. A., and E. Z. Vogt (eds) (1972), *Reader in Comparative Religion* (New York: Harper & Row).

Levin, A. J. (1951), "The fiction of the death instinct," *Psychiatric Quarterly*, 25: 257–81.

Levine, R. A., and D. T. Campbell (1972), *Ethnocentrism: Theories of Conflict, Ethnic Attitudes, and Group Behavior* (New York: Wiley).
Levins, R. (1968), *Evolution in Changing Environments: Some Theoretical Explorations* (Princeton, NJ: Princeton University Press).
Levins, R. (1970), "Extinction," pp. 77–107 in *Some Mathematical Questions in Biology*, Lectures on Mathematics in the Life Sciences, Vol. 2, ed. M. Gerstenhaber (Providence, RI: American Mathematical Society).
Lévi-Strauss, C. (1949), *Les Structures élémentaires de la parenté* (Paris: Presses Universitaires de France).
Lévi-Strauss, C. (1950), "Introduction à l'oeuvre de Marcel Mauss," Marcel Mauss, *Sociologie et Anthropologie* (Paris: Presses Universitaires de France).
Lévi-Strauss, C. (1963, orig. 1962), *Totemism* (Boston, Mass.: Beacon Press).
Lévi-Strauss, C. (1969), *The Elementary Structures of Kinship* (Boston, Mass.: Beacon Press).
Lewinson, P. (1932), *Race, Class, and Party: A History of Negro Suffrage and White Politics in the South* (New York: Grosset & Dunlap).
Lewis, L. S., and J. Lopreato (1962), "Arationality, ignorance, and perceived danger in medical practices," *American Sociological Review*, 27: 509–14.
Lewis, O. (1966), "The culture of poverty," *Scientific American*, 215: 19–25.
Lewis, S. (1924), *Arrowsmith* (New York: Designer Publishing).
Lewontin, R. C. (1978), "Adaptation," *Scientific American*, 239: 212–30.
Lewontin, R. C. (1979), "Sociobiology as an adaptationist program," *Behavioral Science*, 24: 5–14.
Lewontin, R., D. Kirk, and J. Crow (1968), "Selective mating, assortative mating, and inbreeding: definitions and implications," *Eugenics Quarterly*, 15: 141–3.
Lewy, G. (1974), *Religion and Revolution* (London: Oxford University Press).
Lifton, R., and E. Olson (1974), *Living and Dying* (New York: Praeger).
Lin, Y. (1947), *The Lolo of Liang-shan* (Shanghai: Commercial Press).
Lindzey, G. (1967), "Some remarks concerning incest, the incest taboo, the psychoanalytic theory," *American Psychologist*, 22: 1051–9.
Lipset, S. M. (ed.) (1967), *Student Politics* (New York: Basic Books).
Lipset, S. M., and R. Bendix (1959), *Social Mobility in Industrial Society* (Berkeley, Calif.: University of California Press).
Loehlin, J. C., and R. C. Nichols (1976), *Heredity, Environment, and Personality* (Austin, Texas: University of Texas Press).
Lopreato, J. (1967), *Peasants No More: Social Class and Social Change in an Underdeveloped Society* (San Francisco: Chandler).
Lopreato, J. (1970), *Italian Americans* (New York: Random House).
Lopreato, J. (1971), "The concept of equilibrium: sociological tantalizer," pp. 309–43 in *Institutions and Social Exchange: The Sociologies of Talcott Parsons and G. C. Homans*, ed. H. Turk and R. L. Simpson (Indianapolis, Ind.: Bobbs-Merrill).
Lopreato, J. (1977), *La Stratificazione Sociale negli USA: Fatti e Teorie (1945–1975)* (Torino: Edizioni della Fondazione).
Lopreato, J. (1981), "Toward a theory of genuine altruism in Homo sapiens," *Ethology and Sociobiology*, 2: 113–26.
Lopreato, J., and L. Alston (1970), "Ideal types and the idealization strategy," *American Sociological Review*, 35: 88–96.
Lopreato, J., and J. Saltzman Chafetz (1970), "The political orientation of skidders: a middle-range theory," *American Sociological Review*, 35: 440–51.
Lopreato, J., and J. Saltzman Chafetz (1979), "Social integration, regulation of needs, and suicide: an emendation of Durkheim's theory," *Revue Européenne des Sciences Sociales* 17: 115–33.

Lopreato, J., and L. E. Hazelrigg (1972), *Class, Conflict, and Mobility: Theories and Studies of Class Structure* (San Francisco: Chandler).
Lorenz, K. (1958), "The evolution of behavior," *Scientific American*, 199: 67–8.
Lorenz, K. (1965), *Evolution and Modification of Behavior* (Chicago: University of Chicago Press).
Lorenz, K. (1966), *On Aggression* (New York: Harcourt Brace Jovanovich).
Lorenz, K. (1970), *Studies in Animal and Human Behavior* (Cambridge, Mass.: Harvard University Press), Vol. 1.
Lorenz, K. (1973), *Civilized Man's Eight Deadly Sins* (New York: Harcourt Brace Jovanovich).
Lorenz, K. (1974), "Analogy as a source of knowledge," *Science*, 185: 229–34.
Lorenz, K. (1977), *Behind the Mirror* (New York: Harcourt Brace Jovanovich).
Lorenz, K., and P. Leyhausen (1973), *Motivation of Human and Animal Behavior: An Ethological View* (New York: Van Nostrand Reinhold).
Lotka, A. (1922), "Contribution to the energetics of evolution," *Proceedings of the National Academy of Sciences*, 8: 147–51.
Lotka, A. (1925), *Elements of Physical Biology* (Baltimore, Md: Williams & Wilkins).
Lowie, R. (1937), *History of Ethnological Theory* (New York: Farrar & Rinehart).
Luhman, R., and S. Gilman (1980), *Race and Ethnic Relations: The Social and Political Experience of Minority Groups* (Belmont, Calif.: Wadsworth).
Lumsden, C. J., and E. O. Wilson (1981), *Genes, Mind, and Culture: The Coevolutionary Process* (Cambridge, Mass.: Harvard University Press).
McBride, G. (1971), "The nature-nurture problem in social evolution," pp. 35–56 in *Man and Beast: Comparative Social Behavior*, ed. J. F. Eisenberg and W. S. Dillon (Washington, DC: Smithsonian Institution Press).
McClelland, D. C. (1961), *The Achieving Society* (New York: Van Nostrand).
McGill, M. J. (1942), "Scheler's theory of sympathy and love," *Philosophy and Phenomenological Research*, 2: 273–91.
McKusick, V. A. (1975), *Mendelian Inheritance in Man* (Baltimore, Md: Johns Hopkins University Press).
McKusick, V. A., and F. H. Ruddle (1977), "The status of the gene map of the human chromosome," *Science*, 196: 390–405.
McLemore, S. D. (1980), *Racial and Ethnic Relations in America* (Boston, Mass.: Allyn & Bacon).
Malinowski, B. (1923), *Argonauts of the Western Pacific* (New York: Dutton).
Malinowski, B. (1926a), "Anthropology," pp. 131–9 in *Encyclopaedia Britannica*, First Supplementary Volume (New York: Encyclopaedia Britannica).
Malinowski, B. (1926b), *Crime and Customs in Savage Society* (New York: Dutton).
Malinowski, B. (1927), *Sex and Repression in Savage Society* (London: Kegan Paul).
Malinowski, B. (1935), *The Foundations of Faith and Morals: An Anthropological Analysis of Primitive Beliefs and Conduct with Special Reference to the Fundamental Problems of Religion and Ethics* (Oxford: Oxford University Press).
Malinowski, B. (1944), *A Scientific Theory of Culture* (Chapel Hill, NC: University of North Carolina Press).
Malinowski, B. (1948), *Magic, Science and Religion and Other Essays* (New York: The Free Press).
Malthus, T. R. (1817, orig. 1798), *An Essay on the Principle of Population* (London: Murray).
Mannheim, K. (1936), *Ideology and Utopia* (New York: Harcourt, Brace).
Mariéjol, J. H. (1961), *The Spain of Ferdinand and Isabella* (New Brunswick, NJ: Rutgers University Press).
Marler, P. R., and M. Tamura (1964), "Culturally transmitted patterns of vocal behavior in sparrows," *Science*, 146: 1483–6.

Marris, P. (1974), *Loss and Change* (London: Routledge & Kegan Paul).
Marsden, H. M. (1968), "Agonistic behaviour of young rhesus monkeys after changes induced in social rank of their mothers," *Animal Behaviour*, 16: 38–44.
Marshall, T. H. (1950), *Citizenship and Social Class, and Other Essays* (Cambridge: Cambridge University Press).
Martin, M. K. (1974), *The Foraging Adaptation: Uniformity or Diversity* (Reading, Mass.: Addison-Wesley).
Marx, K. (1961, orig. 1844), *Economic and Philosophical Manuscripts*, pp. 90–196 in *Marx's Concept of Man*, ed. E. Fromm (New York: Ungar).
Marx, K. (1963, orig. 1847), *The Poverty of Philosophy* (New York: International Publishers).
Marx, K. (1934, orig. 1850), *The Class Struggles in France (1848–1850)* (New York: International Publishers).
Marx, K. (1967, orig. 1867), *Capital* (New York: International Publishers).
Marx, K. (1959, orig. 1875), *Critique of the Gotha Program*, pp. 112–32 in *Marx and Engels: Basic Writings on Politics and Philosophy*, ed. L. S. Feuer (Garden City, NY: Doubleday).
Marx, K., and F. Engels (1947, orig. 1845–6), *The German Ideology* (New York: International Publishers).
Marx, K., and F. Engels (1955, orig. 1848), *The Communist Manifesto*, ed. S. H. Beer (New York: Appleton-Century-Crofts).
Marx, K., and F. Engels (1935), *The Correspondence of Marx and Engels* (New York: International Publishers).
Marx, K., and F. Engels (1955), "Wage labour and capital," in *Selected Works in Two Volumes* (Moscow: Foreign Languages Publishing House).
Maslow, A. H. (1940), "Dominance-quality and social behavior in infra-human primates," *Journal of Social Psychology*, 11: 313–24.
Maslow, A. H. (1954), *Motivation and Personality* (New York: Harper & Row).
Maslow, A. H. (1968), *Toward a Psychology of Being* (New York: Van Nostrand).
Mason, K. O. (1974), *Women's Labor Force Participation and Fertility* (Research Triangle Park, NC: Research Triangle Institute).
Matthews, D. R. (1954), "United States senators and the class structure," *Public Opinion Quarterly*, 18: 5–22.
Matthews, D. R. (1960), *U.S. Senators and their World* (Chapel Hill, NC: University of North Carolina Press).
Mauss, M. (1954, orig. 1925), *The Gift* (London: Cohen & West).
Mauss, M. (1972, orig. 1950a), *A General Theory of Magic* (London: Routledge & Kegan Paul).
Mauss, M. (1950b), *Sociologie et anthropologie* (Paris: Presses Universitaires de France).
Maynard Smith, J. (1958), *The Theory of Evolution* (New York: Penguin).
Maynard Smith, J. (1964), "Group selection and kin selection," *Nature*, 201: 1145–7.
Maynard Smith, J. (1976), "Evolution and the theory of games," *Scientific American*, 64: 41–5.
Maynard Smith, J. (1978), "The evolution of behavior," *Scientific American*, 239: 176–92.
Maynard Smith, J., and G. A. Parker (1976), "The logic of asymmetric contests," *Animal Behavior*, 24: 159–75.
Maynard Smith, J., and G. R. Price (1973), "The logic of animal conflicts," *Nature*, 246: 15–18.
Mayr, E. (1942), *Systematics and the Origin of Species* (New York: Columbia University Press).

Mayr, E. (1978), "Evolution," *Scientific American*, 239: 47–55.
Mazlish, B. (1965), *The Railroad and the Space Program: An Exploration in Historical Analogy* (Cambridge, Mass.: MIT Press).
Mazur, A. (1973), "A cross-species comparison of status in small established groups," *American Sociological Review*, 38: 513–30.
Mazur, A. (1976), "On Wilson's sociobiology," *American Journal of Sociology*, 82: 697–700.
Mead, G. H. (1934), *Mind, Self, and Society* (Chicago: University of Chicago Press).
Menaker, E. (1979), *Masochism and the Emergent Ego: Selected Papers of Esther Menaker*, ed. L. Lerner (New York: Human Sciences Press).
Menzel, R., and J. Erber (1978), "Learning and memory in bees," *Scientific American*, 239: 102–10.
Merton, R. K. (1941), "Intermarriage and the social structure: fact and theory," *Psychiatry*, 4: 361–74.
Merton, R. K. (1968), *Social Theory and Social Structure* (New York: The Free Press), enlarged edition.
Métraux, A. (1946), pp. 400–1 in *Handbook of South American Indians*, ed. J. Steward (Washington, DC: Smithsonian Institution, Bureau of American Ethnology), Vol. 5.
Michels, R. (1959, orig. 1914), *Political Parties* (Glencoe, Ill.: The Free Press).
Middleton, J. (1960), *Lugbara Religion: Ritual and Authority among an East African People* (London: Oxford University Press).
Milgram, S. (1974), *Obedience to Authority* (New York: Harper & Row).
Miller, E. J. (1954), "Caste and territory in Malabar," *American Anthropologist*, 56: 410–20.
Miller, H. P. (1968), "The poor are still here," pp. 5–49 in *Aspects of Poverty*, ed. B. B. Seligman (New York: Crowell).
Miller, S. M. (1960), "Comparative social mobility," *Current Sociology*, 9: 1–89.
Miller, S. M., and F. Reissman (1968), *Social Class and Social Policy* (New York: Basic Books).
Mills, C. W. (1956), *The Power Elite* (New York: Oxford University Press).
Missakian, E. A. (1972), "Genealogical and cross-genealogical dominance relations in a group of free-ranging rhesus monkeys (*Macaca mulatta*) on Cayo Santiago," *Primates*, 13: 169–80.
Montagu, M. F. A. (ed.) (1968), *Man and Aggression* (London: Oxford University Press).
Monteil, C. V. (1924), *The Bambara of Segón and Kaarta: An Historical, Ethnographical and Literary Study of a People of the French Sudan* (Paris: LaRose).
Moody, R. (1978), *Reflections on Life After Death* (New York: Bantam Books).
Morgan, L. H. (1877), *Ancient City* (New York: Holt).
Mosca, G. (1939, orig. 1896), *The Ruling Class* (New York: McGraw-Hill).
Moskos, C. C., Jr. (1970), *The American Enlisted Man* (New York: Russell Sage Foundation).
Mueller, H. C. (1971), "Oddity and specific search image more important than conspicuousness in prey selection," *Nature* (London), 233: 345–6.
Muggeridge, M. (1971), *Something Beautiful for God* (New York: Harper & Row).
Mulkay, M. J. (1971), *Functionalism, Exchange and Theoretical Strategy* (New York: Schocken Books).
Murdock, G. P. (1940), "The cross-cultural survey," *American Sociological Review*, 5: 361–70.
Murdock, G. P. (1945), "The common denominator of culture," pp. 124–42 in *The Science of Man in the World Crisis*, ed. R. Linton (New York: Columbia University Press).

Murdock, G. P. (1949), *Social Structure* (New York: Macmillan).
Murdock, G. P. (1957), "World ethnographic sample," *American Anthropologist*, 59: 644–87.
Murdock, G. P. (1967), *Ethnographic Atlas* (Pittsburgh, Pa: University of Pittsburgh Press).
Murphy, R. F. (1957), "Intergroup hostility and social cohesion," *American Anthropologist*, 59: 1018–35.
Murray, B. G. (1971), "The ecological consequences of interspecific territorial behavior in birds," *Ecology*, 52: 414–23.
Murstein, B. I. (1976), *Who Will Marry Whom?* (New York: Springer).
Myrdal, G. (1944), *An American Dilemma* (New York: Harper).
Nagel, E. (1961), *The Structure of Science* (New York: Harcourt, Brace & World).
Nagel, S. S. (1970), "The tipped scales of American justice," pp. 114–27 in *The Politics of Local Justice*, ed. J. R. Klonoski and R. I. Mendelsohn (Boston, Mass.: Little, Brown).
Nagel, T. (1970), *The Possibility of Altruism* (Oxford: Clarendon Press).
National Catholic Almanac (1980) (Huntington, Ind.: Our Sunday Visitor).
Neel, J. V. (1970), "Lessons form a primitive people," *Science*, 170: 815–22.
Neel, J. V. (1978), "The population structure of an Amerindian tribe, the Yanomama," *Annual Review of Genetics*, 12: 365–413.
Newby, I. A. (1965), *Jim Crow's Defense* (Baton Rouge, La: Louisiana State University Press).
Newman, W. M. (1973), *American Pluralism: A Study of Minority Groups and Social Theory* (New York: Harper & Row).
Newsweek (1978), "A vote for the Sasquatch," 26 June.
Nielsen, J. (1964), "Mental disorder in married couples (assortative mating)," *British Journal of Psychiatry*, 110: 683–97.
Nisbet, R. A. (1969), *Social Change and History* (London: Oxford University Press).
Noble, G. K. (1939), "The role of dominance in the life of birds," *Auk*, 56: 263–73.
Nordskog, E. (ed.) (1960), *Social Change* (New York: McGraw-Hill).
Novak, M. (1971), *The Rise of the Unmeltable Ethnics* (New York: Macmillan).
Noyes, R., and R. Kletti (1976), "Depersonalization in the face of life-threatening danger: a description," *Psychiatry*, 39: 19–27.
O'Dea, T. F. (1966), *The Sociology of Religion* (Englewood Cliffs, NJ: Prentice-Hall).
Ogburn, W. F. (1950, orig. 1922), *Social Change: With Respect to Culture and Original Nature* (New York: Viking Press).
Ogburn, W. F. (1964, orig. 1942), "Inventions, population, and history," pp. 62–77 in W. F. Ogburn, *On Culture and Social Change*, ed. O. D. Duncan (Chicago: University of Chicago Press).
Olmsted, F. L. (1959), *The Slave States* (New York: Putnam).
Opler, M. E. (1936), "An interpretation of ambivalence of two American Indian tribes," *Journal of Social Psychology*, VII: 82–115.
Oster, G. F., and E. O. Wilson (1978), *Caste and Ecology in the Social Insects* (Princeton, NJ: Princeton University Press).
Otterbein, K. F. (1970), *The Evolution of War* (New Haven, Conn.: HRAF Press).
Padover, S. K. (1978), *Karl Marx: An Intimate Biography* (New York: McGraw-Hill).
Pareto, V. (1971, orig. 1906), *Manual of Political Economy* (New York: Kelly).
Pareto, V. (1963, orig. 1916), *A Treatise on General Sociology* (also known as *The Mind and Society*) (New York: Dover Publications), 4 vols.
Park, R. E. (1941), "The social function of war: observations and notes," *American Journal of Sociology*, 46: 551–70.

Park, R. E. (1950), *Race and Culture* (New York: The Free Press).
Park, R. E. (1967), *On Social Control and Collective Behavior*, ed. R. H. Turner (Chicago: University of Chicago Press).
Park, R. E., and H. A. Miller (1921), *Old World Traits Transplanted* (New York: Harper & Row).
Parker, S. (1976), "The precultural basis of the incest taboo: toward a biosocial theory," *American Anthropologist*, 78: 285–305.
Parsons, T. (1937), *The Structure of Social Action* (Glencoe, Ill.: The Free Press).
Parsons, T. (1951), *The Social System* (New York: The Free Press).
Parsons, T. (1954), *Essays in Sociological Theory* (Glencoe: Ill.: The Free Press).
Parsons, T. (1961), "An outline of the social system," pp. 30–79 in *Theories of Society*, ed. T. Parsons, E. Shils, K. D. Naegele, and J. R. Pitts (New York: The Free Press), Vol. I.
Parsons, T. (1964), "Evolutionary universals in society," *American Sociological Review*, 29: 339–57.
Parsons, T. (1966), *Societies: Evolutionary and Comparative Perspectives* (Englewood Cliffs, NJ: Prentice-Hall).
Parsons, T. (1971), *The System of Modern Societies* (Englewood Cliffs, NJ: Prentice-Hall).
Parsons, T., and R. F. Bales (1955), *Family, Socialization and Interaction Process* (New York: The Free Press).
Pattee, H. H. (ed.) (1973), *Hierarchy Theory: The Challenge of Complex Systems* (New York: Braziller).
Pearson, K., and A. Lee (1903), "On the laws of inheritance in man," *Biometrika*, 2: 357–462.
Penrose, L. S. (1944), "Mental illness in husband and wife: a contribution to the study of assortative mating in man," *Psychiatric Quarterly Supplement*, 18: 161–6.
Percy, W. (1958), "Symbol, consciousness and intersubjectivity," *Journal of Philosophy*, 55: 631–41.
Percy, W. (1961), "The symbolic structure of interpersonal process," *Psychiatry*, 24: 39–52.
Perrins, C. M. (1965), "Population fluctuations and clutch size in the great tit. *Parus major*," *Journal of Animal Ecology*, 34: 601–47.
Perry, R. H., C. H. Chilton, and S. D. Kirkpatrick (1963), *Chemical Engineer's Handbook* (New York: McGraw-Hill).
Perry, W. J. (1923), *The Children of the Sun* (London: Methuen).
Peterson, N. (1975), "Hunter-gatherer territoriality: the perspective from Australia," *American Anthropologist*, 77: 53–68.
Pettigrew, T. F. (1978), "Three issues in ethnicity: boundaries, deprivations, and perceptions," pp. 25–49 in *Major Social Issues: A Multidisciplinary View*, ed. J. M. Yinger and S. J. Cutler (New York: The Free Press).
Pfeiffer, J. E. (1969), *The Emergence of Man* (New York: Harper & Row).
Pfeiffer, J. E. (1977), *The Emergence of Society* (New York: McGraw-Hill).
Phelps, E. S. (ed.) (1975), *Altruism, Morality, and Economic Theory* (New York: Russell Sage Foundation).
Piaget, J. (1951), *Play, Dreams, and Imitation in Childhood* (New York: Norton).
Piaget, J. (1952), *The Origins of Intelligence in Children* (New York: International University Press).
Piaget, J. (1954), *The Construction of Reality in the Child* (New York: Basic Books).
Piaget, J. (1970), *Genetic Epistemology* (New York: Columbia University Press).
Piaget, J. (1971), *Biology and Knowledge* (Chicago: University of Chicago Press).
Piaget, J. (1976), *Behavior and Evolution* (New York: Pantheon).
Pirenne, H. (1937, orig. 1933), *Economic and Social History of Medieval Europe* (New York: Harcourt, Brace & World).

Pitt-Rivers, A. L.-F. (1906), *The Evolution of Culture and Other Essays*, ed. J. L. Myres (Oxford: Clarendon Press).
Plomin, R., J. C. DeFries, and J. C. Loehlin (1977), "Genotype environment interaction and correlation in the analysis of human behavior," *Psychological Bulletin*, 84: 309–22.
Plomin, R., J. C. DeFries, and G. E. McClearn (1980), *Behavioral Genetics: A Primer* (San Francisco: Freeman).
Polanyi, M. and H. Prosch (1975), *Meaning* (Chicago: University of Chicago Press).
Popov, A. (1933), "Consecration ritual for a blacksmith novice among the Yakuts," *Journal of American Folk-Lore*, 46: 257–71.
Porter, J. (1965), *The Vertical Mosaic* (Toronto: University of Toronto Press).
Poston, D. L., Jr (1976), "Characteristics of voluntarily and involuntarily childless wives," *Social Biology*, 23: 198–209.
Power, H. W. (1975), "Mountain bluebirds: Experimental evidence against altruism," *Science*, 189: 142–3.
Premack, D. (1970), "A functional analysis of language," *Journal of the Experimental Analysis of Behavior*, 14: 107–25.
Presser, H. B. (1971), "The timing of the first birth, female roles and black fertility," *Milbank Memorial Fund Quarterly*, 49: 329–59.
Price, D. de Solla (1975), *Science since Babylon* (New Haven, Conn.: Yale University Press).
Price, W. J., and L. W. Bass (1969), "Scientific research and the innovative process," *Science*, 164: 802–6.
Prigogine, I., G. Nicolis, and A. Babloyantz (1972), "Thermodynamics of evolution," *Physics Today* (November and December): 23–8, 38–44.
Pulliam, H. R., and C. Dunford (1980), *Programmed to Learn: An Essay in the Evolution of Culture* (New York: Columbia University Press).
Pusey, A. E. (1980), "Inbreeding avoidance in chimpanzees," *Animal Behaviour*, 28: 543–52.
Quadagno, J. S. (1979), "Paradigms in evolutionary theory: the sociobiological model of natural selection," *American Sociological Review*, 44: 100–9.
Quillian, W. S. (1945), *The Moral Theory of Evolutionary Naturalism* (New Haven, Conn.: Yale University Press).
Quinton, A. (1966), "Ethics and the theory of evolution," pp. 107–30 in *Biology and Personality*, ed. I. T. Ramsey (Oxford: Blackwell).
Radcliffe-Brown, A. R. (1922), *The Andaman Islanders* (Cambridge: Cambridge University Press).
Radcliffe-Brown, A. R. (1930), "The social organization of Australian tribes," *Oceania* 1: 34–63.
Radcliffe-Brown, A. R. (1951), "The comparative method in social anthropology," *Journal of the Royal Anthropological Institute*, 81: 15–22.
Rainwater, L. (1970), "The problem of lower class culture," *Journal of Social Issues*, 26: 133–48.
Rao, P. S. S., and S. G. Inbaraj (1979), "Trends in human reproductive wastage in relation to long-term practice of inbreeding," *Annals of Human Genetics*, 42: 401–13.
Reed, E. W. (1971), "Mental retardation and fertility," *Social Biology*, 18: 42–9.
Reid, R. M. (1976), "Effects of consanguineous marriage and inbreeding on couple fertility and offspring mortality in rural Sri Lanka," *Human Biology*, 48: 139–46.
Reik, T. (1941), *Masochism in Modern Man* (New York: Farar & Rinehart).
Reiss, I. L. (1980), *Family Systems in America* (New York: Holt, Rinehart & Winston).
Reynolds, V. (1966), "Open groups in hominid evolution," *Man*, 1: 441–52.

Richards, R. J. (1974), "The innate and the learned: the evolution of Konrad Lorenz's theory of instinct," *Philosophy of the Social Sciences*, 4: 111–33.
Richerson, P. J., and R. Boyd (1978), "A dual inheritance model of the human evolutionary process: I, basic postulates and a simple model," *Journal of Social and Biological Structures*, 1: 127–54.
Rigby, P. (1968), "Some Gogo rituals of 'purification': an essay on social and moral categories," pp. 153–78 in *Dialectic in Practical Religion* (Cambridge: Cambridge University Press).
Rindfus, R. R. and J. A. Sweet (1977), *Postwar Fertility Trends and Differentials in the United States* (New York: Academic Press).
Rivers, W. H. R. (1906), *The Todas* (New York: Macmillan).
Rivers, W. H. R. (1911), "The ethnological analyses of culture," *Report of the British Association for the Advancement of Science*, 81: 490–9.
Rogin, M. P. (1974), "Liberal society and the Indian question," pp. 9–52 in *The Politics and Society Reader*, ed. I. Katznelson, G. Adams, P. Brenner, and A. Wolfe (New York: McKay).
Rosaldo, M. Z., and L. Lamphere (eds) (1974), *Women, Culture, and Society* (Stanford, Calif.: Stanford University Press).
Rosaldo, R. I., Jr (1972), "Metaphors of hierarchy in a Mayan ritual," pp. 359–69 in *Reader in Comparative Religion*, ed. W. A. Lessa and E. Z. Vogt (New York: Harper & Row).
Rosch, E., and B. B. Lloyd (eds) (1978), *Cognition and Categorization* (Hillsdale, NJ: Lawrence Earlbaum Associates).
Roscoe, J. (1923a), *The Banyankole* (Cambridge: Cambridge University Press).
Roscoe, J. (1923b), *The Bakitara* (Cambridge: Cambridge University Press).
Rosenfeld, E. (1951), "Social stratification in a 'classless' society," *American Sociological Review*, 16: 766–74.
Rossi, A. S. (1968) "Transition to parenthood," *Journal of Marriage and the Family*, 30: 26–39.
Rowe, J. H. (1946), "Inca culture at the time of the Spanish conquest," pp. 183–330 in *Handbook of South American Indians*, ed. J. Steward (Washington, DC: Bureau of American Ethnology).
Rubin, Z. (1968), "Do American women marry up?" *American Sociological Review*, 33: 750–60.
Ruse, M. (1979), *Sociobiology: Sense or Nonsense?* (Boston, Mass.: Reidel).
Ruyle, E. E. (1973), "Genetic and cultural pools: some suggestions for a unified theory of biocultural evolution," *Human Ecology*, 1: 201–15.
Ryan, W. (1976), *Blaming the Victim* (New York: Vintage Books).
Sade, D. S. (1967), "Determinants of dominance in a group of free ranging rhesus monkeys," pp. 99–114 in *Social Communication among Primates*, ed. S. A. Altmann (Chicago: University of Chicago Press).
Sade, D. S. (1968), "Inhibition of son-mother mating among free-ranging rhesus monkeys," *Scientific Psychoanalysis*, 12: 18–37.
Sade, D. S. (1975), "The evolution of sociality," *Science*, 190 (17 October): 261–3.
Sahagún, B. de. (1958, orig. circa 1530), "Aztec sacrifice," pp. 442–5 in *Reader in Comparative Religion*, eds W. A. Lessa and E. Z. Vogt (New York: Row, Peterson).
Sahlins, M. D. (1976), *The Use and Abuse of Biology: An Anthropological Critique of Sociobiology* (Ann Arbor, Mich.: University of Michigan Press).
Sahlins, M. D., and E. R. Service (1960), *Evolution and Culture* (Ann Arbor, Mich.: University of Michigan Press).
Samuelson, K. (1961, orig. 1957), *Religion and Economic Action: A Critique of Max Weber* (New York: Harper & Row).

Sanford, C. (1961), *The Quest for Paradise: Europe and the American Moral Imagination* (Champaign, Ill.: University of Illinois Press).
Sapir, E. (1949, orig. 1929), "The status of linguistics as a science," pp. 160–6 in *Selected Writings of Edward Sapir*, ed. D. Mandlebaum (Berkeley, Calif.: University of California Press).
Sawyer, J. (1966), "The altruism scale: a measure co-operative, individualistic, and competitive interpersonal orientation," *American Journal of Sociology*, 71: 407–16.
Scarr, S., and R. A. Weinberg (1978), "The influence of 'family background' on intellectual attainment," *American Sociological Review*, 43: 674–92.
Schacht, R. (1978), "On power and powerlessness," pp. 425–38 in *Major Social Issues*, ed. J. M. Yinger and S. J. Cutler (New York: The Free Press).
Schacter, S. (1951), "Deviation, rejection, and communication," *Journal of Abnormal and Social Psychology*, 46: 190–207.
Schaller, G. B. (1965), "The behavior of the mountain gorilla," pp. 324–67 in *Primate Behavior*, ed. I. DeVore (New York: Holt, Rinehart & Winston).
Schaller, G. B. (1972), *The Serengeti Lion: A Study of Predator-Prey Relations* (Chicago: University of Chicago Press).
Schapera, I. (1956), *Government and Politics in Tribal Societies* (London: Watts).
Scheler, M. (1970, orig. 1926), "The sociology of knowledge: formal problems," pp. 170–86 in *The Sociology of Knowledge: A Reader*, ed. J. E. Curtis and J. W. Petras (New York: Praeger).
Schneider, D. M. (1968), *American Kinship* (Englewood Cliffs, NJ: Prentice-Hall).
Schneider, H. K. (1977), "Prehistoric transpacific contact and the theory of culture change," *American Anthropologist*, 79: 9–25.
Schneirla, T. C. (1972), "Interrelationship of the 'innate' and the 'acquired' in instinctive behavior," pp. 48–100 in *Function and Evolution of Behavior*, ed. P. H. Klopfer and J. P. Hailman (Reading, Mass.: Addison-Wesley).
Schram, L. M. J. (1961), *The Monguors of the Kansu-Tibetan Frontier. Part III. Records of the Monguors Clans: History of the Monguors in Huangchung and the Chronicles of the Lu Family* (Philadelphia, Pa: American Philosophical Society).
Schull, W. J., and J. V. Neel (1965), *The Effects of Inbreeding on Japanese Children* (New York: Harper & Row).
Schulman, S. R. (1978), "Kin selection, reciprocal altruism, and the principle of maximization," *Quarterly Review of Biology*, 53: 283–6.
Schumpeter, J. (1951), *Imperialism and Social Classes* (New York: Kelley).
Scott, J. P. (1958), *Aggression* (Chicago: University of Chicago Press).
Scott, J. P. (1964), "The effects of early experience on social behavior and organization," pp. 231–55 in *Social Behavior and Organization among Vertebrates*, ed. W. Etkin (Chicago: University of Chicago Press).
Sebeok, T., and D. J. Umiker-Sebeok (eds) (1980), *Speaking of Apes* (New York: Plenum).
Seemanová, E. (1971), "A study of children of incestuous matings," *Human Heredity*, 21: 108–28.
Seligman, M. (1971), "Phobias and preparedness," *Behavior Therapy*, 2: 307–20.
Service, E. R. (1962), *Primitive Social Organization* (New York: Random House).
Sharp, R., and A. Green (1975), *Education and Social Control: A Study in Progressive Primary Education* (London: Routledge & Kegan Paul).
Shepher, J. (1971), "Mate selection among second-generation kibbutz adolescents and adults: incest avoidance and negative imprinting," *Archives of Sexual Behavior*, 1: 293–307.
Shepher, J. (1979), *Incest: The Biosocial View* (Cambridge, Mass.: Harvard University Press).
Shepherd, G. M. (1974), *The Synaptic Organization of the Brain: An Introduction* (London: Oxford University Press).

Shields, J., and I. I. Gottesman (1971), *Man, Mind, and Heredity* (Baltimore, Md: Johns Hopkins University Press).
Short, J. F., and F. I. Nye (1957), "Reported behavior as a criterion of deviant behavior," *Social Problems*, 5: 207–13.
Shryock, H. S., and J. S. Siegel (1971), *The Methods and Materials of Demography* (Washington, DC: Government Printing Office).
Siegel, R. K. (1980), "The psychology of life after death," *American Psychologist*, 35: 911–31.
Siegel, R. K. (1981), "Accounting for 'after-life' experiences," *Psychology Today*, 15: 65–75.
Siegel, R. K., and L. J. West (eds) (1975), *Hallucinations: Behavior, Experience, and Theory* (New York: Wiley).
Simmel, G. (1950), *The Sociology of George Simmel*, ed. K. H. Wolff (Glencoe, Ill.: The Free Press).
Simmel, G. (1955), *Conflict and the Web of Group-Affiliations* (New York: The Free Press).
Simmons, L. W. (1942), *Sun Chief* (New Haven, Conn.: Yale University Press).
Simon, H. A. (1979), *Models of Thought* (New Haven, Conn.: Yale University Press).
Simpson, G. G. (1944), *Tempo and Mode in Evolution* (New York: Columbia University Press).
Simpson, G. G. (1967, orig. 1949), *The Meaning of Evolution* (New Haven, Conn.: Yale University Press).
Simpson, G. G. (1953), *The Major Features of Evolution* (New York: Columbia University Press).
Singelmann, P. (1981), *Structures of Domination and Peasant Movements in Latin America* (Columbia, Mo: University of Missouri Press).
Skinner, B. F. (1965, orig. 1953), *Science and Human Behavior* (New York: The Free Press).
Skinner, B. F. (1966), "The phylogeny and ontogeny of behavior," *Science*, 153: 1205–13.
Skinner, B. F. (1974), *About Behaviorism* (New York: Knopf).
Skocpol, T. (1979), *States and Social Revolutions* (Cambridge: Cambridge University Press).
Skolnick, J. H. (1969), *The Politics of Protest* (New York: Simon & Schuster).
Slobodkin, L. B., and A. Rapoport (1974), "An optimal strategy of evolution," *Quarterly Review of Biology*, 49: 181–200.
Smith, G. E. (1928), *In the Beginning: The Origin of Civilization* (New York: Morrow).
Sociobiology Study Group of Science for the People (SSGSP) (1978, orig. 1976), "Sociobiology: another biological determinism," pp. 280–90 in *The Sociobiology Debate*, ed. A. L. Caplan (New York: Harper & Row).
Solomon, M. (1977), *Beethoven* (New York: Schirmer).
Solzhenitsyn, A. (1973), *The Gulag Archipelago 1918–1956* (New York: Harper & Row), 2 vols.
Sorokin, P. A. (1941), *The Crisis of Our Age* (New York: Dutton).
Sorokin, P. A. (1950a), *Altruistic Love* (Boston, Mass.: Beacon Press).
Sorokin, P. A. (ed.) (1950b), *Explorations in Altruistic Love* (Boston, Mass.: Beacon Press).
Sorokin, P. A. (1954a), *Forms and Techniques of Altruistic and Spiritual Growth* (Boston, Mass.: Beacon Press).
Sorokin, P. A. (1954b), *The Ways and Power of Love: Types, Factors, and Techniques of Moral Transformation* (Boston, Mass.: Beacon Press).

Sorokin, P. A. (1957), *Social and Cultural Dynamics* (Boston, Mass.: Porter Sargent).
Sparks, R. F., H. G. Genn, and D. J. Dodd (1977), *Surveying Victims* (New York: Wiley).
Spencer, H. (1915, orig. 1857a), *Essays: Scientific, Political and Speculative* (New York: Appleton-Century-Crofts).
Spencer, H. (1857b), "Progress: its laws and causes," *Westminster Review*, 67: 445–85.
Spencer, H. (1864), *Social Statics* (New York: Appleton).
Spencer, H. (1895–8), *The Principles of Ethics* (New York: Appleton), 2 vols.
Spencer, H. (1896–9), *Principles of Psychology* (New York: Appleton).
Spencer, H. (1897), *The Principles of Sociology* (New York: Appleton).
Spiro, M. E. (1952), "Ghosts, Ifaluk, and teleological functionalism," *American Anthropologist*, 54: 497–503.
Spuhler, J. N. (1967), "Behavior and mating patterns in human populations," pp. 241–68 in *Genetic Diversity and Human Behavior*, ed. J. N. Spuhler (Chicago: Aldine).
Spuhler, J. N. (1968), "Assortative mating with respect to physical characteristics," *Eugenics Quarterly*, 15: 128–40.
Stanley, S. M. (1981), *The New Evolutionary Time Table* (New York: Basic Books).
Starcke, C. N. (1901, orig. 1889), *The Primitive Family in its Origins and Development* (New York: Appleton).
Stearns, S. C. (1976), "Life-history tactics: a review of the ideas," *Quarterly Review of Biology*, 51: 3–47.
Stein, M. I. (1975), *Stimulating Creativity* (New York: Academic Press).
Stern, C. (1973), *Principles of Human Genetics* (San Francisco: Freeman).
Stevens, C. F. (1979), "The neuron," *Scientific American*, 214: 54–65.
Steward, J. H. (1938), *Basin-Plateau Aboriginal Sociopolitical Groups* (Washington, DC: Smithsonian Institution. Bureau of American Ethnology, Bulletin 120).
Steward, J. H. (1955), *Theory of Culture Change* (Urbana, Ill.: University of Illinois Press).
Stokes, A. W. (ed.) (1974), *Territory* (Stroudsburg, Pa: Dowden, Hutchinson, & Ross).
Stone, L. (1960–1), "Marriage among the English nobility in the sixteenth and seventeenth centuries," *Comparative Studies in Society and History*, 3: 182–206.
Stouffer, S. A., et al. (1949), *The American Soldier* (Princeton, NJ: Princeton University Press), 2 vols.
Stowe, H. B. (1948, orig. 1852), *Uncle Tom's Cabin* (New York: Random House).
Sumner, W. G. (1883), *What Social Classes Owe to Each Other* (New York: Harper).
Sumner, W. G. (1906), *Folkways* (New York: Ginn).
Sumner, W. G. (1911), *War and Other Essays* (New Haven, Conn.: Yale University Press).
Sumner, W. G. (1914), *The Challenge of Facts and Other Essays* (New Haven, Conn.: Yale University Press).
Sumner, W. G. (1919), *The Forgotten Man and Other Essays* (New Haven, Conn.: Yale University Press).
Sumner, W. G. (1963), *Social Darwinism: Selected Essays of William Graham Sumner*, ed. S. Persons (Englewood Cliffs, NJ: Prentice-Hall).
Susanne, C. (1977), "Heritability of anthropological characters," *Human Biology*, 49: 573–80.
Sutherland, E. (1949), *White Collar Crime* (New York: Holt).
Swanson, G. E. (1960), *The Birth of the Gods* (Ann Arbor, Mich.: University of Michigan Press).
Sztompka, P. (1974), *System and Function: Toward a Theory of Society* (New York: Academic Press).

Takakura, S. (1960), *The Ainu of Northern Japan: A Study in Conquest and Acculturation* (Philadelphia, Pa: American Philosophical Society).

Talmon, Y. (1964), "Mate selection in collective settlements," *American Sociological Review*, 29: 491–508.

Tawney, R. H. (1926), *Religion and the Rise of Capitalism* (New York: Harcourt, Brace).

Tax, M. (1980), *The Rising of the Women: Feminist Solidarity and Class Conflict, 1880–1917* (New York: Monthly Review Press).

Tax, S. (ed.) (1960), *Evolution after Darwin* (Chicago: University of Chicago Press), Vol. II.

Taylor, I. A., and J. W. Getzels (eds) (1975), *Perspectives in Creativity* (Chicago: Aldine).

Terrace, H. (1979), *Nim* (New York: Knopf).

Thibaut, J. W., and H. H. Kelley (1959), *The Social Psychology of Groups* (New York: Wiley).

Thiessen, D. D., and B. Gregg (1980), "Human assortative mating and genetic equilibrium: an evolutionary perspective," *Ethology and Sociobiology*, 1: 111–40.

Thomas, W. I. (1923), *The Unadjusted Girl* (Boston, Mass.: Little, Brown).

Thomas, W. I., and F. Znaniecki (1918), *The Polish Peasant in Europe and America* (Chicago: University of Chicago Press), 2 vols.

Thorpe, W. H. (1956), *Learning and Instinct in Animals* (Cambridge, Mass.: Harvard University Press).

Thorpe, W. H. (1963), *Learning and Instinct in Animals*, 2nd edn (London: Methuen).

Thorpe, W. H. (1974), *Animal Nature and Human Nature* (Garden City, NY: Anchor Books).

Thrupp, S. (1962), *The Merchant Class of Medieval London* (Ann Arbor, Mich.: University of Michigan Press).

Tiger, L. (1970), *Men in Groups* (New York: Random House).

Tiger, L., and R. Fox (1971), *The Imperial Animal* (New York: Holt, Rinehart & Winston).

Tiger, L., and J. Shepher (1976), *Women in the Kibbutz* (New York: Harcourt Brace Jovanovich).

Tilly, C. (1964), *The Vendée* (Cambridge, Mass.: Harvard University Press).

Time, 15 December 1980.

Tinbergen, N. (1951), *The Study of Instinct* (Oxford: Clarendon).

Tinbergen, N. (1953), *Social Behavior in Animals* (London: Methuen).

Tinbergen, N. (1960), "Behavior, systematics, and natural selection," pp. 381–402 in *Evolution after Darwin*, Vol. I, ed. S. Tax (Chicago: University of Chicago Press).

Tiryakian, E. A. (1976), "Biosocial man, *sic et non*," *American Journal of Sociology*, 82: 701–6.

Toffler, A. (1970), *Future Shock* (London: Pan Books).

Tokuda, K. (1961–2), "A study of sexual behavior in a Japanese monkey troop," *Primates*, 3: 1–40.

Tokuda, K., and G. D. Jensen (1968), "The leader's role in controlling aggressive behavior in a monkey group," *Primates*, 9: 319–22.

Tolstoy, L. N. (1898), *What is Art?* (Philadelphia, Pa: Altemus).

Toynbee, A., *et al.* (eds) (1976), *Life after Death* (New York: McGraw-Hill).

Trilling, L. (1972), *Sincerity and Authenticity* (Cambridge, Mass.: Harvard University Press).

Trivers, R. L. (1971), "The evolution of reciprocal altruism," *Quarterly Review of Biology*, 46: 35–47.

Trivers, R. L. (1974), "Parent-offspring conflict," *American Zoologist*, 14: 249–64.

Trivers, R. L., and D. E. Willard (1973), "Natural selection of parental ability to vary the sex ratio of offspring," *Science*, 179: 90–2.
Tsui, A. O., and D. J. Bogue (1978), "Declining world fertility: trends, causes, implications," *Population Bulletin*, 33: 1–36.
Turnbull, C. M. (1963), "The lesson of the pygmies," *Scientific American*, 208: 28–37.
Turnbull, C. M. (1972), *The Mountain People* (New York: Simon & Schuster).
Twain, M. (1885), *The Adventures of Huckleberry Finn* (New York: Webster).
Tylor, E. B. (1958, orig. 1871), *Primitive Culture* (New York: Harper).
Tylor, E. B. (1889), "On a method of investigating the development of institutions; applied to laws of marriage and descent," *Journal of the Royal Anthropological Institute*, 18: 245–69.
Ulč, O. (1978), "Some aspects of Czechoslovak society since 1968," *Social Forces*, 57: 419–35.
United Nations (1974), *Demographic Yearbook* (New York: United Nations Publications).
US National Center for Health Statistics (1980), *Vital Statistics of the United States* (Washington, DC).
US National Center for Social Statistics (1980), *Adoptions in 1970, Adoptions in 1975*, Report E-10 (Washington, DC).
Vaihinger, H. (1924, orig. 1911), *The Philosophy of As If: A System of the Theoretical, Practical and Religious Fictions of Mankind* (London: Routledge & Kegan Paul).
Vaillant, G. C. (1941), *Aztecs of Mexico* (Garden City, NY: Doubleday).
Valentine, C. A. (1968), *Culture and Poverty: Critique and Counter-Proposals* (Chicago: University of Chicago Press).
Vandenberg, S. G. (1972), "Assortative mating, or who marries whom?," *Behavior Genetics*, 2: 127–58.
van den Berghe, P. L. (1965), *South Africa: A Study in Conflict* (Middletown, Conn.: Wesleyan University Press).
van den Berghe, P. L. (1974), "Bringing beasts back in," *American Sociological Review*, 39: 777–88.
van den Berghe, P. L. (1978a), "Bridging the paradigms: biology and the social sciences," pp. 32–52 in *Sociobiology and Human Nature*, ed. M. S Gregory, A. Silvers, and D. Sutch (San Francisco: Jossey-Bass).
van den Berghe, P. L. (1978b), *Man in Society: A Biosocial View* (New York: Elsevier).
van den Berghe, P. L. (1980), "Incest and exogamy: a sociobiological reconsideration," *Ethology and Sociobiology*, 1: 151–62.
van den Berghe, P. L., and D. P. Barash (1977), "Inclusive fitness and human family structure," *American Anthropologist*, 79: 809–23.
van der Leeuw, G. (1963, orig. 1937), *Sacred and Profane Beauty* (New York: Holt, Rinehart & Winston).
van Gennep, A. (1960, orig. 1908), *The Rites of Passage* (London: Routledge & Kegan Paul).
van Lawick-Goodall, J. (1971), *In the Shadow of Man* (Boston, Mass.: Houghton Mifflin).
Veblen, T. (1899), *The Theory of the Leisure Class* (New York: Macmillan).
Veevers, J. E. (1972), "Factors in the incidence of childlessness in Canada: an analysis of census data," *Social Biology*, 19: 266–74.
Ventura, S. J. (1982), "Trends in first births to older mothers, 1970–79," *Monthly Vital Statistics Report 31* (Supplement 2), 27 May (Washington, DC: National Center for Health Statistics).

Verner, J. (1964), "Evolution of polygyny in the long-billed marsh wren," *Evolution*, 18: 252–61.
Vines, K. N. (1964), "Federal district judges and race relations cases in the South," *Journal of Politics*, 26: 337–57.
Volgyes, I. (1978), "Modernization, stratification and elite development in Hungary," *Social Forces*, 57: 500–21.
Wach, J. (1951), *Types of Religious Experience, Christian and Non-Christian* (Chicago: University of Chicago Press).
Waddington, C. H. (1957), *The Strategy of the Genes* (London: Allen & Unwin).
Waddington, C. H. (1960), *The Ethical Animal* (London: Allen & Unwin).
Waddington, C. H. (1975), "Mindless societies," *New York Review of Books*, 22 (7 August): 30–2.
Wallace, A. F. C. (1966), *Religion: An Anthropological View* (New York: Random House).
Wallace, R. A. (1973), *The Ecology and Evolution of Animal Behavior* (Pacific Palisades, Calif.: Goodyear).
Ward, L. (1893), *The Psychic Factors of Civilization* (Boston, Mass.: Ginn).
Ward, L. (1913–18), *Glimpses of the Cosmos* (New York: Putnam's Sons), 6 vols.
Washburn, S. L., and E. R. McCown (eds) (1978), *Human Evolution: Biosocial Perspectives* (Menlo Park, Calif.: Benjamin-Cummings).
Watanabe, H. (1964), *The Ainu: A Study of Ecology between Man and Nature in Relation to Group Structure* (Tokyo: University of Tokyo Press).
Watts, C. R., and A. W. Stokes (1971), "The social order of turkeys," *Scientific American*, 224: 112–18.
Weber, M. (1958, orig. 1904–5), *The Protestant Ethic and the Spirit of Capitalism* (New York: Charles Scribner's Sons).
Weber, M. (1968, orig. 1922), *Economy and Society* (New York: Bedminster Press), 3 vols.
Weber, M. (1927), *General Economic History* (New York: Greenberg).
Weber, M. (1946), *From Max Weber: Essays in Sociology*, ed. H. H. Gerth and C. W. Mills (New York: Oxford University Press).
Weber, M. (1947), *The Theory of Social and Economic Organization* (New York: The Free Press).
Weber, M. (1949), *The Methodology of the Social Sciences* (Glencoe, Ill.: The Free Press).
Weiner, A. B. (1976), *Women of Value, Men of Renown: New Perspectives in Trobriand Exchange* (Austin, Texas: University of Texas Press).
Weisbord, R. (1975), *Genocide: Birth Control and the Black American* (Westport, Conn.: Greenwood Press).
West Eberhard, M. J. (1975), "The evolution of social behavior by kin selection," *Quarterly Review of Biology*, 50: 1–33.
Westermarck, E. A. (1891), *The History of Human Marriage* (London: Macmillan).
White, L. A. (1948), "The definition and prohibition of incest," *American Anthropologist*, 50: 416–35.
White, L. A. (1949), *The Science of Culture* (New York: Farrar, Straus & Giroux).
White, L. A. (1959), *The Evolution of Culture* (New York: McGraw-Hill).
Whitehead, A. N. (1917), *The Organisation of Thought, Educational and Scientific* (London: Williams & Norgate).
Whiting, J. W. M. (1941), *Becoming a Kwoma* (New Haven, Conn.: Yale University Press).
Whorf, B. L. (1956), *Language, Thought, and Reality: Selected Writings* (Cambridge, Mass.: Technology Press of MIT).
Whyte, W. F. (1943), *Street Corner Society* (Chicago: University of Chicago Press).

Wickelgren, W. A. (1979), *Cognitive Psychology* (Englewood Cliffs, NJ: Prentice-Hall).
Wiley, R. H. (1973), "Territoriality and non-random mating in sage grouse, *Centrocercus urophasianus*," *Animal Behaviour Monographs*, 6: 85–169.
Wilkie, J. S. (1950), "The problem of the temporal relation of cause and effect," *British Journal for the Philosophy of Science*, 1: 211–29.
Williams, B. J. (1974), "A model of band society" (Washington, DC: Society for American Archaeology), Memoir No. 29.
Williams, F. E. (1930), *Orokaiva Society* (London: Oxford University Press).
Williams, G. C. (1966), *Adaptation and Natural Selection* (Princeton, NJ: Princeton University Press).
Williams, G. C. (ed.) (1971), *Group Selection* (Chicago: Aldine-Atherton).
Williams, T. (1975), "Family resemblance in abilities: the Wechsler scales," *Behavior Genetics*, 5: 405–9.
Wilson, D. S. (1980), "A theory of group selection," pp. 104–10 in *Selected Readings in Sociobiology*, ed. J. H. Hunt (New York: McGraw-Hill).
Wilson, E. O. (1971a), "Competitive and aggressive behavior," pp. 183–217 in *Man and Beast: Comparative Social Behavior*, ed. J. F. Eisenberg and W. Dillon (Washington, DC: Smithsonian Institution Press).
Wilson, E. O. (1971b), *The Insect Societies* (Cambridge, Mass.: Harvard University Press).
Wilson, E. O. (1975a), "Human decency is animal," *New York Times Magazine*, 12 October, pp. 38–50.
Wilson, E. O. (1975b), *Sociobiology: The New Synthesis* (Cambridge, Mass.: Harvard University Press).
Wilson, E. O. (1976), "Academic vigilantism and the political significance of sociobiology," *BioScience*, 26: 183, 187–90.
Wilson, E. O. (1977), "Preface" to D. P. Barash, *Sociobiology and Behavior* (New York: Elsevier).
Wilson, E. O. (1978), *On Human Nature* (Cambridge, Mass.: Harvard University Press).
Wilson, E. O. (1979), "Biology and anthropology: a mutual transformation?," pp. 519–21 in *Evolutionary Biology and Human Social Behavior: An Anthropological Perspective*, ed. N. A. Chagnon and W. Irons (North Scituate, Mass.: Duxbury Press).
Wilson, M. H. (1951), *Good Company: A Study of Nyakyusa Age-Villages* (London: Oxford University Press).
Wilson, M. H. (1957), *Rituals of Kinship among the Nyakyusa* (London: Oxford University Press).
Wilson, R. S. (1978), "Synchronies in mental development: an epigenetic perspective," *Science*, 202: 939–48.
Winch, R. F. (1962), "The theory of complementary needs in mate-selection: final results on the test of the general hypothesis," *American Sociological Review*, 27: 552–5.
Wispé, L. (ed.) (1978), *Altruism, Sympathy, and Helping* (New York: Academic Press).
Witkin, H. A., S. A. Mednick, F. Schulsinger, E. Bakkestrom, K. O. Christiansen, D. R. Goodenough, K. Hirschhorn, C. Lundsteen, D. R. Owen, J. Philip, D. B. Rubin, and M. Stocking (1976), "Criminality in XYY and XXY men," *Science*, 193: 547–55.
Wittke, C. (1939), *We Who Built America: The Saga of the Immigrant* (Englewood Cliffs, NJ: Prentice-Hall).
Wolf, A. P. (1966), "Childhood association, sexual attraction and the incest taboo," *American Anthropologist*, 68: 883–98.

Wolf, A. P., and C. Huang (1980), *Marriage and Adoption in China, 1845–1945* (Stanford, Calif.: Stanford University Press).
Wolfenden, G. E. (1975), "Florida scrub jay helpers at the nest," *Auk*, 92: 1–15.
Wright, S. (1921), "Systems of mating III: assortative mating based on somatic resemblance," *Genetics*, 6: 144–61.
Wright, S. (1931), "Evolution in Mendelian populations," *Genetics*, 16: 97–159.
Wrong, D. (1961), "The oversocialized conception of man in modern sociology," *American Sociological Review*, 26: 184–98.
Wuthnow, R. (1976), "Recent patterns of secularization: a problem of generations," *American Sociological Review*, 41: 850–67.
Wuthnow, R. (1978), *Experimentation in American Religion: The New Mysticisms and their Implications for the Churches* (Berkeley, Calif.: University of California Press).
Wuthnow, R., and C. Y. Glock (1973), "Religious loyalty, defection, and experimentation among college youth," *Journal for the Scientific Study of Religion*, 12: 157–80.
Wynne-Edwards, V. C. (1962), *Animal Dispersion in Relation to Social Behaviour* (Edinburgh: Oliver & Boyd).
Wynne-Edwards, V. C. (1963), "Intergroup selection in the evolution of social systems," *Nature*, 200: 623–6.
Wynne-Edwards, V. C. (1972), "Ecology and the evolution of social ethics," pp. 61–9 in *Biology and the Human Sciences*, ed. J. W. S. Pringle (New York: Oxford University Press).
Yancey, W. L., E. P. Ericksen, and R. N. Juliani (1976), "Emergent ethnicity: a review and reformulation," *American Sociological Review*, 41: 391–403.
Yang, M. (1945), *A Chinese Village: Taitou, Shantung Province* (New York: Columbia University Press).
Yinger, J. M. (1976), "Ethnicity in complex societies: structural, cultural, and characterological factors," pp. 197–216 in *The Uses of Controversy in Sociology*, ed. L. A. Coser and O. N. Larsen (New York: The Free Press).
Zaltman, G., R. Duncan, and J. Holbek (1973), *Innovations and Organizations* (New York: Wiley).
Zeitlin, M., K. Lutterman, and J. Russell (1974), "Death in Vietnam: class, poverty, and the risks of war," pp. 53–68 in *The Politics and Society Reader*, ed. I. Katznelson, G. Adams, P. Brenner, and A. Wolfe (New York: McKay).
Zetterberg, H. L. (1963), *On Theory and Verification in Sociology* (Totowa, NJ: Bedminster Press).
Ziegenhagen, E. A. (1977), *Victims, Crime, and Social Control* (New York: Praeger).
Zilboorg, G. (1943), "Fear of death," *Psychoanalytic Quarterly*, 12: 465–75.

Index of Names

Abramson, J. H. 332
Adam, H 137
Adams, M. 316
Adams, R. N. 114, 177, 263
Adler, A. 111
Ajzen, I. 305, 332
Alba, R. D. 334
Alcock, J. 17, .18
Alexander, B. K. 319
Alexander, K. L. 131
Alexander, R. D. 18, 94, 165, 201, 202, 208, 224
Alland, A., Jr 94
Allee, W. C. 170, 206
Allen, E. *et al.* 22–3
Allport, G. W. 333
Alston, L. 235
Alstrom, C. H. 313
Altmann, S. A. 173, 286
Ammerman, A. J. 238
Amory, C. 173
Anderson, E. 148
Anderson, J. R. 223
Anderson, R. C. 79
Arens, W. 43
Arensberg, C. M. 327
Argyris, C. 150n
Arieti, S. 238
Aristotle 160, 258
Aron, R. 224
Asimov, I. 270
Ayala, F. J. 4, 5, 67, 80

Babloyantz, A. 223
Back, K. W. 184, 187
Badillo, G. 139
Bahr, H. M. 332
Bakkestrom, E. 50
Bales, R. F. 315
Baltzell, E. D. 73
Bandura, A. 49, 70, 73, 79, 178, 192, 212, 238, 244, 265, 300
Banfield, E. C. 158
Barash, D. P. 17, 18, 20, 69, 99, 106, 108, 111, 134, 166, 172, 174, 181, 201, 205, 207, 208, 217, 221, 252, 325, 328
Barab, P. G. 178
Barber, B. 310
Barber, E. G. 327
Barnett, H. G. 238, 241n, 253
Barth, F. 331
Barzun, J. 2
Bass, L. W. 243, 247

Becker, E. 276
Beckman, L. 205, 311
Beecher, H. K. 45
Beit-Hallahmi, B. 317
Bell, C. 26n
Bell, D. 176
Bellah, R. N. 230, 231, 280, 285, 292–3
Bellamy, E. 13
Bendix, R. 88, 163, 327
Benedict, R. 155, 161
Bentler, P. M. 312
Berent, J. 327
Berlyne, D. E. 261n
Bernstein, I. S. 170
Bersheid, E. 312
Bierstedt, R. 114
Bischof, N. 318
Bishop, C A. 124
Bitterman, M. 68
Blake, J. 327
Blassingame, J. W. 137
Blau, P. M. 75, 114, 130, 132–3, 163, 186, 210, 329
Bloch, M. 158, 327
Block, N. 69
Blumberg, B. S. 108
Blute, M. 21, 238, 265
Boehm, C. 248
Bogue, D. J. 216
Bonham, G. S. 220n
Boorman, S. A. 202
Bottomore, T. B. 309
Bowers, W. J. 135
Bowles, S. 131
Boyd, R. 264
Brandon, S. G. F. 92, 229–33
Braudel, F. 88
Brinton, C. 99, 224
Brown, J. L. 17, 18, 70, 120, 123, 127, 170
Bruce, H. M. 207
Bullock, T. H. 239, 253
Bumpass, L. L. 217
Bunzel, R. 279
Burchinal, L. G. 108, 328
Burgess, E. W. 327
Burton, F. D. 77
Bury, J. B. 2
Busino, G. 57
Butler, J. S. 139
Butler, S. 275

Campbell, B. 178

Campbell, D. T. 26, 27, 178, 203–4, 208, 225, 247, 264, 333, 338
Caplan, A. L. 18
Carlin, J. E. 136
Carpenter, C. R. 120
Caso, A. 141
Cavalli-Sforza, L. L. 238, 265
Chadwick, B. A. 332
Chafetz Saltzman, J. 84, 118
Chagnon, N. A. 18, 108, 138
Chance, M. R. A 211
Charnov, E. L. 199
Cherns, A. 168
Childe, V. G. 264
Chilton, C. H. 223
Chomsky, N. 39, 95, 243–4
Choron, J. 276
Christiansen, K. O. 50
Clark, P. J. 311
Clarke, D. L. 64
Cloak, F. T., Jr 264
Cochrane, S. H. 109
Cohen, E. 121
Cohen, J. 245
Cohen, J. 88
Cohen, R. 264
Cohen, Y. 315
Coles, R. 134
Collias, N. E. 206
Collins, R. 35, 36, 174–5
Colson, C. W. 149, 191
Colson, E. 279
Combs, A. W. 254, 305–6
Comte, A. 34, 111, 196, 242, 243, 246
Condorcet 2
Cook, M. 131
Cooley, C. H. 227
Coser, L. A. 316
Count, E. W. 33–4, 82, 164, 172, 174
Coursey, C. K. 109
Coursey, D. G. 109
Cowan, W. M. 239
Cowell, F. R. 327
Cox, H. 180
Crippen, T. 190, 283
Crockett, H. J., Jr 163
Crognier, E. 313
Crook, J. H. 121, 239, 325
Crow, J. F. 311
Curry, C. D. 139
Curtis, C. P., Jr 57

Dahrendorf, R. 101, 176, 241, 309
Damas, D. 123
Dante, A. 142
Darlington, C. D. 163
Darwin, C. 5–8, ch. 1 passim, 94, 242, 247, 264–6
Davie, M. R. 94, 99

Davies, J. C. 99
Davis, K. 88, 175, 176, 309
Dawidowicz, L. S. 141
Dawkins, R. 18, 24, 74, 108, 127, 201, 207, 233, 264, 328
Dawson Scanzoni, L. 322
Deacon, A. B. 155
DeFries, J. C. 51, 162, 206
Deutsch, H. 273–4
DeVore, B. I. 121, 123, 153, 162, 164, 170
Dewey, J. 268
Dickemann, M. 327
Dickens, C. 114, 149, 217, 226
Dinnerstein, L. 332
Djilas, M. 116–17, 166–7
Dobzhansky, T. 2, 3, 4, 5, 16, 17, 67, 71, 80, 100, 243, 249
Dodd, D. J. 135
Dohrenwend, B. P. 134
Dohrenwend, B. S. 134
Douglas, F. 137, 188, 193
Douglas, M. 244
Driedger, L. 334
Dumond, D. 98
Dumont, L. 173, 305
Duncan, O. D. 130, 132–3, 163, 329
Duncan, R. 238
Dunford, C. 70, 179, 253
Durham, W. H. 26, 94, 106–7, 108–9, 122, 123, 138, 203, 252, 264
Durkheim, E. 4, 5, 35–6, 42–3, 45, 46, 53, 58, 84–5, 122, 142–3, 160, 179, 182, 186, 190, 193, 196, 210–11, 214–15, 219, 228–30, 274, 278, 281, 282–5, 289, 293, 297–8, 301–3, 307, 309, 322
Dyson-Hudson, R. 123–5

Eaton, G. G. 170
Eckland, B. K. 22, 24, 79, 205, 311
Edelman, G. M. 239
Ehrman, L. 311
Eibl-Eibesfeldt, I. 82, 253, 318
Eisenberg, J. F. 170, 224, 318, 325
Eisenstadt, S. N. 224, 264n
Elkin, A. P. 230
Ember, M. 315, 318
Emlen, S. T. 264
Engels, F. 3, 13, 99, 116, 166, 175–6
Erber, J. 73
Ericksen, E. P. 335
Erikson, E. 338
Errington, P. L. 206
Ervin, F. R. 70
Esser, A. H. 121
Essock-Vitale, S. M. 222
Evans-Pritchard, E. E. 144, 193, 230

Farb, P. 271
Fast, H. 129

Feagin, C. B. 338
Feagin, J. R. 133, 175, 338
Feder, H. M. 153
Feifel, H. 276
Feldman, M. W. 238, 265
Felsenstein, J. 311
Fenn, R. K. 299
Festinger, L. 184, 186–7, 223
Feuer, L. S. 77, 97
Firth, R. W. 136, 140
Fischer, M. M. J. 184
Fischer-Galati, S. 166
Fishbein, M. 305, 332
Fisher, G. J. 170
Fisher, R. A. 16
Flannery, R. 213
Flew, A. 11, 197, 230
Flink, J. J. 261
Fodor, J. A. 69
Forbes, E. 147
Fortes, M. 315
Fortune, R. F. 230, 315
Foster, G. M. 238
Foster, J. 327
Fox, R. 75, 211–12, 314, 315
Francis, E. K. 330
Frazer, Sir J. G. 43–4, 55, 277, 278
Freedman, D. G. 108
Freud, S. 35, 43, 151, 215, 221, 223, 227, 276, 303
Friedl, E. 327
Friedrichs, R. W. 196, 210
Fromm, E. 150n

Gadgil, M. 202
Gallup, G., Jr 50, 92
Gamson, W. A. 114
Gans, H. J. 175
Garcia, J. 70
Gardner, B. T. 73
Gardner, R. A. 73
Gartlan, J. C. 80
Gazzaniga, M. S. 239
Gebhard, P. H. 218
Geertz, C. 30–3, 57, 81–2, 281–2, 284, 287–8
Geiger, T. 176
Geis, G. 136
Geist, V. 163, 206
Gelfand, D. E. 334
Genn, H. G. 135
George, H. 13
Geschwender, J. A. 332
Geschwind, N. 239
Getzels, J. W. 238
Ghiselin, M. T. 208
Gibson, M. 238, 241n
Giddens, A. 176
Giddings, F. H 111
Gillespie, C. C. 1

Gilman, S. 332
Gintis, H. 131
Girard, R. 226
Glazer, N. 99, 165, 332
Glock, C. Y. 280, 291
Gluckman, M. 144
Goffman, E. 58, 174–5, 301
Goldman, N. 135
Goode, W. J. 313, 327, 328
Goodenough, D. R. 50
Gordon, M. M. 165, 332, 334
Gottesman, I. I. 311
Gould, M. S. 134
Gould, S. J. 16, 24–7
Gouldner, A. W. 15, 156–60, 196, 210
Grace, J. 137
Granovetter, M. 264n
Greeley, A. M. 99, 165, 332, 333, 334
Green, A. 130
Gregg, B. 205, 310–14, 315
Gregory, M. S. 34
Greenberg, J. H. 264
Greene, G. 272
Grinnell, A. 239, 253
Grobstein, C. 169
Grossberg, S. 239, 253
Guhl, A. M. 170, 206, 224
Gulick, J. 238, 241n
Gusinde, M. 125n

Habbakuk, H. J. 327
Haggett, P. 238
Haldane, J. B. S. 16
Halley, P. 220
Hamblin, R. L. 238
Hamilton, W. D. 18, 152–3, 199–201, 217
Handlin, O. 99
Hardin, G. 207, 309
Hardy, A. 226, 243
Harris, M. 43, 53, 54, 56, 57, 141, 264, 315
Hawthorne, N. 281
Haydn, F. J. 147–8, 259
Hayes, C. 73
Hayes, K. J. 73
Hazelrigg, L. E. 28, 118, 133, 163–4, 166, 176, 277, 327, 329–30
Hediger, H. P. 178
Hegel, G. W. F 261n
Henderson, L. J. 57, 310
Henderson, N. D. 51
Herskovits, M. J. 315n
Hertz, R. 43, 140, 193, 229, 272, 276, 278
Hesiod 43, 308–9
Hesser, J. E. 108
Hill, C. T. 312
Hill, J. L. 318, 319
Hinde, R. A. 70, 82, 178
Hirschhorn, K. 50
Hobbes, T. 152, 197, 256

Hodge, R. W. 118
Hogbin, H. I. 144
Hofstadter, R. 8, 9, 13
Hogg, G. 43, 46
Holbek, J. 238
Hollingshead, A. B. 130, 133–4, 311
Holmes, G. A. 327
Homans, G. C. 57, 75–6, 88, 114, 156–60, 184, 186–7, 201, 252
Homer 146, 324
Honigmann, J. J. 238, 241n
Hooglund, E. J. 145
Horn, J. M. 51
Houghton, J. 88
Howard, J. 136
Howe, F. 131
Howitt, A. W. 144
Hrdy, S. B. 286
Hsu, F. L. K. 327
Huang, C. 315, 318
Hubel, D. H. 239
Huber, J. 322
Hubert, H. 43, 140, 193, 230
Hull, D. L. 33
Hullum, J. R. 104
Hume, D. 34
Hunt, J. H. 33
Hutter, M. 322
Huxley, A. 299
Huxley, J. 16
Huxley, T. H. 223, 270

Inbaraj, S. G. 313
Inkeles, A. 131
Irons, W. 18, 70, 107–8, 201, 203

Jackson, E. F. 163
Jacobson, M. 239
Jakobi, L. 313
James, J. J. 144
James, W. 71n
Jellinck, E. M. 45
Jensen, G. D. 170
Jolly, A. 318
Jones, R. A. 44–5
Jones, T. A. 116, 117
Jones, W. 138
Juliani, R. N. 335

Kaffman, M. 317
Kahl, J. 173, 187
Kamazani, R. K. 145
Kandel, E. R. 239
Kanowitz, L. 322
Kant, I. 152, 261n
Karlin, S. 205
Kawai, M. 164, 178
Keats, A. S. 45
Keller, A. G. 247, 264

Keller, S. 100
Kelley, H. H. 186–7
Keyes, C. F. 330
Kharuzin, N. 321
Kimball, S. T. 327
King, G. E. 122
Kirk, D. 311
Kirkpatrick, S. D. 223
Kletti, R. 279
Knox, I. 261n
Koelling, R. 70
Kohlberg, L. 185
Konishi, M. 73
Krebs, J. R. 199
Kreitman, N. 311
Krober, A. 213
Kropotkin, P. 12, 196, 210, 223
Kübler-Ross, E. 279
Kuehn, R. E. 170, 224
Kuffer, S. W. 239
Kuhn, T. S. 1, 2, 4, 14, 147, 238, 243, 248, 258–9, 268–9
Kummer, H. 50, 170, 286
Kunkel, J. H. 238
Kurtén, B. 287
Kushner, G. 238, 241n

Lack, D. 207–8, 325
Lamanna, M. A. 322
Lamarck, J. B. de 5
Lamphere, L. 321
Lang, O. 157–8
Langer, S. 287, 301
Langton, J. 241n, 247, 256, 264–5
Laqueur, W. 140–1
Lasagna, L. 45
Laumann, E. O. 328
Lauter, P. 131
Lavandera, B. 244
Leach, E. 244
Leach, W. 322
Leavitt, G. 94
LeBoeuf, B. J. 163, 206
Ledoux, J. E. 239
Lee, A. 311
Lee, R. B. 123, 153, 164
Lee, R. D. 334
Lehrman, D. S. 70
Lenski, G. E. 94, 98, 112, 116, 138, 164–5, 167, 175, 241n, 264, 320, 325–6
Lenski, J. 94, 98, 138, 164, 241n, 264, 320, 325–6
Lessa, W. A. 230, 277–9, 307
Levin, A. J. 276
Levine, R. A. 333
Levins, R. 202
Lévi-Strauss, C. 57, 155, 210, 230, 305, 315, 316
Levitt, P. R. 202

Lewinson, P. 137
Lewis, L. S. 245
Lewis, O. 175
Lewontin, R. C. 20–2, 208, 252, 311, 335
Lewy, G. 293
Leyhausen, P. 249
Lifton, R. 141
Lin, Y. 137, 144, 316
Lindzey, G. 315
Link, B. 134
Lipset, S. M. 148, 163, 327
Lloyd, B. B. 289
Locke, H. J. 327
Locke, J. 197, 288
Loehlin, J. C. 51
Lopreato, J. 27, 28, 63, 88, 97, 98, 106, 115, 118, 133, 158, 163–4, 165, 166, 176, 196, 204, 235, 241, 244, 277, 310, 327, 329–30
Lorenz, K. 17, 47, 63, 67, 69, 70, 178, 189, 249, 250, 299–300, 318, 319, 338
Lotka, A. 263
Lowie, R. 53
Luhman, R. 332
Lumsden, C. J. 18–20, 23, 27, 33, 34, 57, 60, 63–5, 69, 70, 80, 105–6, 108, 210, 238, 252, 254–5, 264, 265–6
Lundsteen, C. 50
Lutterman, K. 139

McBride, G. 70
McClearn, G. E. 162, 206
McClelland, D. C. 100
McCown, E. R. 18
McDill, E. L. 131
McGill, M. J. 149
McGuire, M. T. 222
McKusick, V. A. 235, 313
McLemore, S. D. 99, 165, 332, 334–6
Malinowski, B. 35, 144, 154–6, 210, 230, 281, 301, 315
Malthus, T. R. 6–7
Mannheim, K. 166
Mariéjol, J. H. 327
Marler, P. R. 73
Marquer, P. 313
Marris, P. 243, 246, 272
Marsden, H. M. 164
Marshall, T. H. 98
Martin, M. K. 123
Marx, K. 2–3, 13, 28, 77, 88, 91, 99, 101, 115–19, 150n, 156, 166–7, 175–6, 233, 302, 316
Maslow, A. H. 35, 105, 146, 150, 173, 238
Mason, K. O. 217
Matthews, D. R. 163
Mauss, M. 43, 57, 140, 154–6, 193, 230, 301
Maynard Smith, J. 18, 200, 233, 255
Mayr, E. 16
Mazlish, B. 261

Mazur, A. 24, 170
Mead, G. H. 35, 77, 227, 285–8, 303
Mednick, S. A. 50
Meier, R. F. 136
Menaker, E. 213
Mendel, G. 8, 18
Menzel, R. 73
Merton, R. K. 83–8, 187, 309, 328
Messinger, S. L. 136
Métraux, A. 138
Michels, R. 110, 116, 172
Middleton, J. 279
Milgram, S. 223
Miller, E. J. 327
Miller, H. A. 99
Miller, H. P. 175
Miller, J. L. L. 238
Miller, S. M. 163
Mills, C. W. 173, 175, 224, 309
Missakian, E. A. 164
Montagu, M. F. A. 121
Monteil, C. V. 140
Montesquieu 2
Moody, R. 279–80
Moore, W. E. 175, 176
Morgan, L. H. 53, 315
Mosca, G. 110, 119, 224
Moskos, C. C., Jr 139
Mosteller, F. 45
Mountcastle, V. B. 239
Moynihan, D. P. 99, 165, 332
Mueller, H. C. 178
Muggeridge, M. 209
Mulkay, M. J. 105
Murdock, G. P. 27, 57, 92, 136, 164, 229, 257, 325–6
Murphy, R. F. 138
Murray, B. G. 121
Murstein, B. I. 205, 311, 327

Nagel, E. 235
Nagel, S. S. 135–6
Nagel, T. 196, 198
Neel, J. V. 163, 313, 317
Neugebauer, R. 134
Newby, I. A. 137
Newcomb, M. D. 312
Newman, W. M. 332, 334
Newton, Sir I. 259
Nicholls, J. G. 239
Nicolis, G. 223
Nielsen, J. 311
Nietzsche, F. W. 149
Nisbet, R. A. 264n
Noble, G. K. 120
Nonas, R. 238, 241n
Nordskog, E. 264n
Novak, M. 165, 311, 334, 335
Noyes, R. 279

Nye, F. I. 135

O'Dea, T. F. 301
Ogburn, W. F. 34, 253, 260
Olmsted, F. L. 137
Olson, E. 141
Opler, M. E. 279
Orkand, R. 239, 253
Orwell, G. 245
Oster, G. F. 21
Otterbein, K. F. 94
Owen, D. R. 50

Padover, S. K. 308
Pareto, V. 35, 38–9, 41–2, 47, 54, 56, 57–9, 63, 97, 99, 101–2, 108, 110–13, 122, 146, 170, 172, 175, 179, 182, 186, 189–90, 210–11, 215–16, 224, 228, 242, 246, 249, 268, 276–7, 282, 289–91, 295, 298–9, 305, 308, 309
Park, R. E. 99, 122, 184, 299
Parker, G. A. 255
Parker, S. 315, 318
Parsons, T. 31, 45, 57, 75, 108, 169, 241, 264, 309, 315
Pattee, H. H. 169
Pearson, K. 311
Penrose, L. S. 311
Peplau, L. A. 312
Percy, W. 287, 288
Perrin, C. M. 207–8
Perry, R. H. 223
Perry, W. J. 56
Peterson, N. 120, 123
Pettigrew, T. F. 331, 334, 338
Pfeiffer, J. E. 178, 319, 320
Phelps, E. S. 196
Philip, J. 50
Piaget, J. 76–7, 81, 178, 289
Pirenne, H. 158
Pitt-Rivers, A. L.-F. 52
Plato 238
Plomin, R. 51
Polanyi, M. 244
Popov, A. 322
Porter, J. 130–2
Poston, D. L. 216–17
Power, H. W. 207
Premack, D. 73
Presser, H. B. 217
Price, D. de Solla 241n
Price, G. R. 255
Price, W. J. 243, 247
Prigogine, I. 223
Prosch, H. 244
Pulliam, H. R. 70, 179, 253
Pusey, A. E. 320

Quadagno, J. S. 21

Quillian, W. S. 197
Quinton, A. 9

Rabin, A. I. 317
Radcliffe-Brown, A. R. 122, 301, 305
Rainwater, L. 175
Rao, P. S. S. 313
Rapoport, A. 106
Redlich, F. C. 133–4
Reed, E. W. 311
Reid, R. M. 313
Reik, T. 213
Reimers, D. M. 332
Reiss, I. L. 322
Reissman, F. 175
Reynolds, V. 121
Richards, A. 254, 305–6
Richards, F. 254, 305–6
Richards, R. J. 71
Richerson, P. J. 264
Riedman, A. 322
Rigby, P. 192
Rindfus, R. R. 109
Rivers, W. H. R. 56, 194
Rogin, M. P. 139
Rosaldo, M. Z. 321
Rosaldo, R. I., Jr 168
Roscoe, J. 315n
Rosch, E. 289
Rosenfeld, E. 167–8
Rossi, A. S. 217
Rousseau, J. J. 259
Rowe, J. H. 315n
Rubin, D. B. 312
Rubin, Z. 328, 329
Ruse, M. 69, 70
Russell, J. 139
Ruyle, E. E. 238, 256, 264
Ryan, W. 130, 136

Sade, D. S. 164, 168, 318
Sahagún, B. de 56, 230
Sahlins, M. D. 28, 31–3, 81–2, 106, 151–2, 238, 264, 274
Samuelson, K. 88
Sanford, C. 304–5
Sapir, E. 287–288
Sawyer, J. 196
Saxton, D. E. 238
Scanzoni, J. 322
Scarr, S. 51
Schacht, R. 114
Schacter, S. 184, 187
Schaller, G. B. 163, 170, 207
Schapera, I. 164
Scheler, M. 149, 269–70
Schneider, D. M. 315
Schneider, H. K. 247
Schneirla, T. C. 71

Schram, L. M. J. 125n
Schull, W. J. 313, 317
Schulman, S. R. 106, 152
Schulsinger, F. 50
Schumpeter, J. 327
Scott, J. P. 170, 318, 319
Sebeok, T. 74
Seemanová, E. 317
Seligman, M. 69
Service, E. R. 123, 238, 264
Shakespeare, W. 142
Sharp, R. 130
Sharpe, L. G. 170
Shepher, J. 315, 317, 323
Shepherd, G. M. 239
Shields, J. 311
Short, J. F. 135
Shryock, H. S. 217
Siegel, J. S. 217
Siegel, R. K. 278–80
Silvers, A. 34
Simmel, G. 45, 170, 316
Simmons, L. W. 279, 309
Simon, H. A. 239
Simpson, G. G. 16
Singelmann, P. 98
Skinner, B. F. 265
Skocpol, T. 224
Skolnick, J. H. 148
Slobodkin, L. B. 106
Smith, E. A. 123–5
Smith, G. E. 56
Solomon, M. 147, 259
Solzhenitsyn, A. 166
Sorokin, P. A. 94, 147, 196, 210, 258, 260
Sparks, R. F. 135
Spencer, H. 5, 8–9, 34, 53–4, 59, 146, 172, 197, 261
Spinoza, B. 197
Spiro, M. E. 279
Spuhler, J. N. 311, 313
Stanley, S. M. 16
Starcke, C. N. 315
Strauss, J. H. 332
Stearns, S. C. 106
Stebbins, G. L. 4, 5, 67, 80
Stein, M. I. 238
Stern, C. 317
Stevens, C. F. 239
Steward, J. H. 124, 264
Stocking, M. 50
Stokes, A. W. 121, 206
Stone, L. 327
Stouffer, S. A. 184
Stowe, H. B. 137
Struhsaker, T. T. 80
Such, D. 34
Sumner, W. G. 9–12, 35, 39, 94, 179, 184, 189–90, 246, 247, 248, 270, 305, 333

Susanne, C. 311
Sutherland, E. 136
Swanson, G. E. 5, 43, 229–30, 274
Sweet, J. A. 109
Sztompka, P. 310

Takakura, S. 125n
Talmon, Y. 317
Tamura, M. 73
Tawney, R. H. 88
Tax, M. 322
Tax, S. 264
Taylor, I. A. 238
Terrace, H. 74
Thibaut, J. W. 186
Thiessen, D. D. 205, 310–14, 315
Thomas, W. I. 35, 146, 151, 186–7, 242–3, 246
Thorndike, A. H. 68–9
Thorpe, W. H. 72–3
Thrupp, S. 327
Tiger, L. 75, 323
Tilly, C. 327
Tinbergen, N. 17, 69, 70–1, 127, 300
Tiryakian, E. A. 24
Toffler, A. 260
Tokuda, K. 170, 318
Tolstoy, L. N. 14, 47, 261n
Toynbee, A. 279
Treiman, D. J. 118
Trilling, L. 226
Trivers, R. L. 27, 77, 108, 152–3, 159, 201–4, 222–4, 323–4, 328
Tsui, A. O. 216
Turnbull, C. M. 164
Twain, M. 233
Tylor, E. B. 122, 230, 315

Ulé, O. 117
Umiker-Sebeok, D. J. 74

Vaihinger, H. 58
Vaillant, G. C. 141
Valentine, C. A. 175
Valentine, J. W. 4, 5, 67, 80
van Beethoven, L. 147–8, 259
Vandenberg, S. G. 205, 311
van den Berghe, P. L. 111, 122, 130, 137, 162, 171, 201, 209, 316, 318, 319–20, 328
van der Leeuw, G. 261n
van Gennep, A. 214, 273, 301
van Lawick-Goodall, J. 28, 50, 164, 173, 178, 286, 318
Veblen, T. 35, 163
Veevers, J. E. 217
Verner, J. 325
Vico, G. 2
Vines, K. N. 134–5
Virgil 290

Vogt, E. Z. 230, 277–9, 307
Volgyes, I. 117
Volterra, V. 263
von Felsinger, J. M. 45
von Frisch 17

Wach, J. 301
Waddington, C. H. 17, 184, 249
Wallace, A. F. C. 188, 193, 219, 229–30, 307
Wallace, A. R. 6n
Walster, E. 312
Ward, L. 11–12
Washburn, S. L. 18, 170
Watanabe, H. 125n
Watts, C. R. 206
Weber, M. 9, 30–2, 88–93, 114, 119, 161, 173, 175, 176, 182, 190, 211, 213, 261, 268, 275, 294–5, 309, 327
Weinberg, R. A. 51
Weiner, A. B. 157
Weisbord, R. 141
West, L. J. 279
West Eberhard, M. J. 201
Westermarck, E. A. 315, 317
Westoff, C. F. 217
White, L. A. 264, 287–8, 315
Whitehead, A. N. 239
Whiting, J. W. M. 138
Whorf, B. L. 95
Whyte, W. F. 81, 99, 165
Wickelgren, W. A. 239, 253
Wiley, R. H. 163, 206
Wilkie, J. S. 63
Willard, D. E. 108, 322–3, 328
Willerman, L. 51
Williams, B. J. 123
Williams, G. C. 18, 202

Williams, F. E. 187, 278
Williams, T. 51
Wilson, D. S. 202
Wilson, E. O. 17–20, 20–2, 22–33, 34, 40, 45, 47, 50, 51, 55, 57, 59–60, 63–5, 69, 70, 71n, 72, 80–1, 94, 105–6, 107, 108, 111, 120–1, 126, 128, 134, 137, 162, 164, 168, 170, 173, 177–8, 185, 195–235 *passim*, 238, 249, 252, 253, 254–5, 260, 264, 265–6, 282–3, 320–1, 323, 324–5, 328
Wilson, M. H. 192, 193, 194
Wilson, R. S. 51
Winch, R. F. 311
Wispé, L. 196, 210
Witkin, H. A. 50
Wittke, C. 99
Wolf, A. P. 315, 318
Wolfenden, G. E. 217
Wright, S. 16, 208, 311, 313
Wrong, D. 75
Wunsch-Hitzig, R. 134
Wuthnow, R. 180, 291–2
Wynne-Edwards, V. C. 34, 127, 162, 164, 168–9, 202, 224

Yancey, W. L. 335
Yang, M. 221
Yinger, J. M. 332

Zaltman, G. 238
Zavalloni, M. 100
Zetterberg, H. L. 88
Zeitlin, M. 139
Ziegenhagen, E. A. 144
Zilboorg, G. 276
Znaniecki, F. 146, 242

Index of Subjects

Adaptation: in Darwin 6–8; and the struggle for existence 7; and fitness 7, 251–5; defined 20, 252; and the maximization principle 20–2, 106–10; as a problem of sociobiology 20–2; impediments to 21–2; as a relational (comparative) term 21–2; and the nature of learning 69–78; and the theory of anomie 83–8; and inclusive fitness 106–10; and culture 106–10; and education 109–10; and the climbing maneuver 110–20; and territoriality 120–9; and the urge to victimize 129–41; and the need for vengeance 141–6; and the need for recognition 146–50; and reciprocation 151–61; and predispositions of domination and deference 161–76; and the need for conformity 177–86; and the need for social approval 186–8; and the need for self-purification 188–94; and altruism 195–235; and group selection 202–4; and asceticism 211–28; and self-deception 225–35; and religion 228–35, 271–303; and the soul ch. 6 *passim*, 271–4; and self-deception 222–35; and biocultural evolution 236–65; and the explanatory urge 267–74; and relational self-identification 267–74; and the denial of death 274–81; and symbolization 281–93; and reification 281–93; and the susceptibility to charisma 294–8; and the imperative to act 298–303; and the need for ritual consumption 298–303; and homologous affiliation 304–14; and heterologous affiliation 304–14, and assortative mating 310–14; and incest avoidance 314–21; and family relations 321–30; and heterologous contraposition 330–6

Altruism: definition of 20, 198, 205–7; as a central problem of sociobiology 20; and natural selection 20; and the maximization principle 20; among Eskimos 25–7; and civilization 59–60; and group selection 195, 202–4; and kin selection 199–201; and social insects 199–200; reciprocal 201–2, 222–5; as favoritism 204–5; ascetic 205–22; and *Homo sapiens* 207–10; and suicide 210; and asceticism 211–16; and avoidance of parenthood 216–18; and heroism 219; and ritual chastity 220; and adoption 221; and the theory of ascetic altruism 222–35; and conformity 222–5; and the dominance order 222–5; and social approval 222–5; and self-deception 225–8; and the soul 225–35; and the social self 227–8; and religion 228–33; and Buddhism 232–3.

Analogy: and cultural universals 65–8; defined 65; and their heuristic value 66–7; and the generalizing tendency 308; and the production of knowledge 308

Anomie: and biocultural science 83–8; and Durkheim's theory of suicide 84–5; and Merton's theory 84–8; and deviance, 84–8

Anthropocentrism: defined 1; and the rationalistic bias 40; and reification 281–93

Asceticism: defined 211; and hominid evolution 211–12; and deferred gratification 211–12; and religion 212–14; and rites of expiation 212–14; and variants 212–14; and masochism 213; and prohibitionism 213–14; and fasting 214; and sex 215–16; and altruism 216–28; and Buddhism 232–3

Assortative mating: defined 310; and the coefficient of relationship 310; and Mendel's First Law 310; in the literature 310–14; and marital stability 312; and inbreeding depression 313; and kinship classifications 313–14

Behavior: and morphology 17–18; and homology 17–18, 52; as pacemaker in evolution 18; and meaning 29–33, 88–92; and variability 31–3, 82–103; and the normative tendency 34; and biocultural interplay 37–49; and cross-species comparisons 50; and direct genetic evidence 50–1; and identical twins 50–1; and adopted children 51; and deception 58–60; and explanation in social science 68; nature and nurture 68–78; and genes 74–5; and anomie 83–8; and religion 88–92, 267–303; and status in the dominance order 172–5; and ascetic altruism 195–235; and the predispositions of variation and selection 238–51; and the evolution of the brain 238–40; and the criteria of sociocultural selection 251–61; and the predispositions underlying religious behavior 267–303; and the predispositions underlying family and ethnicity 304–40; see Behavioral predispositions, Cultural universals, Culture, Variants

Behavioral predispositions: defined 33, 60–1, 63–4; and human nature 33; methods of inferring them 37–49; and the deep structure in linguistics 39; appearing in compound form 48–9, 105; conceptual discussion of 49–56; and genetic evidence 50–1; and early sociocultural evolutionists 52–6; and cultural universals 56–65; and variants 56–65; and epigenetic rules 64, 265–6; and the nature-nurture debate 68–78; as teaching devices 69–70, and behavior 70–8; and their varying intensity 79–92, 277–81; and their distribution 92–5; and their modifiers 95–103; and symbols 95–6; and great events 96–7; and migration 99; and social mobility 100–2; and the problem of classification 105; classification of 105; the climbing maneuver 110–20; territoriality 120–9; the urge to victimize 129–41; the need for vengeance 141–6; the need for recognition 146–50; reciprocation 151–61; domination and deference 161–76; the need for conformity 177–86; the need for social approval 186–8; the need for self-purification 188–94; asceticism 211–16; and the forces of variation and selection 238–51; the explanatory urge 267–74; relational self-identification 267–74; the denial of death 274–81; symbolization 281–93; reification 281–93; susceptibility to charisma 294–8; the imperative to act 298–303; the need for ritual consumption 298–303; homologous affiliation 304–14; heterologous affiliation 304–14; incest avoidance 314–21; heterologous contraposition 330–6

Behavioral scaling: see intensity

Behaviorism: and the nature vs. nurture problem 68–70

Biocultural evolution: defined 237; and diffusion 238; and creativity 238; and the process of variation and selection 238–42; and the brain 239–40; and the forces of variation 239–46; and socialization 241; and the forces of selection 246–51; and the fluctuating nature of change 250, 277–81; and the criteria of selection 251–61; and the problem of adaptation 251–5; and the leash principle 254–5; and group selection 262–3; and the model of variation and selective retention 264–6; and religious behavior 267–303; and the evolutionary foundations of family and ethnicity 304–40; and assortative mating 310–14; and incest avoidance 314–21; and family relations 321–30; and heterologous contraposition 330–6; see Biocultural science, Sociobiology

Biocultural science: definition of 18–19; and the comparative method 18; and the problem of determinism 22–8; and general principles 25–33; and inter-specific behavioral traits 27; and human nature 33, 37–56; and etiological ordering 41–2, 46–9, 60–5; and behavioral predispositions 49–56; and the case of XYY men 50; and identical twins 50–1; and adopted children 51–2; and cultural universals 56–65; and variants 56–65; and the case of baptism 63; and the problems of homology and analogy 65–8; and the nature–nurture debate 68–78; and intensity of causal units 79–92; and the theory of anomie 83–8; and Weber's theory of Protestantism and capitalism 88–92; and the distribution of causal units 92–5; and the modifiers of behavioral predispositions 95–103; and the maximization principle 106–10, 234–5; and the adaptiveness of culture 106–10; and inclusive fitness 106–10; and the biocultural principle 107; and the climbing maneuver 110–20; and territoriality 120–9; and the urge to victimize 129–41; and the need for vengeance 141–6; and the need for recognition 146–50; and reciprocation 151–61; and the predispositions of domination and deference 161–76; and the need for conformity 177–86; and the need for social approval 186–8; and the need for self-purification 188–94; and ascetic altruism 195–235; and conceptualization of altruism 196–207; and asceticism 211–16; and the idea of the soul ch. 6 *passim*; and religion 228–35, 267–303; and biocultural evolution 236–66; and combiners 242–6; and selectors 246–51; and criteria of sociocultural selection 251–61; and group selection 262–3; and the explanatory urge 267–74; and relational self-identification 267–74; and the denial of death 274–81; and symbolization 281–93; and reification 281–93; and the susceptibility to charisma 294–8; and the imperative to act 298–303; and the need for ritual consumption 298–303; and homologous affiliation 304–14; and heterologous affiliation 304–14; and assortative mating 310–14; and incest avoidance 314–21; and family relations 321–30; and heterologous contraposition 330–6; and social science 386–40; see Biocultural evolution, Sociobiology

Biogram: see Human nature

Brain: and self-deception 225–8; and neurological activity 239; and the process of biocultural variation and selective retention 238–42, 253–4; see Mind

Capitalism: and religion 88–92

Causation: see Ultimate causation, Proximate causation, General principles, Determinism

Change: see Biocultural evolution
Charisma: the predisposition of susceptibility to 294–8; and the dominance order 294; and the predisposition of deference 294; and Weber's types of authority 294–5; defined 294–5; and charismatic leaders 295–6; in relation to symbolization and reification 296–8; and deities 296–7; evolution of 297; and power 297
Cheating: and the theory of reciprocal altruism 151–3, 222–5; and reciprocal behavior 158–61; forms of 159; and the dominance order 159–76
Class consciousness: and the climbing maneuver 112, 115–20; and Marxism 115–20; as action in process 117–18; as action in retrospect 117–18; and kinship 118–19
Class struggle: see Class consciousness
Climbing maneuver: and selfish behavior 110–20; and deception 110–20; and social mobility 110–20; and politics 111–12; and capitalism 111–12; and class consciousness 112, 115–20; and the democracy-aristocracy paradox 112–13; and exploitation 113; and hypocrisy 114; and power 114–15; and Marxism 115–20; and group selection 117–18; and nonevolutionary theories of social stratification 118–19; and the Industrial Revolution 119–20; and the potlatch 115–16; and domination and deference 162–76
Coevolution: see Biocultural evolution
Combinations: as the novelties of biocultural evolution 240; and the combiners 242–6; see Combiners, Selections
Combiners: as forces of biocultural variation 242–6; in social theory 242–3; and symbol systems 243; and scientific innovation 243; and creativity in language 243–4; and model learning 244; and magical behavior 244–5; and gambling 245–6; active and passive sides of 246; in relation to selectors 249–50; and personality types 250–1; the explanatory urge 267–74; symbolization 281–93; the imperative to act 298–303; see Selectors
Competition: see Biocultural evolution
Conformity: and deviance 84–8; and social mobility 101; and the need for conformity 177–88; and stages of personal development 184–5; and relational self-identification 267–74
Conservatism: and sociobiology 28–9; and the futility of ideological interpretations of scientific theories 28–9
Copernican revolution: 1–5, ch. 1 *passim*
Creativity: and the theory of biocultural evolution 236–461; in language 243–4; and the explanatory urge 267–74; and symbolization 281–93; see Combiners
Creature comforts: and sociocultural selection 256–7
Cultural universals: defined 33, 56–7, 63; behavioral predispositions 33, 37–56; as inferential bases of behavioral predispositions 37–59; and causation 46–9, 60–5; and the problem of error in classification 48–9, 105; and alternative explanations thereof 51–2; and early evolutionary theory 52–6; and a dispute over them 57–8; and variants 56–65; and socialization 62; and culturgens 63–4; and the problems of homology and analogy 65–8; and their varying intensity 79–92, 277–81; and their distribution 92–5; and natural selection 92–5; and the problem of classification 105; and the climbing maneuver 110–20; and territoriality 120–9; and the urge to victimize 129–41; and the need for vengeance 141–6; and the need for recognition 146–50; and reciprocation 151–6; and predispositions of domination and deference 161–76; and the need for conformity 177–86; and the need for social approval 186–8; and the need for self-purification 188–94; and ascetic altruism 195–235; and asceticism 211–28; and the idea of the soul ch. 6 *passim*, 271–4; and biocultural evolution 236–65; and the explanatory urge 267–74; and relational self-identification 267–74; and the denial of death 274–81; and symbolization 281–93; and reification 281–93; and the susceptibility to charisma 294–8; and the imperative to act 298–303; and the need for ritual consumption 298–303; and religion 267–303; and homologous affiliation 304–14; and incest avoidance 314–21; and family relations 321–30; and heterologous contraposition 330–6
Culture: as an adaptive mechanism 22, 106–10; and the mind 23; and the problem of meaning 29–31; and the problem of variability 31–3; and genetics 50–1; and early evolutionists 52–6; and cultural universals 56–65; and variants 56–65; and the nature-nurture debate 68–78; and the answer to the Hobbesian question 75–8; in causal relation to biology 79–95; in reciprocal relation with natural selection 92–103; and the maximization principle 106–10; and inclusive fitness 106–10; and the biocultural principle 107; and predispositions of self-enhancement 104–50; and predispositions of sociality 151–235; and altruism 195–235; and asceticism 211–28; and self-deception 222–35; and religion 228–35, 267–303; and the soul ch. 6 *passim*, 271–4; and the definition of

biocultural evolution 237–8; and the forces of variation and selection 238–51; and combiners 242–6; and selectors 246–51; and criteria of selection 251–61; and the genetic leash 254–5; and the explanatory urge 267–74; and relational self-identification 267–74; and the denial of death 274–81; and symbolization 281–93; and reification 281–93; and the susceptibility to charisma 294–8; and the imperative to act 298–303; and the need for ritual consumption 298–303; and homologous affiliation 304–14; and heterologous affiliation 304–14; and the assortative mating 310–14; and incest avoidance 314–21; and family relations 321–30; and heterologous contraposition 330–6

Deception: and Weber's theory of capitalism 88–93; and self-deception 225–8; and the idea of the soul 225–8, 271–4; see Altruism, Denial of death, Explanatory urge, Need for social approval, Reciprocation, Variants

Denial of death: as a selector of the soul concept 274–81; and Weber's theory of capitalism 274–5; and relational self-identification 275; and genetic mimicking 275–6; and sex 276; and relations between the living and the dead 276–81; and its varying intensity 277–81; and apparitions 278–81; and ancestral spirits 278–9; and solidarity 279; and after-life experiences 279–80; and fear of death 280–1; and suicide 281; and age 281; and religious belief 281

Deprivation, relative: and reciprocal behavior 160

Determinism: and sociobiology 23–8; and the leash principle 23; and biocultural interplay 23–8; and the problems of homology and analogy 65–8; and the nature vs. nurture debate 68–78; economic, and Weber's theory of religion 88–92

Deviance: and the theory of anomie 83–8; and conformity 177–86; and the need for self-purification 188–94; and altruism ch. 6 *passim*

Diffusion: as a possible source of cultural universals 51–2, 56; and sociocultural evolution 238; and systemic immanence 259

Discrimination: defined 333; see Prejudice

Distribution: defined 79; as a causal property 92–5; in relation to culture and natural selection 92–103; and the modifiers of behavioral predispositions 95–103

Distributive justice: and reciprocal behavior 160; see Altruism

Dominance order: and power 114–15; and the climbing maneuver 115; and class consciousness 115–20; and Marxism 115–20, 166–8; and communist society 116–17, 166–85; and nonevolutionary theory 118–19; and territoriality 122; and victimization 120–9; and education 130–3; and mental illness 133–4; and slavery 137; and reciprocity 151–61; and cheating 159–76; and symbolism 168; among animals 162–74; in relation to conflict and solidarity 170–2; and current theories 175–6; defined 177; and the need for social approval 186–8; and ascetic altruism 222–8; and biocultural evolution 253

Domination and Deference: defined 161; and power 162; and the climbing maneuver 162–76; and territoriality 162–6; and fitness 162–3; among animals 162–5; and status inheritance 164; in pre-agrarian society 164–5; and the street corner gang 165; and the classless society 166–8; and attempts at destratification 166–8; and the Israeli kibbutz 167–8; and symbolism 168; and solidarity 170–2; and politics 171–2; and the behavior of dominants and subordinates 172–5; and their varying intensity 172–5; entertainment stars and beauty contests 173–4; and current theories of stratification 175–6; and the need for conformity 177–86; and ascetic altruism 222–8

Education: and exploitation 130–3; and the track system 130–1; and social stratification 130–3; and the family 131–3; and barriers to equality of opportunity 132–3; and the vicious circle 133

Environment: and behaviour predispositions 72–6; and the plasticity of behavior 74; and social scientists' focus on it 75–8; see Culture

Ethnicity: and the family 330–2; defined 330–2; and homologous affiliation 330–2; and heterologous contraposition 332–40; and prejudice and discrimination 333–40; and religion 333–4; and acculturation 334–6; and assimilation 334–6; in the USA 334–40; and the need for recognition 335; and the struggle for existence 335–40

Ethnocentrism: defined 184, 333; and the need for conformity 184; and group selection 204; and heterologous contraposition 330–40; and ethnicity 330–40; regional 336

Ethology: and sociobiolgy 17–18; definition of 17; and the problem of nature vs.

nurture 69–78

Evolution: see Biocultural evolution

Evolutionary theory: and Darwin 5–8; and natural selection according to Darwin 6–8; and variation according to Darwin 7–8; and Lester Ward 11–12; and K. Marx 13; and its misreading in social science 14; and the problem of free will 14–15; and the *tabula rasa* assumption 15; and the assumption of socialization 15; and slow beginnings 16–17; and human nature 33–6; and behavioral predispositions 49–56; among early social scientists 52–6; and adaptive radiation 54–6; and parallel evolution 55, 66; and sociocultural universals 56–65; and variants 56–65; and the problems of homology and analogy 65–8; and the nature-nurture debate 68–78; and the intensity of causal units 79–92; and the theory of anomie 83–8; and Weber's theory of Protestantism and capitalism 88–93; and the distribution of causal units 92–5; and group selection 93–4; and the modifiers of behavioral predispositions 95–103; and social mobility 101–2; and the shape of sociocultural change 102–3; and the maximization principle 106–10; and inclusive fitness 106–10; and the adaptiveness of culture 106–10; and predispositions of self-enhancement 106–50, 255–6; and the biocultural principle 107; and predispositions of sociality 151–235, 255–6; and altruism 195–235; and kin selection 199–201; and group selection 202–4, 262–3; and asceticism 211–16; and ascetic altruism 216–35; and religion 228–33, 267–303; and the definition of biocultural evolution 237–8; and the process of variation and selection 238–42; and the brain 239–40; and the forces of variation 242–6, 267–303; and the forces of selection 246–51, 267–308; and fitness enhancement 251–5; and creature comforts 256–7; and systemic immanence 258–9; and the multiplier effect 260–1; and homologous affiliation 304–14; and heterologous affiliation 304–14; and assortative mating 310–14; and incest avoidance 314–21; and family relations 321–30; and heterologous contraposition 330–6

Exchange, social: see Reciprocation

Exogamy: see Incest avoidance

Expiation: rites of 188–94; and the need for self-purification 188–94; and acts demanding it 192–3

Explanation: see Biocultural science, General principles, Ultimate causation, Proximate causation

Explanatory urge: and knowledge 267–74; and relational self-identification 267–74; and religion 268–74; and magic 268–74; and exploratory behavior 268; and science 267–74; and self-deception 269–70; and the sociology of science 269–70; myth and legend 269–70; and talk shows 271; and the soul 271–4; and death 272–4; and mourning 272–4

Exploitation: and the climbing maneuver 113; and reciprocation 151–61; and the dominance order 161–76; and the need for social approval 186–8, 222–5

Extinction: and group selection 202–4, 262–3; and selective retention 241n; and incest 321

Family: and assortative mating 310–14; and incest avoidance 314–21; and incest taboo 314–21; and exogamy 314–21; and the kibbutz 317–18; and the status of women 321–3; and sex roles 320–30; and the relative work of the genders 322–3; and mating patterns in humans and animals 323–30; and conditions of monogamy 324–8; and conditions of polygamy 324–8; and the bride price 325–8; and heterogamy 325–30; and theories of hypergamy 328–30

Fertility: and education 109–10; and territoriality 127–8; and asceticism 216–20; and incest avoidance 314–21

Fitness: according to Darwin 7; and the struggle for existence 7; and variation according to Darwin 7–8; and learning 68–78; and education 109–10; and ascetic altruism ch. 6 *passim*; and biocultural evolution 236–65; as a criterion of sociocultural selection 251–5; and religious behavior 267–303; and assortative mating 310–14; and the coefficient of relationship 310; and inbreeding depression 313–21; and mating patterns 323–30

Free will: as a methodological issue 71; as a by-product of genetic noise 71n

General principles: function of 25–33; and the necessity to render them explicit 26–7; and the problems of meaning and cultural variability 31–2; and causation 41–2, 46–9, 60–5; and the maximization principle 106–10; and the biocultural principle 107; and inclusive fitness 106–10; and the theory of ascetic altruism 233–5; and social science 337

Genes: direct and indirect effects of 25, 74–5; and behavior 50–1; and learning 74–5; and Mendel's First Law 77; and altruism 195–235; and kin selection 199–201; and haplodiploidy 199–201; and the soul

225–8, 271–4; and the denial of death 275–6; and assortative mating 310–14; and incest avoidance 314–21

Genetics: see Genes, Population genetics

Genotype: and early sociocultural evolutionists 53; and Mendel's First Law 77

Great events: as modifiers of behavioral predispositions 96–7; and the Great Depression 97; and the Diaspora 97

Group selection: and technology 92–5, 262–3; and social mobility 101–2; and war 109, 263; and the climbing maneuver 117–18; and human sacrifice 139–41; and the need for self-purification 188–94; in relation to hierarchy and territoriality 168–9; and altruism 202–4; defined 203, 262; and biocultural evolution 262–3; and ascetic altruism 262; and religion 263, 303; and incest 321; and ethnicity 330–40

Heredity: see Genes, Behavioral predispositions, Population genetics

Heterologous affiliation: defined 304; in social theory 304–6; evolution of 306; and the generalizing tendency 306–10; and magic 307–9; and language 309–10; and assortative mating 310–14; and incest avoidance 314–21

Heterologous contraposition: and ethnicity 330–40; and ethnocentrism 330–40; defined 332; and homologous affiliation 332; and prejudice and discrimination 333–40

Homogamy: see Assortative mating

Homologous affiliation: defined 304; in social theory 304–6; evolution of 306; and the generalizing tendency 306–10; and the halo effect 307; and magic 307–9; and analogy 308; and psychobiographies 308; and language 309–10; and assortative mating 310–14; and inbreeding depression 313–21

Homology: and cultural universals 52, 65–8; defined 65

Homosexuality: and altruism 218

Human nature: defined 33; and behavioral predispositions 33, 49–56; and biocultural science 33; the concept in social science 34–6; and cultural universals 56–65; and variants 56–65; and nurture 68–78; and predispositions of self-enhancement 104–50; and predispositions of sociality 151–235; and altruism 195–235; and the model of biocultural evolution 236–63; and the predispositions underlying religious behavior 267–303; and the predispositions underlying family and ethnicity 304–40; see Behavioral predispositions

Human sociobiology: see Biocultural science, Sociobiology

Hypertrophy: defined 45, 59; and the evolution of ritual consumption 45; as a source of variants 59; and the need for social approval 222–5

Ideal types: and the classification of meaning 30–1; and the theory of ascetic altruism 233–5

Ideas: see Brain, Culture, Mind, Variants

Imitation: and the need for conformity 178–86; and learning 178–9; and developmental stages 184–5

Imperative to act: and the need for ritual consumption 298–303; and the source of ritual 298–303; and the sanctification of society 298; and collective behavior 299; in relation to the other behavioral predispositions 299–301; and symbols 301; and glossolalia 301–2; and group selection 303

Inbreeding depression: defined 313; and assortative mating 313; and incest 314–21

Incest avoidance: and the incest taboo 314–21; and exogamy 314–21; and inbreeding depression 314–21; and practices of incest 315; and theories of 315–17; and civilization 316; evolution of 316–21; mimicking of, in the kibbutz 317–18, and in China 318; in animals 318–20; and population extinction 321

Incest taboo: see Incest avoidance

Inclusive fitness: and the maximization principle 106–10; and the adaptiveness of culture 106–10; and the climbing maneuver 110–20; and territoriality 120–9; and the urge to victimize 129–41; and the need for vengeance 141–6; and the need for recognition 146–50; and reciprocation 151–61; and predispositions of domination and deference 161–76; and the need for conformity 177–86; and the need for social approval 186–8; and the need for self-purification 188–94; and altruism 194–235; and kin selection 201; and asceticism 211–16; and the theory of ascetic altruism 216–35; and the idea of the soul 222–33, 271–4; as a criterion of sociocultural selection 251–5; and religious behavior 267–303; and assortative mating 310–14; and incest avoidance 314–21

Instinct: and the nature-nurture debate 71–8; conceptions and misconceptions of 71–3; and learning 72–8

Intensity: defined 79; as a causal property 79–92; and environmental factors 80–1; in relation to anomie and deviance 82–8; in relation to religion and capitalism 88–93; and the modifiers of behavioral predispositions 95–103; and the shape of sociocul-

Index of Subjects   395

tural change 102–3; and the dominance order 172–5; and altruism ch. 6 *passim*; and soul concepts 277–81

Kin selection: defined 199–201; and the soul 274; and religious behavior 292–3; and incest avoidance 314–21; see Altruism, Inclusive fitness, Maximization principle, Natural selection

Lamarckian theory: defined 5, 23
Language: as a basis for inferring behavioral predispositions 38–42; as hindrance to scientific research 39–42; and the deep structure 39; and variants 56–65; in animals 73–4; and developmental stages of the mind 76; and symbolization 281–93; and symbols 288–93
Learning: see Imitation, Mind, Need for conformity, Socialization

Mazimization principle: definition of 20, 106; and natural selection 20; and altruism 20; and adaptation 20–2, 106–10; and learning 69–78; and Weber's theory of Protestantism and capitalism 93; and inclusive fitness 106–10; and the adaptiveness of culture 106–10; and the biocultural principle 107; and the climbing maneuver 110–20; and territoriality 120–9; and the urge to victimize 129–41; and the need for vengeance 141–6; and the need for recognition 146–50; and reciprocation 151–61; and predispositions of domination and deference 161–76; and the need for conformity 177–86; and the need for social approval 186–8; and the need for self-purification 188–94; and altruism 195–235; and kin selection 199–201; and reciprocal altruism 201–2; and group selection 202–4; and ascetic altruism 216–35; and self-deception 225–35; and the idea of the soul 225–35, 271–4; and the theory of biocultural evolution 236–63; and the explanatory urge 267–74; and relational self-identification 267–74; and the denial of death 274–81; and symbolization 281–93; and reification 281–93; and the susceptibility to charisma 294–8; and the imperative to act 298–303; and the need for ritual consumption 298–303; and homologous affiliation 304–14; and heterologous affiliation 304–14; and assortative mating 310–14; and incest avoidance 314–21; and family relations 321–30; and heterologous contraposition 330–6
Meaning: see Mind, Symbols, Variants
Mendelian theory: the Modern Synthesis 15–17; and early opposition to Darwinian gradualism 16; and punctuated equilibrium 16; First Law 77, 310; and kin selection 199–201; and assortative mating 310
Migration: as a modifier of behavioral predispositions 99; and ethnic conflict 99; and the dominance order 165
Mind: and the rationalistic bias 40–9; and behavioral predispositions 37–49; and the placebo effect 44–5; and cultural universals 55–65; and variants 56–65; and deception 58–60; and the nature vs. nurture problem 68–78; and the developmental stages 76–7; and Weber's theory of Protestantism and capitalism 88–93; and ascetic altruism 216–35; and the idea of the soul 225–35, 271–4; and self-deception 225–35; and the brain 239–40; and the forces of sociocultural variation 242–6; and the forces of sociocultural selection 246–51; and the explanatory urge 267–74; and relational self-identification 267–74; and the denial of death 274–81; and symbolization 281–93; and reification 281–93; and the susceptibility to charisma 294–8; and the imperative to act 298–303; and ritual consumption 298–303
Modern Synthesis: 15–17; its fundamental proposition 17; and sociobiology 17–20; and kin selection 199–201; and the theory of ascetic altruism ch. 6 *passim*; and the model of biocultural evolution 236–65
Monogamy: see Family
Mourning: and the explanatory urge 267–74; and relational self-identification 267–74; and the social self 271–4; and the soul 272–4; defined 273
Multiplier effect: as a criterion of sociocultural selection 260–1; and autocatalytic reaction 260–1; and cultural reversals 260–1; and counterforces 260–1; and Spencer's law of evolution 261
Mutation: and early Mendelians 16; and early opposition to gradualism 16; and punctuated equilibrium 16; as a problem of adaptation 21; definition of 21; and ascetic altruism 235

Natural selection: and Darwin's definition 6–8; and horizontal evolution 6; and T. Malthus's influence 6–7; and adaptation in Darwin 6–8; and variation in Darwin 7–8; and the Mendelian problem 7–8, 16; and the survival of the fittest 8; and the Modern Synthesis 16–17; and the problems of analogy and homology 66–8; as a cause of culture 75; in reciprocal relation with culture 92–103; and the modifiers of behavioral predispositions 95–103; and the maximization principle 106–10; and inclusive fitness 106–10; and the adaptiveness of culture 106–10; and predispositions of self-enhancement 104–50; and the biocultural principle 107; and the climbing maneuver 110–20; and territoriality 120–9; and the urge to victimize 129–41; and the need for vengeance 141–6; and the need for recognition 146–50; and reciprocation 151–61; and predispositions of domination and deference 161–76; and the need for conformity 177–86; and the need for social approval 186–8; and the need for self-purification 188–94; and altruism 195–235; and kin selection 199–201; and group selection 202–4; 262–3; and the definition of altruism 205–7; and asceticism 211–28; and the theory of ascetic altruism 216–35; and self-deception 222–35; and the idea of the soul 222–35, 271–4; and religion 222–35, 267–303; and biocultural evolution 236–63; and the forces of sociocultural variation and selection 238–51; and fitness enhancement 251–5; and the explanatory urge 267–74; and relational self-identification 267–74; and the denial of death 274–81; and symbolization 281–93; and the susceptibility to charisma 294–8; and the imperative to act 298–303; and the need for ritual consumption 298–303; and homologous affiliation 304–14; and heterologous affiliation 304–14; and assortative mating 304–14; and incest avoidance 314–21; and family relations 321–30; and heterologous contraposition 330–6

Need for conformity: defined 177; and predators' preference for deviates 177–8; and imitation 178–86; and socialization 178–86; and model learning 179; and conflict 179; and religious conflict 179–82; and the use of force 179–86; and intrasocietal cleavages 182–6; and international conflict 183; and societal solidarity 183–4; and ethnocentrism 184; in small groups 184; and stages of personal development 184–5; and reciprocal altruism 201–2; and the theory of ascetic altruism 222–8; see Domination and Deference, Reciprocation

Need for recognition: defined 146–7; and the history of science 147; and the history of music 147–8; and the search for roots 48; and social movements 148–9; and the USA immigrant 148–9; in literature 149; and the affairs of state 149; and the hierarchy of needs 150

Need for ritual consumption: see Ritual consumption

Need for self-purification: defined 188; and expiation 188–94; and reciprocation 188; and cultural deviance 189; and animal behavior 189; and the baptism 189–91; and the idea of original sin 190; and born-again religion 190–1; and the idea of the soul 191–2; and the concept of Purgatory 191–2; and rites of expiation 192–3; and the maximization principle 193–4

Need for social approval: defined 186; in social theory 186; and the need for recognition 186; and the social self 186–7; and social stratification 187–8; as a symbolic reward 187; and the theory of ascetic altruism 222–8

Need for vengeance: and the problem of reciprocation 141–2, 151–61; in Durkheim's sociology 142–3; as a family affair 143–4; and territorial disputes 144; and modern man 145; and the case of Iran 145–6; against imaginary as well as real offenders 146; and reciprocal altruism 201–2

New Synthesis: see Sociobiology

Optimality principle: see Maximization principle

Parochialism: and religion 182–6; and heterologous contraposition 330–40; and ethnicity 330–40; see Kin selection
Personification: see Reification
Phenotype: see Behavior, Culture, Assortative mating
Pleiotrophy: defined 21
Polygamy: see Family
Population genetics: and the marriage of Mendelian and Darwinian laws 16–17; and Mendel's First Law 77; and the modifiers of behavioral predispositions 95–103; and migration 99; and population control among animals 168–9; and kin selection 199–201; and group selection 202–4, 262–3; and incest avoidance 314–21
Power: defined 114–15; and the climbing maneuver 114–20; and territoriality 120–9; and the urge to victimize 129–41; and the need for vengeance 141–6; and the need for recognition 146–50; and reciprocal behavior 160
Prejudice: and pre-evolutionist assumptions 3–5; and Social Darwinism 8–11; and anti-evolutionist assumptions in social science 14; and ethnicity 333–40; and social class 336; institutional 338–40; its distribution 336–40; and deception 336–40
Proximate causation: and biocultural interplay 47; and the rationalistic bias 40–9; and Great Man theories 47–8; and the property

of intensity 80–103; and the reach of explanation 82–3; and the influence of culture and natural selection 92–103; and symbols 95–6; and great events 96–7; and socioeconomic changes 97–9; and migration 99; and social mobility 100–2; and criteria of biocultural selection 241, 251–61

Rationality: and irrational behavior 15; and meaning 29–31; and human nature 34–6; and the rationalistic bias 40–9, 52–6; and proximate causation 47; and Great Man theories 47–8; and variants 56–65; and the explanatory urge 267–74; and symbolization 281–93

Reciprocation: and vengeance 141–2; and the theory of reciprocal altruism 151–3, 159–60; and theories of reciprocity in philosophy 152; and theories of reciprocity in sociocultural science 154–61; and the gift 154–5; and the potlatch 155–6; and Gouldner's principle of moral absolutism 156–7; and reciprocity in China 157–8; and cheating 158–61; and feudal society 158; and exchange theory 159–60; and the dominance order 159–76; and relative deprivation 160; and distributive justice 160–1; and power 160; and reciprocal altruism 199–201; and altruism ch. 6 *passim*; and the theory of ascetic altruism 222–8; and the soul 274

Reciprocity: see Reciprocation

Reification: and symbolization 281–93; and its natural selection 285–90; and relational self-identification 285–8; and the social self 285–7; and the hunt 285–7; and the game 285–7; degrees of 289–93; and language 289–93; and personification 290–3

Relational self-identification: and knowledge 267–74; and the explanatory urge 267–74; and the soul 271–4; and the social self 272–4; and death 272–4; and mourning 272–4; and the denial of death 275; and relations between the living and the dead 277–81

Religion: and early evolutionists 53–4; and Weber's theory of capitalism 88–93; born-again 92; and the urge to victimize 140; and the need for conformity 179–82; and conflict 179–82; 291–3; and intrasocietal conflict 182–6; and the need for self-purification 188–94; and asceticism 211–16; and the theory of ascetic altruism 222–8; and the soul 225–9, 271–4; types of 229–33; and the explanatory urge 267–74; and relational self-identification 267–74; and the denial of death 274–81; and relations between the living and the dead 277–81; and symbolization 281–93; and reification 281–93; and its definition 281–4; and a mechanism of group selection 283, ch. 8 *passim*; and the idea of the sacred 284–93; and the clan 284–93; Durkheim's theory of 282–5, 298–303; and the social self 285–7; and secularization 291–3; in USA history 292–3; and kin selection 292–3; and ethnicity 292–3; and intrasocietal boundaries 293; and revolution 293; and the susceptibility to charisma 294–8; and group immortality 297–8; and the need for ritual consumption 298–303; and the imperative to act 298–303; and ethnicity 333–4

Ritual: the tivah 278; the corrobbori 284–5; and the imperative to act 298–303; and ritual consumption 290–303; as reinforcer of behavioral predispositions 301–3; and the evolutionary viability of societies 301; see Ritual consumption

Ritual consumption: in mythology and ethnography 42–6; and medical practices 44–6; and the placebo effect 44–5; and the Christian communion 45–6; and communion 303; and socialization 303; and the soul 303; and group selection 303

Sacrifice, human: and head-hunting 138; and war 138–9; and population density 138; and technology 138–9; and scalping 138; and cannibalism 138–9; and the American Indian 139–40

Selections: defined 240; as the enduring novelties in biocultural evolution 240; and universals 240; and variants 240–1; and selectors 246–51; and the criteria of selection 251–61; and religious behavior 267–303

Selectors: as forces of the selective retention of combinations 240–1, 246–51; and the conservative mentality 246–8; in social theory 246–7; and scientific innovation 247–8; and war 249; in relation to combiners 249–50; and personality types 250–1; and the criteria of selection 251–61; relational self-identification 267–74; the denial of death 274–81; reification 281–93; susceptibility to charisma 294–8; the need for ritual consumption 298–303; see Combiners

Self, social: and the need for social approval 186–7, 222–5; and the idea of the soul 225–8, 272–4; and the explanatory urge 267–74; and relational self-identification 267–74; and death 272–4; and mourning 272–4; and persistence of relations between the living and the dead 277–81; and symbolization 285–7; and reification

285–7; and the hunt 285–7; its evolution 286–7; and the game 286–7
Sex: and asceticism 215–16; and the denial of death 276; see Fitness, Inclusive fitness, Maximization principle
Slavery: and victimization 136–7; in time and place 136–7
Social Darwinism: 8–11; and Spencer's theory 8–9; and capitalist ideology 8–13; and Puritanism 9; and the role of the state 9–11; in the USA 9–11; and Sumner's theory 9–10; and laissez-faire economics 10–11; and misconception of Darwinism 10–15; and the reaction 11–15; and artificial selection 12; and Kropotkin on mutual aid 12; and Marxism 13
Socialization: and evolutionary theory 15; and the case of XYY men 50; and identical twins 50–1; and adopted children 51; defined 62, 75; and cultural universals 62; nature and nurture 68–78; and adaptation 69; and the nature of learning 69–78; and types of learning 71–3; and learning in animals 72–8; and imitation 73–8; and genes 74–5; and environmental variation 74; and learning as a genetic fact 74; and the sterility of social theory 75–8; and developmental stages 76–7; and the theory of anomie 87–8; and imitation 178; and developmental stages 184–5; as a mechanism of biocultural evolution 241–4
Social mobility: and the theory of social structure and anomie 82–8; as a modifier of behavioral predispositions 100–2; and the perception of legitimacy 100–1; and class conflict 101; and conformity 101; and revolution 101; and the climbing maneuver 110–20; and the potlatch 155–6; and the dominance order 161–76; and altruism 222–5; and hypergamy 325–30
Social stratification: see Dominance order, Domination and Deference
Sociobiology: 17–20; as offshoot of Modern Synthesis 17; definition of 17; and ethology 17–18; and the challenge to sociocultural science 17–20; and coevolution 18–20; and the maximization principle 20, 106–10; and altruism 20; criticisms of 20–3; and the problem of adaptation 20–2; 106–10; and the problem of determinism 22–8; and the problem of conservatism 28–9; and the problem of meaning 29–33; and the problem of cultural variability 30, 31–3; and human nature 33–6; and behavioral predispositions 49–56; and behavior genetics 50–1; and the problems of homology and analogy 65–8; and the nature vs. nurture problem 68–78; and the intensity of causal units 79–92; and the distribution of causal units 92–5; and the modifiers of behavorial predispositions 95–103; and the maximization principle 106–10, 233–5; and inclusive fitness 106–10; and the adaptiveness of culture 106–10; and the biocultural principle 107; and the climbing maneuver 110–20; and territoriality 120–9; and the urge to victimize 129–41; and the need for vengeance 141–6; and the need for recognition 146–50; and reciprocation 151–61; and predispositions of domination and deference 161–76; and the need for conformity 177–86; and the need for social approval 186–8; and the need for self-purification 188–94; and altruism 195–235; and kin selection 199–201; and reciprocal altruism 201–2; and group selection 202–4, 262–3; and types of altruism 204–7; and asceticism 211–16; and the theory of ascetic altruism 216–35; and religion 225–33, 267–333; and the model of biocultural evolution 236–65; and the definition of biocultural evolution 237–8; and the forces of variation and selective retention 238–51; and the criteria of selection 251–61; and the explanatory urge 267–74; and relational self-identification 267–74; and the denial of death 274–81; and symbolization 281–93; and reification 281–93; and the susceptibility to charisma 294–8; and the imperative to act 298–303; and the need for ritual consumption 298–303; and homologous affiliation 304–14; and heterologous affiliation 304–14; and assortative mating 310–14; and incest avoidance 314–21; and family relations 321–30; and heterologous contraposition 330–6; and the social sciences 337–40
Socioeconomic change: as modifier of behavioral predispositions 97–9; and technology 97–9; and the Industrial Revolution 98–9; and sociopolitical revolutions 98–9; its fluctuating nature 102–3
Soul, idea of: and the need for self-purification 191–2; and self-deception 225–8; and ascetic altruism 225–8; and the gene 225–8; and the social self 227–8, 272–4; defined 228–9; and the idea of immortality 271–4; and death 272–4; and mourning 272–4; and kin selection 274; and civilization 274; and the denial of death 274–81; and relations between the living and the dead 277–81
Speciation: and Darwin's theory 7–8; and punctuated equilibrium 16
Struggle for existence: in Darwin 7; and fitness 7; and natural selection 6–8; and

Social Darwinism 8–11; and Kropotkin on mutual aid 12
Suicide: and biocultural science 83–8
Survival of the fittest: and natural selection in Darwin 8; and Social Darwinism 8–11
Susceptibility to charisma: see Charisma
Symbolization: and reification 281–93; and the symbol 281–93; and its natural selection 285–90; and relational self-identification 285–8; and the social self 285–7; and the hung 285–7; and the game 286–8; and language 288–93; its evolution 289
Symbols: and religion 58, 225–33, 267–303; and Weber's theory of Protestantism and capitalism 82–8; as modifiers of behavioral predispositions 95–6; and silence 96; and reiteration 96; and persecution 96; and the dominance order 168; and the theory of ascetic altruism 225–35; and the soul 225–8, 271–4; and sociocultural variation and selection 238–51; and the explanatory urge 267–74; and relational self-identification 267–74; and symbolization 281–93; and the social self 285–7; and the hunt 285–7; and the game 286–7; defined 287–9; and language 288–93; and ritual 301–3
Synthetic Theory: see Modern Synthesis
System: and degrees of causation 63; and Weber's theory of religion and capitalism 90–1; and hierarchy theory 169; and emergence 169–70; and the problem of defining altruism 207; and systemic immanence 258–9
Systemic immanence: and the principle of immanent change 258; as a criterion of sociocultural selection 258–9; and scientific discovery 258–9; and artistic innovation 259

Technology: and group selection 92–5, 262–3; and socioeconomic changes 97–9; and the Industrial Revolution 98–9; and population 98; and sociopolitical revolutions 98–9; and human sacrifice 138–9; and the forces of variation and selective retention 238–51; and creature comforts 256–7; and systemic immanence 258–9; and the multiplier effect 260–1
Territoriality: defined 120; and territory 120; and aggression 120–9; and territorial behavior 121–5; and hierarchy 122; and the intensity property 123–9 and cost-benefit analysis 123–5; and the defensibility of territory 123–5; and the predictability of resources 123–5; and the density of resources 123–5; and international conflict 125–6; and the invincible center 126–7; and population control 127–8; and ethnic conflict 128; as emotional attachment 128–9; and earthquakes 128–9
Territory: defined 120; and the invincible center 126–7; and imigration 127–8; see Territoriality
Totemism: and evolutionary theory 4–5; and salvation of the soul 230–1; and the idea of the sacred 284–93; see Religion

Ultimate causation: and biocultural interplay 47; and the reach of explanation 82–3; and natural selection 92–103; and criteria of biocultural selection 241, 251–61

Variants: and causation 49–9, 60–5; and behavioral predispositions 49–65; 95–102; and cultural universals 56–65; and religious symbols 58; and deception 58–60; and their work of disguise 58–65; as products of hypertrophy 59; and the rise of civilization 59–60; and adaptation 59–60; defined 60, 62–3; and culturgens 63–4; and their varying intensity 79–92; as releasens (intensifiers, mitigators) of behavioral predispositions 81–3; and the theory of anomie 83–8; and Weber's theory of Protestantism and capitalism 88–92; as modifiers of behavioral predispositions 95–103; as symbols 95–6; and deception in the climbing maneuver 110–20; and territoriality 120–9; and the urge to victimize 129–41; and the need for vengeance 141–6; and the need for recognition 146–50; and reciprocation 151–61; and predispositions of domination and deference 161–76; and the need for conformity 177–86; and the need for social approval 186–8; and the need for self-purification 188–94; and altruism 195–235; and asceticism 211–16; and the theory of ascetic altruism 222–35; and the soul 222–9, 271–4; and religion 229–33; and biocultural evolution 236–63; and religious behavior 267–303; and the explanatory urge 267–74; and relational self-identification 267–74; and the denial of death 274–81; and symbolization 281–93; and reification 281–93; and the susceptibility to charisma 294–8; and the imperative to act 298–303; and the need for ritual consumption 298–303; and homologous affiliation 304–14; and heterologous affiliation 304–14; and assortative mating 310–14; and incest avoidance 314–21; and family relations 321–30; and heterologous contraposition 330–6
Variation: and Darwin's natural selection

7–8; and Gregor Mendel 8; and biocultural evolution 236–63
Variations: see Selections
Victimization: defined 129; and exploitation 120–34; and the education system 130–3; and the perceived causes of poverty 133; and mental illness 133–4; and the legal system 134–6; and racial relations 134–5; and nepotistic favoritism 135; and criminal behavior 135–6; and slavery 136–7; and the forms of human sacrifice 137–41; and the American Indian 139–40; and religion 140; and ideology 140–1; in the 20th century 141